Localized Corrosion in
Complex Environments

Localized Corrosion in Complex Environments

Mike Yongjun Tan
Deakin University, Australia

WILEY

This edition first published 2023

Registered Office(s)
John Wiley & Sons, Inc., 111 River Street, Hoboken, NJ 07030, USA

For details of our global editorial offices, customer services, and more information about Wiley products visit us at www.wiley.com.

Wiley also publishes its books in a variety of electronic formats and by print-on-demand. Some content that appears in standard print versions of this book may not be available in other formats.

Library of Congress Cataloging-in-Publication Data
Names: Tan, Mike Yongjun, author.
Title: Localized corrosion in complex environments / Mike Yongjun Tan.
Description: Hoboken, NJ : Wiley, 2023.
Identifiers: LCCN 2022039095 (print) | LCCN 2022039096 (ebook) | ISBN 9781119778608 (hardback) | ISBN 9781119778615 (adobe pdf) | ISBN 9781119778622 (epub)
Subjects: LCSH: Corrosion and anti-corrosives.
Classification: LCC TA462 .T27 2023 (print) | LCC TA462 (ebook) | DDC 620.1/1223–dc23/eng/20220919
LC record available at https://lccn.loc.gov/2022039095
LC ebook record available at https://lccn.loc.gov/2022039096

Cover Image: Wiley
Cover Design by Courtesy of Mike Yongjun Tan

Set in 9.5/12.5pt STIXTwoText by Straive, Pondicherry, India

Contents

Preface

Over the past century, significantly improved understanding of materials behavior and performance, extensively developed materials selection standards and software, and various engineering design tools have facilitated the avoidance of "short-term" failure of engineering structures. However, "long-term" materials failure issues, in particular localized forms of corrosion and materials degradation, still remain tenacious threats to the integrity and safety of the huge network of civil and industrial infrastructure assets especially those exposed to complex and varying environmental conditions. Substantial progresses have also been made in corrosion science and engineering, facilitating the effective control of uniform and general corrosion in many industrial structures such as automobile body rusting and radiator corrosion; however, the management and control of localized corrosion in complex engineering environments is still a significant challenge. It is evident by many publically reported catastrophic engineering structure failures and an enormous amount of unreported infrastructure incidents. This problem becomes even more acute when complex forms of localized corrosion occur on "invisible" and highly variable engineering structures such as buried and submerged oil and gas pipelines.

Effective control and management of complex forms of localized corrosion are critical for maintaining the safety and integrity of industrial and civil infrastructures that are vital for the provision of the world's essential services and the maintenance of its economic activities. Although localized corrosion has been widely studied over the past decades, it should be noted that most studies are typically limited to the investigation of specific forms of localized corrosion such as pitting of stainless steels in defined laboratory environments. Conventionally corrosion science research considers a corrosive environment uniform and stable, a simplification of complex and changing industrial environments. Corrosion prediction models, testing methods, and protection measures are mostly developed under such simplification. Unfortunately, most practical engineering structures can be subjected to highly non-uniform corrosion under multiple and dynamically changing environmental conditions. An example is localized corrosion of buried steel pipelines that are affected not only by seasonal changes in soil moisture and oxygen levels, inhomogeneous coating defects and coating disbondment but also by fluctuating stray currents and oscillating mechanical stresses. Another example is localized corrosion and materials degradation on offshore structures such as wind turbines and oil platforms that are affected by multizone and dynamically changing marine environmental conditions. Variable and complex environmental conditions can lead to changes not only in corrosion rates but also in corrosion patterns and mechanisms. Unexpected changes in environment and mechanism could also cause suddenly accelerated localized corrosion damages that are not predictable by conventional corrosion models. Currently, corrosion engineering studies have not sufficiently considered these issues, leaving a major knowledge gap in corrosion science and engineering.

The ultimate goal of corrosion engineering would be to prevent the premature failure of engineering materials and to extend the safe operational life of engineering structures through detecting, mitigating, and minimizing corrosion damage. This could be compared with the protection of human health through detecting, diagnosing, and preventing cancer and other diseases. Human body itself has many "sensors," the eyes, ears, nose, skin, and tongue, that provide disease information through vision, hearing, smell, touch, and taste. Diseases can be further diagnosed and treated through medical testing, doctor's analysis, and the use of various medical treatment. For engineering structures, unfortunately such "sensors" are not naturally available, and therefore it is essential to have artificial devices installed for sensing structural health issues such as complex forms of localized corrosion – the prime threat to the integrity of metallic structures. An "ideal" corrosion sensing and control system should be one that not only provides in situ and site-specific corrosion data required to visualize localized corrosion but also uses such data to inform corrosion prediction and control, for instance, to guide local coating repair and to regulate local cathodic protection potential and corrosion inhibitors injection. In this manner, the threat of localized corrosion to the integrity and safety of engineering structures would be minimized and the safe operational life of infrastructure would be maximized. Considering the variable nature of corrosion environments and mechanisms in practical engineering structures, corrosion management and prevention actions may need to be adjusted smartly and dynamically based on the prevailing corrosion condition and mechanism. A prerequisite for effectively doing so is timely knowledge about the initiation, propagation, and seriousness of localized corrosion occurring over an engineering structure.

Currently, knowledge of localized corrosion is generally from time-based routine inspections using various condition assessment and in-line inspection tools. Although corrosion data from such inspection are useful for identifying longer-term corrosion trends, they often do not have sufficient temporal and spatial resolutions required for predicting dynamic changes in complex localized corrosion. Acquiring corrosion data more frequently is necessary not only for early warning of unanticipated structural failure but also for evaluating the efficiency of corrosion control measures such as cathodic protection, corrosion inhibitors, and protective coatings. Over the past decades variously designed corrosion sensors and probes have been developed for in situ and site-specific corrosion data acquisition. In particular significant progresses have been made in the development and implementation of various forms of electrochemically integrated multi-electrode arrays, often referred to as the wire beam electrode (WBE), which has already been described in an earlier book by the author (Yongjun Tan (2012). *Heterogeneous Electrode Processes and Localized Corrosion*. Wiley). The WBE concept has enabled the design of various forms of novel localized corrosion probes for in situ measurement and monitoring of difficult-to-measure forms of localized corrosion. I believe that future widespread industrial application of effective corrosion monitoring sensors and probes will lead to wider availability of real-time corrosion data that, in conjunction with data analytics and artificial intelligence technologies, will enable much more reliable corrosion prediction, control, and management systems.

This present book aims to document new developments and latest knowledge on localized corrosion in selected industrial environments, with particular focus on those obtained from soil and marine environments through the use of variously designed electrode array corrosion probes. Presentation of new knowledge in this book is mostly based on a range of practical engineering cases associated mostly with important civil and industrial infrastructures, with particular focus on underground and offshore oil and gas pipelines. Case studies frequently include systematic comparison and analysis of results and findings contained in a long list of published articles by coworkers and the author, leading to a clearer view on a range of complex localized corrosion events occurring on buried and submerged steel surfaces exposed to dynamically changing environmental conditions. This new knowledge would assist engineers to enhance infrastructure durability and enable life extension of complex engineering structures through the improved modeling, prediction, and management of complex localized corrosion. Although cases discussed in this book are mainly from steel pipelines, they are believed to be useful references to many other engineering structures although their actual corrosion behavior may vary to some extent with the environmental condition and materials in each engineering structure.

I would like to express my sincere thanks to my colleagues and coworkers who have collaborated with me over the past decade especially, from early to late, Bruce Hinton, Maria Forsyth, Tony Hughes, Facundo (Bob) Varela, Ying Huo, Jianyu Xiong, Fariba Mahdavi, Shyama Ranade, Reza Parvizi, Ke Wang, Aifang

Amy Wei, Mauricio (Max) Leonel Latino, Yunze Xu, Majid Laleh, and Sha Ji. I thank them and my many other colleagues at Deakin University, Australia, who have taught me patiently over the years. I would also thank industry advisers, in particular Alan Bryson, Bruce Ackland, Klaas van Alphen, Craig Bonar, Alan Creffield, Geoff Cope, Brian Martin, Ashley Fletcher, and Alireza Kouklan, for their advice and comments on our applied research work. I also would like to acknowledge that many practical cases discussed in this book are from published work that was funded by the Energy Pipelines Cooperative Research Centre and Future Fuels Cooperative Research Centre, supported through the Australian Government's Cooperative Research Centers Program. Last but not the least, I would like to thank my amazing family for giving me the encouragement, time, and support that anyone could ever wish for throughout these years.

1

Localized Corrosion in Complex Engineering Environments

Although substantial progresses made over the past century in corrosion science and engineering have significantly improved the understanding and control of materials corrosion and degradation of engineering structures, localized corrosion remains a tenacious structural health concern for a wide range of industrial and civil infrastructures such as oil and gas pipelines, piers and ports, water, wastewater and sewer systems, renewal energy production, and transportation facilities. In particular, localized corrosion in complex and variable industrial environments remains a persistent issue that has not been sufficiently managed and controlled. This chapter discusses the characteristics and sources of localized corrosion complexity in practical engineering conditions through the analysis of corrosion cases, from simple to more complex forms. It is shown that localized corrosion complexity is often associated with dynamic changes in corrosion mechanisms and kinetics over an extended period of exposure to variable environments. Localized corrosion complexity makes the prediction, prevention, and management of industrial structural failure a significant challenge.

1.1 Localized Corrosion Complexity

1.1.1 Localized Corrosion and Its Complexity in Engineering Systems

Localized corrosion is usually referred to as corrosion in which there is an intense attack at local sites on a metal surface. A typical example of localized corrosion is pitting corrosion that rapidly attacks a tiny area of an engineering structure such as an oil and gas pipeline, leading to rapid penetration and failure of the whole structure. Localized corrosion is often linked to the local breakdown of a passive or protective film on the surface of a metallic structure, causing accelerated dissolution at localized sites on the structure [1]. In theory, localized corrosion shares

Localized Corrosion in Complex Environments, First Edition. Mike Yongjun Tan.

the same electrochemical principles that are developed to explain uniform or general corrosion by pioneers of corrosion science, among them Evans [2], Fontana [3], Pourbaix [4], and Tomashov [5]. However, localized corrosion has some characteristics that make it significantly more complex to understand and much more difficult to control than uniform or general corrosion. Over the past decades extensive research has been carried out to understand and model localized corrosion, in particular pitting, by many corrosion scientists and engineers. Among them, Frankel [6] provided an overview of pitting processes including the breakdown of passive films, metastable pitting and pit growth, as well as critical factors influencing pitting corrosion such as alloy composition, environment, potential, and temperature. Macdonald [7] presented the point defect model to explain the growth and breakdown of passive films on metal and alloy surfaces in contact with aqueous solutions, and for the development of a deterministic method for predicting localized corrosion damage. Marcus et al. [8] considered diverse mechanisms of passive film breakdown at the oxide grain boundaries. Soltis [9] highlighted that there is a clear separation of the passivity breakdown/pit initiation process from the pit propagation, which can be considered in terms of the well-known pitting localized acidification model [3, 10]. Newman [11] reviewed the use of statistical methods and semi-empirical models, and the fundamental deterministic processes that occur during localized corrosion.

In numerous literatures, pitting of stainless steels is frequently taken as a typical example to describe the characteristics of localized corrosion, probably because of their wide and countless engineering applications. Susceptibility to pitting is well known to be a major weakness of passive alloys including stainless steels and aluminum alloys when they are exposed to some environmental conditions. In order to explain and predict pitting corrosion of stainless steels, many contested pitting models have been proposed over the past decades [6–11], although currently there are still diverse views on pitting initiation and propagation processes. Nevertheless there are some undisputed general observations regarding pitting corrosion characteristics that are often also applicable to other forms of localized corrosion:

- The initiation of pitting corrosion on stainless steels involves a very small pit nucleus that grows over periods of the order of seconds. The cause of the initiation of pitting corrosion is still not entirely clear; however, it is clear that manganese sulfide inclusions play a critical role for stainless steel type 316 [12]. The initiation of pitting is often described as a random or stochastic process with respect to time and location. However, this author has a view that pitting of a specific metal in a particular environment is not an accident, it is a deterministic event determined by the thermodynamic instability of the metal in the environment, regardless of the size and the shape of a metal specimen or an engineering structure [13]. This explains the fact that a small corrosion probe can simulate and detect pitting on a much larger structure surface exposed to the same environment.

- Pits become more stable as they become larger and a local acidic environment is developed in the local cavity area. This is because for very small pits the local acidity in small cavity can be easily neutralized by diffusion into the bulk solution. As the pits grow larger the diffusion distances increase and the cavity become isolated, and it gets harder for the acidity in pits to change through diffusing away. Therefore when pits are established, they are difficult to be stopped; however, solution movements and the introduction of inhibitors could have an effect on this. This fact also suggests that pitting growth would be a three-dimensional event, three-dimensional corrosion models and probes could be necessarily for modelling, simulating and probing such three-dimensional corrosion behavior.
- It is generally believed that pitting requires a passive external surface that can provide a high potential to cause the current to flow into the pit. If the external surface is active, this driving force is not available and therefore pitting would not grow. For this reason, passive metals such as stainless steels are susceptible to pitting because of the passive film of the external surface. Metals such as carbon steel only pit if the solution (e.g. alkaline solutions) passivates it, it would not pit if the solution (e.g. neutral salt solutions) only corrode it generally. Therefore variability in environment (e.g. pH changes) could change carbon steel behavior completely, creating a major uncertainty in corrosion behavior. This characteristic suggests that allowing the formation of local passive and active environments is important for localized corrosion probes. However, it should be noted that if there is an external current source, such as a piece of coupled noble metal, that can behave as cathode to cause the current to flow into the pit, a passive external surface might not be necessary for pitting to occur.
- There is a critical pitting temperature that is considered to provide a robust predictor of the pitting potential [11]. Below the pitting temperature, metal would be unable to maintain active pitting corrosion dissolution at anodes because passivation intervenes, even in the most aggressive possible microenvironment. This characteristic suggests the importance of environmental condition simulation, especially temperature, in localized corrosion testing and monitoring.

Awareness of these characteristics is important not only for predicting and managing pitting but also for understanding many other forms of localized corrosion that have somewhat similar processes and mechanisms as pitting. For instance, crevice corrosion is known to have a similar mechanism as pitting, although it is easier to initiate than pitting because of the pre-existing local environment at the crevice [3, 10]. More extensive discussion on pitting and crevice corrosion characteristics and mechanisms can be found in prime corrosion science and engineering textbooks such as those in references [2, 3, 10]. It should be noted however that although extensive knowledge about localized corrosion has been acquired over the past decades, there is a still major knowledge gap in the field: there is insufficient knowledge of complex localized corrosion in variable engineering environments. Theories and methods discussed in the historical literature are generally limited to specific forms of localized corrosion (e.g. pitting of a stainless steel) in a defined corrosion environment (e.g. a sodium chloride solution). This is because localized corrosion knowledge reported in the historical literature is often developed based on observations and data from simplified and accelerated laboratory experiments where corrosion occurs under controlled environmental and electrochemical polarization conditions. Observations of localized corrosion are usually in relatively small dimensional and short time scales using a range of conventional visual, physical, and electrochemical techniques. Electrochemical methods used for localized corrosion studies are generally designed under steady-state condition hypothesis, which are often ineffective, if not incapable, for probing dynamic and localized interplay between corrosion mechanisms and changing environmental conditions. For instance, most of conventional laboratory corrosion measurement methods such as electrochemical polarization measurement and scanning probe techniques have limitations when applied to practical engineering structures where major variabilities exist in environmental conditions, materials heterogeneities, and local electrochemistry [13].

For these reasons, currently complex forms of localized corrosion, especially those exposed to complicated and variable environmental conditions, are still not well comprehended nor sufficiently characterized and controlled. Although engineers strive to select suitable materials based on available materials property data to design engineering structures that are strong and durable enough to tolerate the service environment, they are usually unable to predict dynamic and localized environment changes and their effects on localized corrosion over long periods of service. If we examine engineering failure records, many major corrosion-induced incidents of engineering structures are due to localized corrosion on invisible structural components exposed to complex environments such as buried, submerged pipelines, and other types of hidden infrastructure. The management and control of such localized corrosion, when it is initiated, remains a very difficult issue although corrosion control technologies such as cathodic protection, coatings, and inhibitors have been developed to mitigate corrosion [14]. The threat of localized corrosion to the safety of engineering structures is evident by widely reported catastrophic infrastructure failures and an enormous amount of unreported infrastructure incidents [15, 16]. Figure 1.1 illustrates localized corrosion-induced major oil and gas pipeline failure incidences reported in the public media.

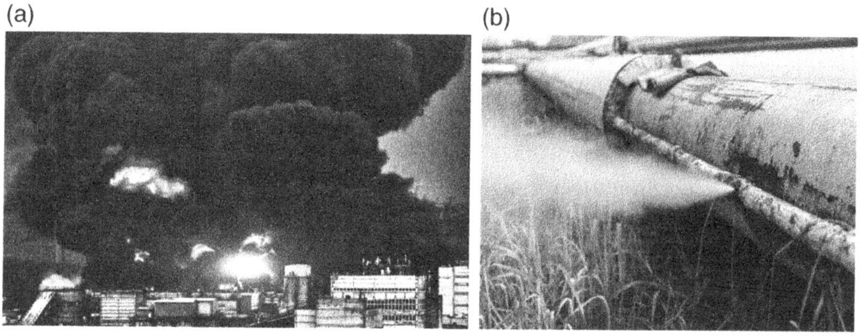

Figure 1.1 Sample localized corrosion-induced infrastructural failure incidences reported in public media. (a) Corrosion induced gas pipeline explosion; (b) Localized corrosion induced pipeline leaking. *Source:* (a) aapsky/Adobe Stock; (b) chitsanupong/Adobe Stock.

1.1.2 Sources of Localized Corrosion Complexity

In practical industrial environments, localized corrosion-induced engineering structural failure may not be due to a single form of localized corrosion, it could be due to multiple and changing forms of localized corrosion whose processes and mechanisms can vary significantly over time and location, leading to localized corrosion complexity. An example is localized corrosion of buried steel pipelines that are affected not only by seasonal changes in soil moisture and oxygen levels but also by fluctuating stray currents and oscillating mechanical stresses. Another example is localized corrosion and materials degradation on offshore structures such as wind turbines and oil platforms that are affected by multi-zone and dynamically changing marine environmental conditions. Such complex environmental conditions can not only lead to changes in corrosion rates but also in corrosion characteristics, mechanisms, and patterns. In some cases other factors such as microbiological activities in soil and ocean could add further complexity to corrosion processes and mechanisms.

One source of localized corrosion complexity in practical engineering structures is major variabilities in environmental conditions. Engineering products and infrastructure often need to function in hostile and changing environments such as corrosive soil and sand, damp air and vapor, flowing water and slurry. For instance, stray currents were found to cause dynamic and rapid localized corrosion namely stray current corrosion of buried and submerged metal structures. The fluctuating stray currents can cause dynamic change in electrochemical corrosion reactions and their thermodynamics and kinetics, adding significant complexity to the protection of underground and marine infrastructures. Over long periods of exposure to service environments, engineering infrastructures can also experience changes in localized corrosion mechanisms. A typical example is a problematic form of localized corrosion on buried and submerged steel pipelines namely corrosion under disbonded coatings (CUDCs). Disbondment of organic coatings can occur on coated pipelines under excessive cathodic protection over extended periods of exposure of a metallic pipeline to soil and seawater. Coating disbondment forms crevices between the pipeline and the disbonded coating layer, leading to changes of local corrosion environmental conditions and changes in the corrosion mechanism from general soil or marine corrosion to highly localized CUDCs. According to a recent survey of the energy pipeline industry, corrosion under coating disbondment is responsible for almost 90% of corrosion-induced damages of buried gas pipelines [17].

Another source of localized corrosion complexity is from the co-existence of many forms of localized corrosion. Examples of corrosion mechanisms that commonly coexist on underground and marine structures include,

- Differential electrochemical potential caused localized corrosion: These include galvanic corrosion (dissimilar-metal corrosion) in which one metal with less noble electrode potential corrodes preferentially when it is in electrical contact with a more noble metal in an electrolyte. Localized corrosion with similar mechanisms include thermogalvanic corrosion, selective leaching, and intergranular corrosion. Stray current corrosion could also be considered to be due to differential potential over different areas of a metal structure.
- Differential aeration cell corrosion (oxygen concentration cell): Localized corrosion are often caused by local environment differences at lap joints, crevices, insulation, as well as debris. Under such conditions, metal areas with less oxygen serves as the anode while areas that are exposed to oxygen usually behave as cathodes of localized corrosion cells. Common examples of this corrosion mechanism include waterline corrosion and filiform corrosion.
- Pitting and crevice corrosion: Pitting and crevice corrosion are considered to have similar mechanisms and characteristics (see descriptions in Section 1.1.1). The main difference is in the initiation and the geometry of the corrosion anodic site. Whereas pitting corrosion occurs on 'weaker' areas over a metal surface, crevice corrosion occurs within a crevice that forms under a fastener, washer or joint, under deposits or under the bottom plate of a storage tank. Pitting and crevice mechanisms could occur simultaneously under some practical engineering conditions. For instance, extended pitting corrosion could generate lots of corrosion products that cover the pits and form crevice, leading to accelerated corrosion as the conditions within the pit become more aggressive.
- Mechanical and velocity-induced localized corrosion occur due to a combination of mechanical factors (e.g. applied and/or residual stresses, cyclic loading, wear) and electrochemical corrosion factors. These usually include stress corrosion cracking, erosion–corrosion, fretting corrosion, corrosion fatigue, abrasion–corrosion, and cavitation corrosion.

These corrosion mechanisms can co-exist and interact with each other, leading to complex forms of localized corrosion to occur on engineering structures. This can be

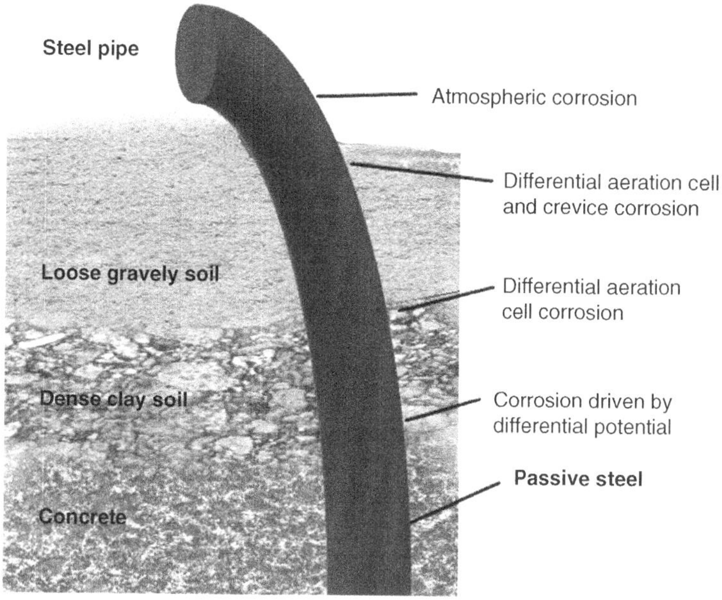

Figure 1.2 Illustrating corrosion mechanisms co-existing on a steel pipe that passes through four environmental zones: atmosphere, loose gravel soil, dense clay soil, and concrete.

illustrated in Figure 1.2 where a steel pipe passes through four environmental zones: atmosphere, the upper loose gravel soil, dense clay soil and concrete. In this structure, there are three areas that can be locally attacked by corrosion. The higher concentration of dissolved oxygen in the loose gravely soil (better aeration) would make the steel surface cathodic to the steel in the dense clay soil, leading to localized corrosion just below the loose and dense soil interface. Dissolved oxygen concentration differences can also be established by crevices existed at the soil and atmosphere interface, leading to corrosion just below the soil surface. Corrosion also occurs in soil at the soil and concrete interface due to a different mechanism: Because of the high pH of concrete, steel is passivated and behaves as cathode because of a higher electrochemical potential. Steel in soil is not passivated and therefore galvanic corrosion would occur to steel in soil. It is similar to dissimilar-metal galvanic corrosion in which one metal with less noble electrode potential corrodes preferentially when it is in electrical contact with a more noble metal in an electrolyte.

Overall, major factors that could contribute to the complexity of localized corrosion include:

- Invisibility (e.g. buried pipes and submerged marine structures)
- Dynamically changing environmental conditions (e.g. soil moisture fluctuations due to wet-dry seasonal changes, stray current-induced potential differences and fluctuations)

- Dynamically changing mechanism (e.g. from general soil corrosion to corrosion under coating disbondment)
- Multi-time and length scale corrosion initiation and propagation (e.g. growth of pitting and crevice corrosion of heterogeneous alloys over extended exposure to corrosion environments)

Other factors can also be added to the list include mechanical stress-induced coating damage and corrosion cracking, hydrogen damage (e.g. hydrogen embrittlement) and microbiological corrosion.

1.2 Corrosion from Simple to Complex

In order to better understand the nature, sources, and characteristics of localized corrosion complexity in engineering structures, let's consider different ways that localized corrosion could become more complex under the effects of various environmental and electrochemical factors.

1.2.1 Corrosion from Uniform to Localized in Engineering Systems

It is well known that aqueous corrosion is the result of a series of electrochemical reactions occurring in electrochemical cells, often referred to as galvanic cells. Galvanic cell is often used as the simplest model for explaining corrosion electrochemical processes. Figure 1.3 shows an ideal textbook galvanic cell where corrosion anodic and cathodic reactions occur in two completely separated half cells. In the galvanic cell formed by dissimilar metals, corrosion concentrates on the anode, leading to metal dissolution "uniformly" on the anode. In this galvanic cell (deoxygenated), an anodic oxidation reaction ($Fe \rightarrow Fe^{2+} + 2e^{-}$) occurs at the iron–solution interface across which iron loses electrons and is corroded, while a cathodic reduction reaction ($Cu^{2+} + 2e^{-} \rightarrow Cu$) occurs "uniformly" at the cathode–solution interface across which cupric ions gain electrons and is deposited as solids. In Figure 1.3, anodic and cathodic reactions can be considered to occur homogeneously over the anode and cathode, respectively, leading to "uniform" corrosion on the anodic iron electrode surface. In this conventional electrochemical model, the mass transfer, chemical reaction, and physical state change are the sequential steps of corrosion processes occurring over two separated anodic and cathodic half cells. Electrochemical corrosion reactions occurring at the electrode–solution interface in the galvanic cell can be described by the most fundamental relationships such as Faraday's law of electrolysis, Nernst equation, and Bulter–Volmer formulation. Through an electrochemical mechanism, corrosion avoids a large activation energy barrier for both oxidation and reduction reactions,

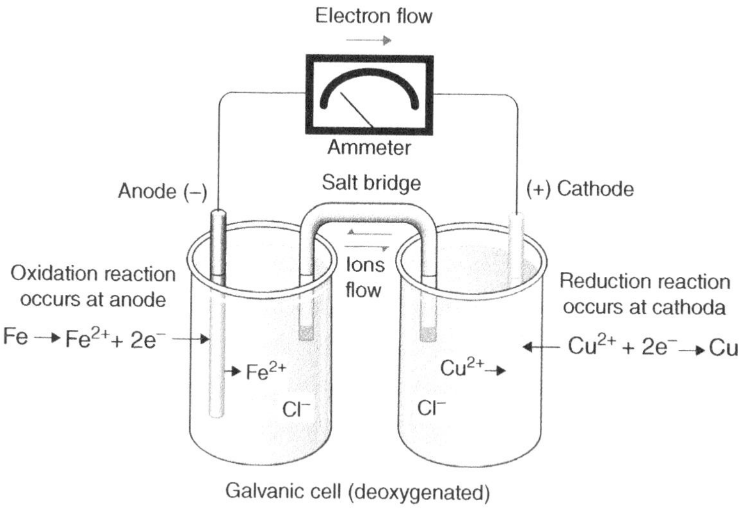

Figure 1.3 A textbook type galvanic corrosion cell model with uniform anodic and cathodic electrochemical reactions occurring on separated electrodes.

leading to much easier corrosion dissolution to occur at the anode. Electrochemical corrosion reactions occurring in the cell are chemical in nature, therefore all factors affecting normal chemical reactions such as the chemical nature of reactants, the concentrations of the reactants, temperature, the ability of reactants to come in contact with each other, and the availability of rate-accelerating or decelerating agents would affect corrosion in the galvanic cell.

A more practical corrosion cell is shown in Figure 1.4 where a porous membrane replaces the salt bridge in Figure 1.3. The cathodic cell in Figure 1.4 is acidified and contains not only cupric ion (Cu^{2+}) but also oxygen (open to the air), creating a more complex corrosion environment. This is a galvanic cell with competing reactions. The oxidation of iron has the lowest reduction potential and is therefore forced to undergo oxidation, $Fe \rightarrow Fe^{2+} + 2e^-$ ($E° = -0.44$ V vs SHE, iron corrosion). The reduction in the cathodic half-cell would include three competing cathodic reactions: $Cu^{2+} + 2e^- \rightarrow Cu$ ($E° = +0.34$, copper plating), $O_2 + 4H^+ + 4e^- \rightarrow 2H_2O$ ($E° = +1.23$, oxygen reduction), and $2H^+ + 2e^- \rightarrow H_2$ ($E° = 0$, hydrogen evolution). In the cathodic half-cell more than one reduction reactions could occur simultaneously over the cathode surface. The major cathodic reaction is the one with the highest reduction potential, i.e. the one with the greater tendency to undergo reduction. This corrosion system is much more complex than that in Figure 1.3, and the application of the fundamental relationships such as the Nernst equation and Bulter–Volmer formulation needs to take multiple cathodic reactions into consideration. Nevertheless anodic and cathodic reactions can still

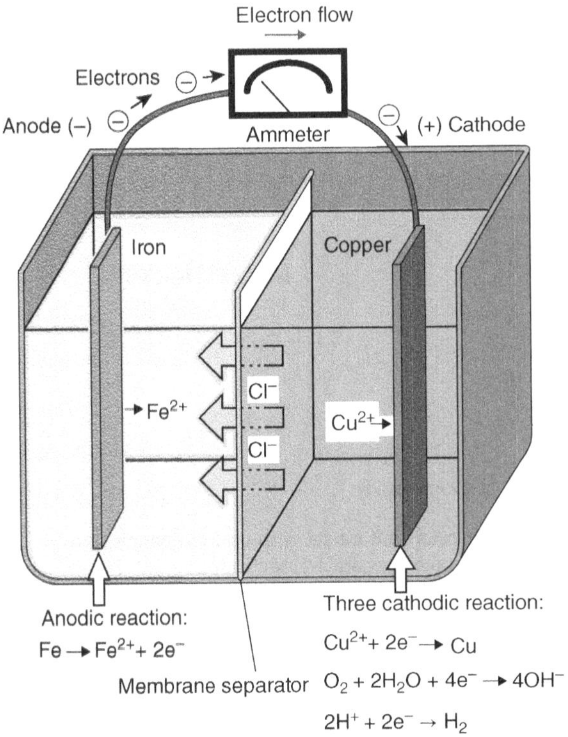

Figure 1.4 A battery form of galvanic corrosion cell.

be considered to occur "uniformly" over the anode and cathode, respectively, and therefore corrosion occurring on the iron electrode surface can still be "uniform."

The most common corrosion cell that could exist in many engineering structures is shown in Figure 1.5 where a galvanic cell is simply made up of electrodes of two dissimilar metals that are electrically connected and exposed to an aqueous solution containing salts and oxygen. This cell simulates galvanic corrosion environments in engineering devices such as a car radiator. In this galvanic cell, iron is forced to undergo oxidation, $Fe \rightarrow Fe^{2+} + 2e^-$ ($E° = -0.44$ V vs SHE, iron corrosion), while oxygen is the dominating reduction cathodic reaction, $O_2 + 2H_2O + 4e^- \rightarrow 4OH^-$ ($E° = +1.23$, oxygen reduction). This corrosion system is more complex than that in both Figures 1.3 and 1.4 since oxygen reduction could occur on both cathode and the anode surfaces, leading to local changes in pH on both surfaces. As shown in Figure 1.5, the iron anode in the cell would not be a uniformly corroding electrode anymore because of the nonuniform distribution of iron oxidation and oxygen reduction reactions, especially at the waterline area where more oxygen exists. Oxygen reduction at the waterline area would lead to a local high pH condition at cathodic areas, leading to local passivity of iron [14, 18]. This

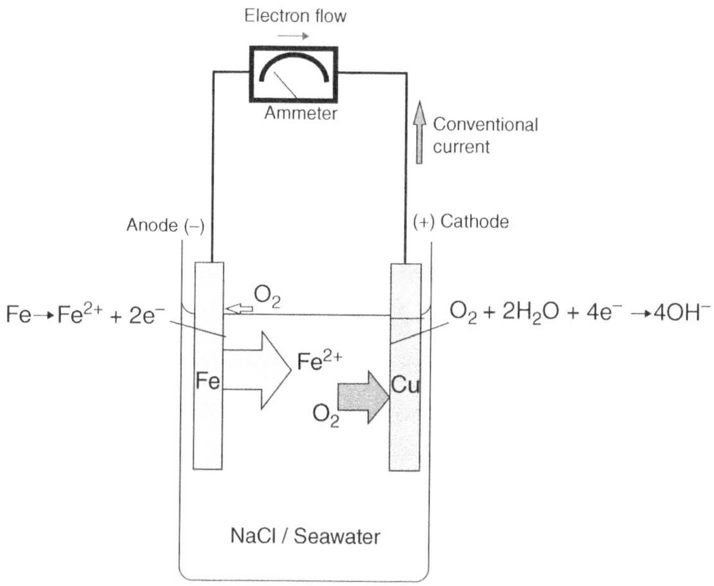

Figure 1.5 Galvanic cell starts to become more complex with localized corrosion starting to appear on the anode.

nonuniform electrode makes the application of the fundamental relationships such as Faraday equation, Nernst equation, and Bulter–Volmer formulation difficult because these relationships are designed for uniform electrodes. The continued operation of the cell would cause significant electrochemical heterogeneity and more complex localized corrosion to occur on the iron electrode as the result of the localization of chemical and electrochemical reactions over the electrode surface.

Corrosion galvanic cells shown in Figures 1.3–1.5 are composed of two dissimilar metals that form separated anode and cathode in the cell; however in practical engineering systems, corrosion usually occurs on the surface of a single metal structure that may be made of the same metallic material or formed by galvanically coupling different metals or alloys. Figure 1.6a shows a simple metallic structure made by joining two copper pieces with a steel bolt exposed to a conductive environment. Galvanic corrosion would occur on the structure under the effect of differential aeration cell corrosion that is often observed on marine structures such as oil and gas production platforms and wind farm tower structures. In the case of Figure 1.6a, iron is the less noble metal and would be the corrosion anode and would be corroded preferentially and nonuniformly because of the existence of the differential aeration cell on its surface. Galvanic corrosion would be concentrated over the small steel bolt, while cathodic reactions occur over the large copper pieces, leading to focused corrosion attack on the iron bolt. This illustrates corrosion due to a dissimilar metal corrosion mechanism in combination with a differential aeration cell corrosion mechanism, leading to more complex corrosion. This structure shown in Figure 1.6a is a poor engineering design since focused corrosion reaction would cause accelerated dissolution of the small iron

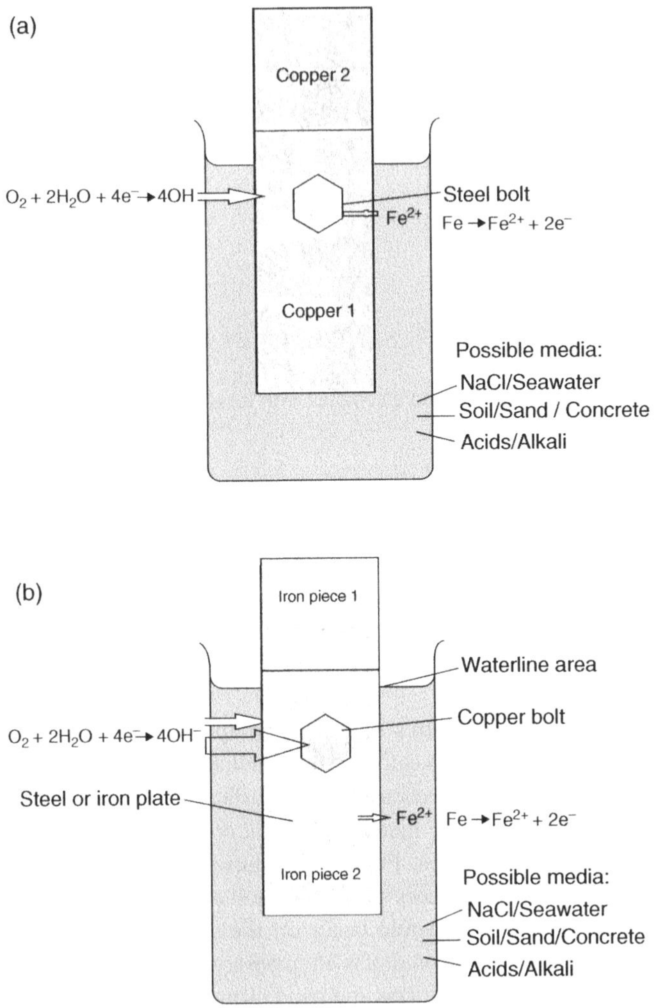

Figure 1.6 A galvanic corrosion cell formed by joining different metal pieces on the same structure. Figure 1.6a illustrates a large cathode and small anode structural design, while figure 1.6b illustrates a small cathode and large anode design.

bolt and rapid failure of the structure. Unfortunately, the structure of the type shown in Figure 1.6a is a rather common engineering design problem that can occur in various engineering systems exposed to fresh and salt water (e.g. desalination plant environments); soil, sand, and concrete (e.g. buried pipeline and metal structures); acids and alkali (e.g. pipeline internal environments). A better design is shown in Figure 1.6b that avoids the situation of a small anode and large cathode and therefore avoids rapid failure of the structure. A simple solution to this galvanic corrosion issue is to use bolts made of copper to avoid the dissimilar metal corrosion situation.

Automobile cooling water system (radiator) is a practical case where multiple dissimilar metals are coupled and exposed to an electrolytic solution, creating a multi-metallic corrosion cell. Cathodic reactions on the more noble metal of the cell (copper) would enhance corrosion of the less noble metal in iron engine, aluminum radiator body and in the radiator' solder etc. To make things more complex, in some cases dissolved noble metals could re-deposit on less noble metal surface, creating a complex galvanic corrosion network. Galvanic corrosion in a radiator is often controlled by maintaining a high resistivity of the coolant and the efficiency of corrosion inhibitors. Please note that although in these cases, copper is nobler than steel and behaves as corrosion cathode, opposite cases can occur in engineering practice. For instance, a copper underground gas pipeline failed due to local attack by galvanic connection with steel in concrete. This was because in a high pH concrete environment, an oxide passive layer existed on steel surface, making steel passive and more noble than copper and causing galvanic corrosion to occur on copper.

1.2.2 General Corrosion in Micro-Electrochemical Cells

In most practical cases, aqueous corrosion reactions occurs over a single piece of metal surface, as shown in Figure 1.7, rather than over separated or jointed pieces of metals or alloys (e.g. in Figures 1.3–1.6). Corrosion on a single piece of metal surface can be in uniform or localized corrosion form depending upon if corrosion cell is a "micro-" or a "macro-" electrochemical cell. The simplest form of corrosion on a metal surface, i.e. uniform or general corrosion, is explained by a "micro-" electrochemical cell model that is based on the following assumptions:

i) Uniform and general corrosion occurs when microscopic local anodic and cathodic sites are sufficiently small and are randomly distributed over a single piece of corroding metal surface. Anodes and cathodes distribute randomly over areas of the metal surface with different electrochemical potentials.
ii) Individual half-reactions occur in these microscopically separated half-cells, causing anode to corrode and electron transfer to cathode through an internal electrical circuit. Ions flow through a conducting solution on metal surface.

iii) Anode and cathode locations change dynamically and a given area on a metal surface could act as both an anode and as a cathode over extended period of time. The averaging effect of these shifting local cells results in a rather uniform attack and general loss of metal and roughening of the surface, usually with the appearance of rusting, as shown in the photo of a corroded pipe of Figure 1.7.

Such uniformly or generally corroding metal surface, as shown in Figure 1.7, is often referred to as a mixed electrode [19] since several different redox reactions with different reduction potentials and kinetics occur simultaneously over the same electrode surface. Wagner–Truad proposed the mixed potential theory to explain the operation of mixed electrode cells operating at a mixed potential [19]. Corrosion is a typical mixed electrode process operating at a mixed potential namely "corrosion potential." Corrosion potential is commonly used in conjunction with the E-pH diagram (often referred to as Pourbaix diagram) [4] that is used as an indicator of corrosion thermodynamic status for predicting if corrosion will occur, for estimating the composition of corrosion products and for predicting environmental changes that would prevent or reduce corrosion attack. The E-pH diagram is a graphical representations of the thermodynamics of common electrochemical and chemical equilibria between metal and water, indicating thermodynamically stable phases as a function of electrode potential and pH. The E-pH diagram visualizes the thermodynamics of corrosion processes and gives information about a metal surface whether it is in a region of immunity where the

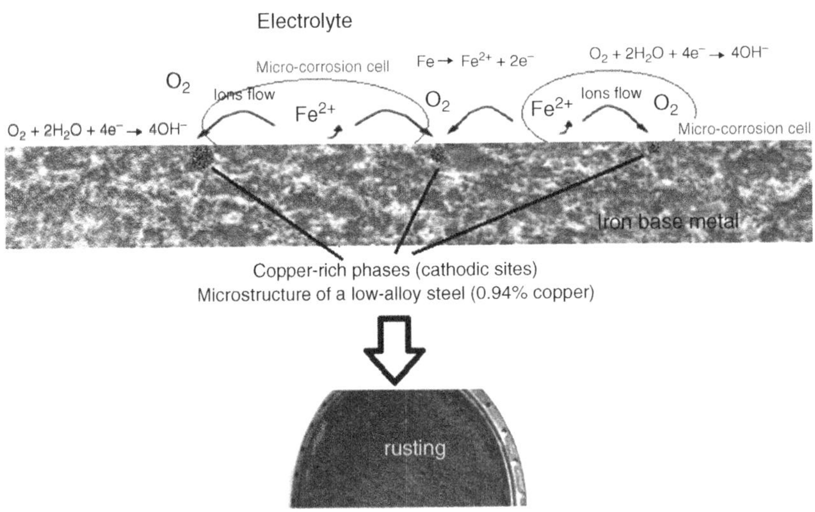

Figure 1.7 Micro-galvanic corrosion cells on a single piece of metal surface.

tendency for corrosion is nil, in a region where the tendency for corrosion is high, or in a region where the tendency for corrosion may still exist but where there is also a tendency and possibility for a protective or passive film to exist. Many E-pH diagrams have been constructed for common materials–environment systems by corrosion scientists including those in Pourbaix's laboratory [2–4]. In many cases, an E-pH diagram can be found from the literature for a particular material–environment combination, although we may need to construct E-pH diagrams for less common systems. As an example, Figure 1.8 shows a typical illustration of the E-pH diagram for the iron–water–soil system at ambient temperature. However, it should be noted that the thermodynamically derived Pourbaix diagrams of the type shown in Figure 1.8 only provides information on corrosion tendency. Like any thermodynamic quantity, the corrosion potential value on its own does not provide information on the rate of corrosion.

The determination of corrosion rates requires the measurement of the kinetics of the corrosion electrochemical process. The rate of a corrosion reaction, for instance $Fe \rightarrow Fe^{2+} + 2e^-$, could be determined if we are able to measure the flows of electrons in the metallic phase or ions in the aqueous phase because the corrosion current, i_{corr}, should be the sum of electron flows. However, unfortunately corrosion electrons flow could not be easily measured from corroding surfaces

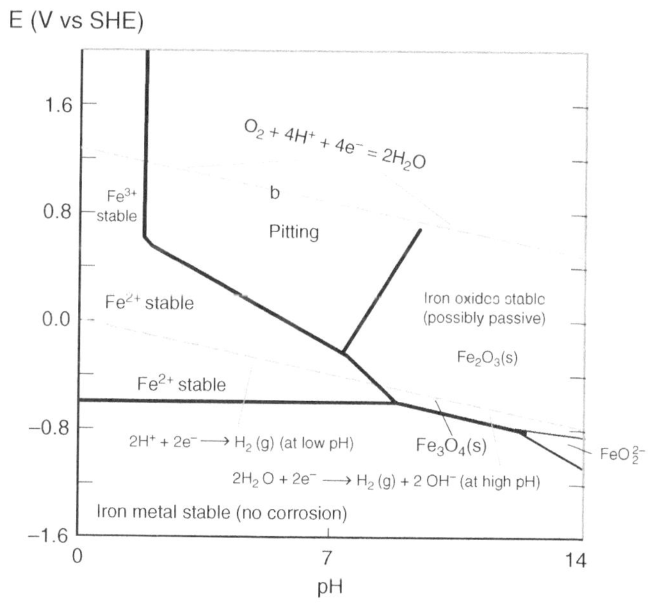

Figure 1.8 A simplified E-pH diagram for iron–water–soil systems. *Source:* Adapted from Pourbaix [4].

because we are unable to directly measure electrons flowing between numerous mini-anodes and mini-cathodes located on the same electrode surface, as illustrated in Figure 1.7. We need to find an indirect way to determine i_{corr} from a corroding electrode surface. A method of analyzing the kinetics of a mixed electrode under dynamic corrosion is a graphical representation of the kinetics of a mixed electrode (often referred to as Evans diagram) [2, 3, 10]. Figure 1.9 illustrates an Evan diagram for iron electrode corrosion in soil. A mixed potential (corrosion potential E_{corr}) is achieved through shifting the potentials of both anodic and cathodic reactions by a corrosion current, $i_{\text{corr,Fe}}$. When oxygen concentration in the soil decreases, the corrosion current values would decrease from point 1 to points 2 and 3, indicating a common oxygen concentration controlled soil corrosion phenomenon. The Evans diagram is very useful tool for analyzing corrosion kinetics and for predicting factors that may affect corrosion reactions; however, it should be noted that the Evans diagram is an imaginative and simplified illustration of electrode "internal" polarization over a corroding metal surface. This diagram is not directly measurable experimentally, and thus it is unable to reveal i_{corr} from a corroding metal surface. Electrochemical polarization techniques such as Tafel polarization, linear polarization, and impedance measurements are traditional methods for measuring i_{corr} from a uniformly corroding metal surface. More details on E-pH and Evans diagram, as well as methods of uniform corrosion rate measurement can be found in classical corrosion textbooks [2–4, 10] and in an earlier book by the author [14].

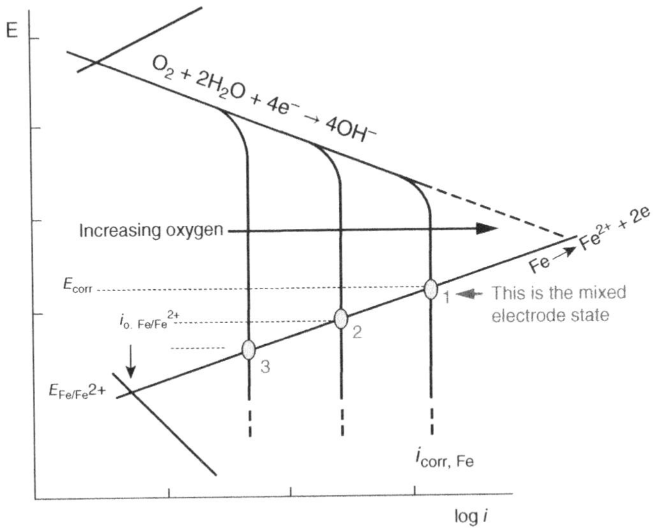

Figure 1.9 An Evan diagram illustrating mixed electrode state of iron corrosion in soil with different oxygen concentration levels. *Source:* Adapted from Evans [2].

1.2.3 Localized Corrosion in Macro-Electrochemical Cells

Uniform or general corrosion such as rusting shown in Figure 1.7 is the least dangerous forms of corrosion because it occurs at predictable rates and it can be effectively controlled by corrosion prevention technologies such as corrosion inhibitors and protective coatings. Unfortunately corrosion occurring on metal surfaces exposed to complex engineering environments most likely has a localized forms such as pitting and crevice corrosion. Under nonuniform corrosion conditions, the micro-electrochemical cell model becomes not applicable because the distribution of anodes and cathodes is no longer random and the locations of anode and cathode become no long changing dynamically. Localized forms of corrosion can be explained by a "macro-" electrochemical cell model illustrated in Figure 1.10. Over extended period of corrosion exposure, as illustrated in Figure 1.10, localized corrosion initiates due to the formation of macroscopic local anodes and cathodes over an electrode surface where local chemistry at an electrode–solution interface is developed.

If the corrosion environment is uniform, localized corrosion on a metal surface would most likely initiate from pre-existing heterogeneities, i.e. spatial, physical, compositional, metallurgical, or chemical nonuniformities existing on electrode surfaces. In industry practice, the most common locations observed for local anodes to form are metal surface metallurgical defects, flaws, scratches, impurities, mill scales, precipitated phases, grain boundaries, dislocation arrays, localized stresses, selective dissolution, and chemicals (such as chloride) induced damages in passive films. The fundamental reason for the formation of macroscopic local anodes and cathodes is that in electrochemical reactions the reactants do not need to be spatially near each other, as normal chemical reactions do, they collide with spatially separated anodes and cathodes. This characteristic permits distinctive separation of electrochemical reactions, leading to a nonuniform

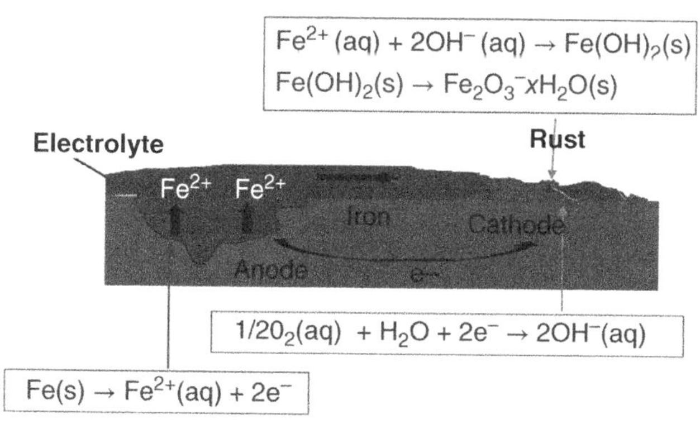

Figure 1.10 Modeling oxygen corrosion of iron or steel in aqueous media.

electrode–solution interface and localization of electrode surface chemistry and physics, i.e. electrochemical heterogeneity [14]. Such heterogeneity can also initiate from an ideally homogeneous surface if it is exposed to a nonuniform environment, and then propagate into extensive separation of corrosion anodes and cathodes, leading to full-scale localized corrosion. Figure 1.11 illustrates localized corrosion on buried pipelines exposed to nonuniform soil environmental condition with and without a disbonded coating film.

Most serious localized corrosion usually occurs on passive metals such as stainless steels and aluminum alloys, or on metal with a damaged surface protective film such as a noble metallic coating with defects. Many metals and alloys rely on a passive film to provide protection, however passive film is not perfect, it can break down under certain environmental conditions. For instance with certain aggressive ions (e.g., Cl^-) in solution, passive film breakdown could be induced at highly localized sites on stainless steel surfaces, often leading to highly localized corrosion such as pitting. In engineering practice, pitting and crevice corrosion are the most destructive form of corrosion that remains the most difficult, yet fundamentally important corrosion issue in corrosion science and engineering. Figure 1.12 shows one of the proposed pitting corrosion model:

- Pitting initiation: Pitting corrosion occurs when a passive film on passive alloys breaks down locally, usually by Cl^- or mechanical impact or erosion. Under natural environment, pitting requires an extended initiation period before visible pits appear (metastable pitting). However the detailed initiation processes and mechanisms of pitting corrosion in natural electrolyte are still not fully understood.

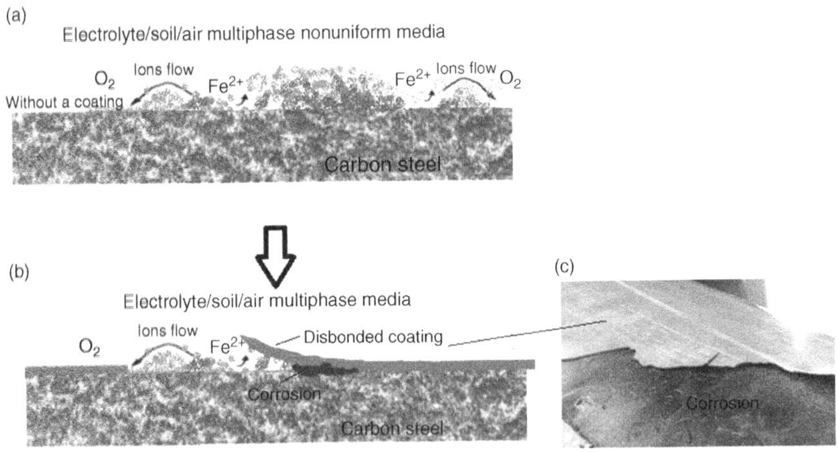

Figure 1.11 Illustrating localized corrosion on buried pipelines exposed to nonuniform soil environmental condition, especially corrosion under disbonded coating. (a) Nonuniform soil corrosion; (b) and (c) localized corrosion under disbonded coatings.

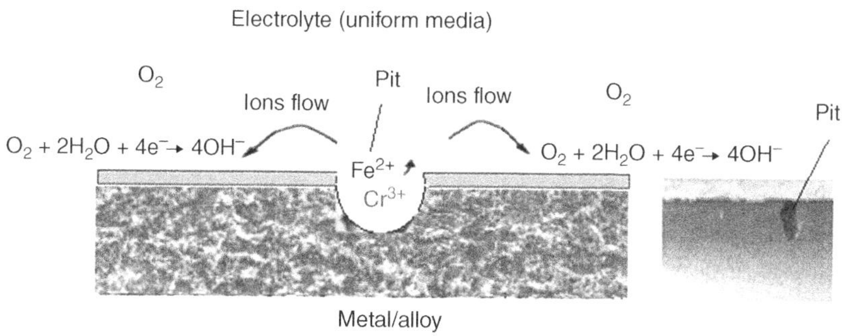

Figure 1.12 Modeling pitting initiation due to the formation of macroscopic local anodes and cathodes over a single piece of metal with damaged surface protective film.

- Pitting propagation: An anode forms where the film has broken, while the unbroken passive film (or protective layer) acts as a cathode. This would accelerate localized attack and pits develop at the anodic spots. Once started, pits penetrate the metal at an increasing rate (pit growth).
- Auto-catalytic pitting grow: The electrolyte inside the growing pit become very aggressive (acidification) which will further accelerate corrosion.
- Repassivation: Under certain condition, pits can die with repassivation.

Major engineering materials such as stainless steels, nickel and aluminum alloys are useful in engineering because of a passive film (an oxidation film) which greatly reduces the rate of corrosion. However if the surface passive conditions cannot be maintained such passive film would be susceptible to localized breakdown, resulting in rapid pitting corrosion. This explains the fact that stainless steels and aluminum alloys have better resistance to localized corrosion in flowing seawater where the supply of oxygen to maintain and repair the passive film varies with the seawater velocity. The higher the seawater velocity, the more sufficient the oxygen and the less the chance of pitting corrosion to occur on stainless steel and aluminum ships and marine structures. However for active metals such as steels, this effect is completely opposite: Increase in the velocity of seawater leads to increase of oxygen supply and thus higher steel corrosion rates. For carbon steel in seawater, corrosion is in general form and its rate-controlling factor is usually the supply of oxygen, therefore corrosion rates of carbon steel in seawater increase with the increase in the velocity of seawater. More details on pitting can be found in classical corrosion textbooks [2–4, 10] and historical reviews of the topic [1, 11].

Similarly another "classical" type of localized corrosion, crevice corrosion, occurs when some sections of a metal are covered by scale or another piece of material forming a crevice, and the flow of oxygen to these sections is restricted. These metal sections may not be able to maintain passive films, and thus become corrosion anodes. An increased corrosion rate at these sections would lead to increased corrosion product residue, further restricting the oxygen flow to worsen the situation. A crevice is an ideal location to become anode due to low oxygen concentration. Crevice corrosion is a contributing mechanism to CUDC shown in Figure 1.11b that occurs frequently on buried pipelines.

Mechanical impact and high velocity fluid can also induce localized corrosion due to the combination of mechanical and erosion factors causing the removal and damage of passive or protective films on metal surfaces. Erosion corrosion usually refers to localized corrosion accelerated by the mechanical effects due to relative movement between a corrosive fluid and the metal surface, e.g. turbulent flow. An example is erosion corrosion that commonly occurs in pipes and pumps. As illustrated in Figures 1.13a and b, localized corrosion dissolution from local anodes that are initiated from local passive/protective surface damage by flow induced mechanical impact. It involves the removal - erosion - of the protective surface passive film, leading to the exposure of bare metal to the corrosive fluid and a rapid corrosion attack. The presence of entrained solids in the fluid would enhance the depassivation process. Erosion corrosion attack usually increases with increasing velocity and varies with the smoothness of the metal surface, degree of turbulence. Figures 1.13b and c show a case of erosion corrosion of carbon steel pipe section with branch in a cooling water system that failed prematurely due to violent turbulence caused by injection of the air into the water-filled pipe, leading to erosion corrosion.

1.3 Cases of Localized Corrosion in Industry

In engineering practice, many structural failure incidents are the results of complex and unexpected localized corrosion due to variabilities in environmental conditions and changes in corrosion mechanisms. Analysis of the causes and solutions of these practical cases could facilitate understanding, management and control of localized corrosion in industry.

1.3.1 Corrosion Due to Unexpected Environmental Changes

A case of localized corrosion failure occurred in a multi-national aerospace components manufacturing company. As illustrated in Figure 1.14, steel components (pistons and slippers) failed due to a rust problem that occurred in a degreaser. Rust changed the dimensions of the precision engineered pistons and slippers, and therefore all rusted components have to be returned for re-processing. This problem occurred only occasionally, and for a long time the root cause of the corrosion problem remained a mystery.

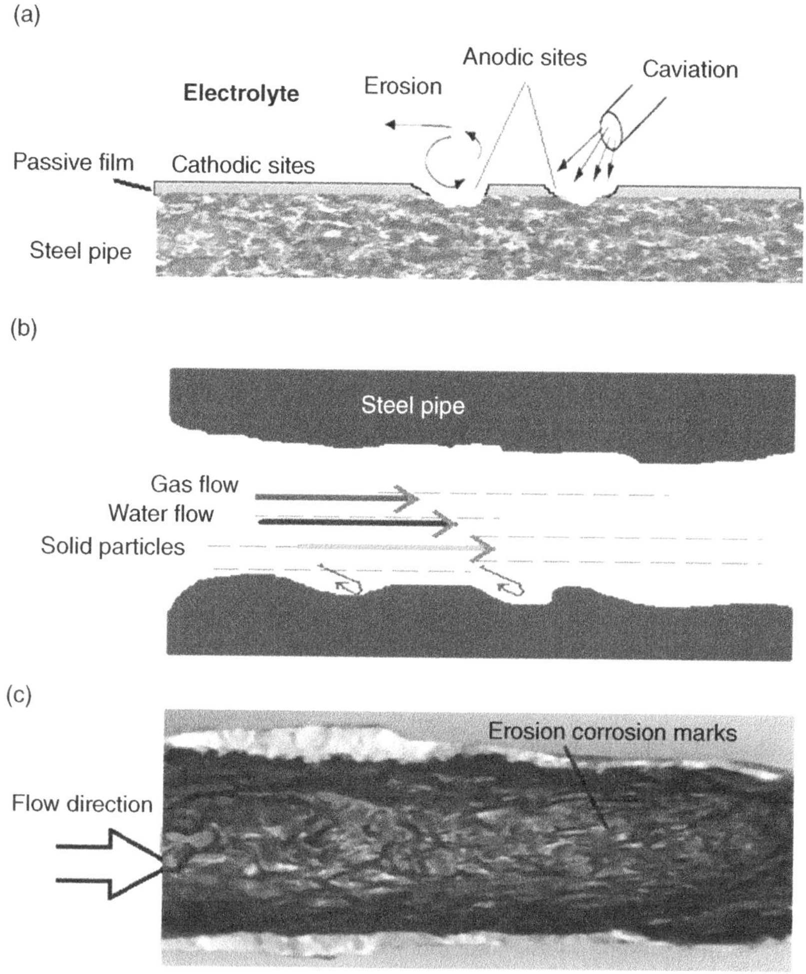

Figure 1.13 (a) Localized corrosion initiates due to erosion induced local anodes; (b) Model and (c) photo of erosion corrosion of carbon steel pipe section in a cooling water system.

Investigation of the problem found that the degreaser was responsible for the rusting problem. As shown in Figure 1.14, an organic solvent, n-propyl bromide, was used in the degreaser. Since this solvent itself is not corrosive, some changes must have occurred in the solvent, leading to changes in corrosion environment. Therefore solvent samples were taken daily for laboratory testing. Slipper samples were molded in epoxy resin to make electrodes for electrochemical corrosion

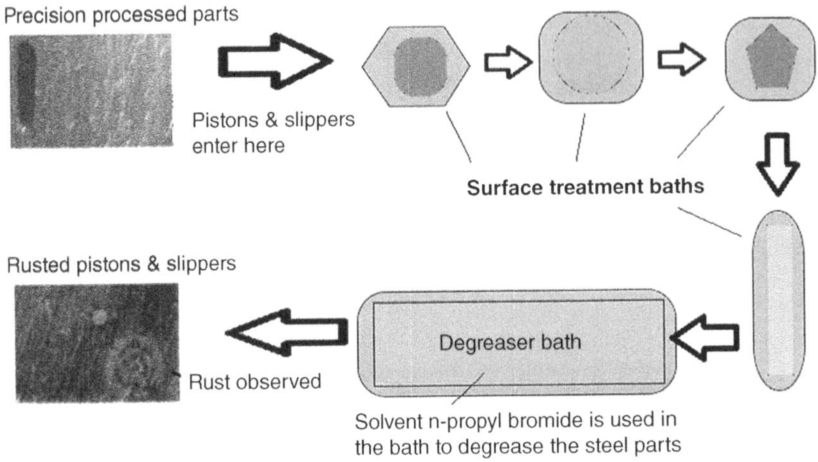

Figure 1.14 Steel components corroded due to a rust problem occurred in a degreaser.

measurements using the polarization measurement technique. Polarization measurements were carried out in each solvent–water mixture (totally 18 solvents mixed with same volume of deionized water). Repeated measurements were carried out in several selected solvent samples in order to understand the change of corrosion rates with time. It was surprised to discover that significant increase in corrosivity and acidity occurred in the solvents after they were mixed with deionized water. There was a typically 10 –30 times jump in corrosivity of the solvents after they were mixed with deionized water for approximately 24 hours. Without exception, corrosion/rusting rates increased significantly with time. This suggests that the solvent itself was not corrosive, it became corrosive only when it was mixed with water. These results suggest that water must have reacted with the solvent, forming acidic substances and contributing remarkably to solvent corrosivity.

However, a question was where did water come from and what happened to the solvent. Careful observation of the degreaser identified two sources of water contamination: (i) moisture condensation from air and (ii) small amount of unremoved water on pistons and slippers from passivating and water rinse tanks. Analysis of the solvent–water chemical reactions suggests that the main component of the solvent *n*-propyl bromide reacted with water forming a very corrosive acid called hydrogen bromide. The process is called "solvolysis" – a nucleophilic substitution reaction, which unambiguously explains the acidity of used solvent and its increased corrosivity.

Analysis of this rust problem has led to the following conclusions on the cause and mechanism of the rust problem: Rusting was due to unexpected environment change and the formation of an aqueous low pH media where rapid acid corrosion occurred together with normal oxygen corrosion. The acid was from a chemical reaction, namely solvolysis, between the solvent *n*-propyl bromide that was used in piston and slipper degreaser, and water that was mainly from atmosphere condensation. According to this analysis, the key to prevent the rust problem is to stop water contamination of solvent in the degreaser. Therefore an essential measure to address this corrosion problem is to control unexpected environmental changes simply by creating a moisture-controlled area for the piston and slipper degreaser by partitioning it off from the rest of the workshop. This would overcome the moisture condensation problem and stop the formation of acid. This is believed to be the simplest and the most economic method of overcoming the problem. Indeed this rust problem was overcome simply through using glass partition and a dehumidizer to create a moisture-controlled area for the piston and slipper degreaser (The humidity of the area should be below 50% RH). This not only helped improving production quality and reliability but also significantly reduced solvent costs because the solvent lasted much longer and frequent change of solvents is no longer necessary. Optional measures to control corrosion in the piston and slipper degreaser could include the addition of corrosion inhibitors to solvents in the tank for improved rustproof capability. Frequent checking and adjustment of pH of water in the tank could also be useful to control corrosion. Timely elimination of acidic water in the tank should also be helpful.

Corrosion due to unexpected environmental changes can occur in many different ways in engineering practice. Another case is acid storage tank corrosion due to significant corrosivity changes of residue liquids after the tank is emptied. For instance, concentrated sulfuric acid is only negligibly corrosive toward a steel storage tank. However, if the tank is incompletely drained and the remaining sulfuric acid liquid is exposed to the air, the acid tends to absorb moisture, resulting in dilution, and rapid corrosion attack would occur in the diluted acid. The key to understand and address such corrosion issues is the identification and monitoring of major environmental changes in engineering structures.

In these cases discussed above, environmental changes can be relatively easily found through visual observation, sampling, and testing. It can be much more difficult for infrastructures that are hidden or not easily visible such as underground pipelines and submerged marine structures.

1.3.2 Pitting Due to Unanticipated Corrosion Mechanism Changes

A multinational company experienced a pitting corrosion problem on its H11 steel aircraft landing gear face, as shown in Figure 1.15(a) and Figure 1.15(b). The H11 gear was bronze plated but this bronze plating layer has to be stripped off using a cyanide based stripper solution. The parts are racked on stainless steel

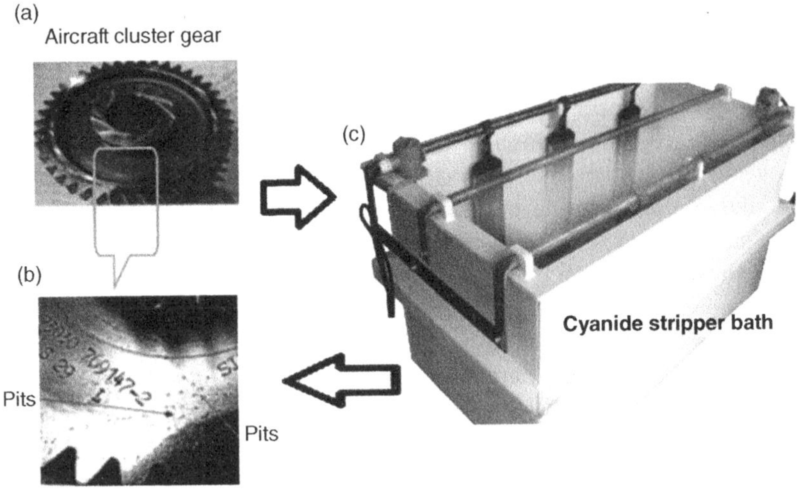

Figure 1.15 (a) and (b) showing pitting on H11 steel gear face due to unexpected environmental variations. (c) Showing the parts racked on stainless steel holders and soaked in the cyanide stripper bath.

holders and soaked in the cyanide stripper bath (Figure 1.15c). The stripping time varies depending on the bronze thickness normally in the range of 1.5–4 hours. Pitting corrosion occurred once in a while to products, and for a long time the root cause of the corrosion problem remained a mystery and there was no solution to it. Analysis of the pitting problem started from understanding of materials perfromance and behaviour. H11 is a chromium containing hot-work tool steel mainly used for die-casting dies, punches, piercing tools, extrusion tooling, forging dies, and high-strength structural components. Its chemical compositions are 0.35% C, 1.50% Mo, 5.00% Cr, and 0.40%V. The corrosion resistance of H11 steel is mainly from its 5% Cr composition that leads to the formation of a very thin protective layer (passive film) on its surface. Under mild atmospheric and aqueous environments, this passivation film can provide sufficient protection. However, since the 5% Cr in H11 is not high enough to form a good quality passivation film (usual stainless steel needs a Cr composition of over 12%), this passivation film can be damaged under more corrosive conditions such as in the cyanide stripper bath. This damage is often localized in nature, leading to localized form of corrosion.

Initiation of pits could be due to local breakdown of passivation film on H11 steel. The passivation film on H11 steel can be damaged locally under corrosive conditions such as in the cyanide stripper bath. Pits can start from surface scratch, dislocation, or other surface defects. These small pits are usually randomly distributed over steel surface. If a small pit is not repaired through repassivation, the pit would grow. Most pitting failures are caused by halide ions such as chloride and chlorine-containing ions. In this case, it is the cyanide ions, which is believed to be similar to chlorides in behavior, that are responsible. Oxidizing metal ions including cupric and ferric halides are also extremely aggressive pitters. In our case, Cu^{2+} and Fe^{3+} may have also contributed to the pitting problem by the following reactions: $Cu^{2+} + 2e \rightarrow Cu$ and $Fe^{3+} + e^- \rightarrow Fe^{2+}$. More importantly, in the bath oxygen concentration could drop occasionally due to uncontrolled activities of the parts racked on stainless steel holders and soaked in the cyanide stripper bath, leading to pitting to occur from time to time.

Based on above analysis, corrosion prevention method was proposed based on the understanding of the corrosion mechanism: For this specific case, the most practical and simplest method was to increase and maintain oxygen concentration in the cyanide-based stripper bath in order to enhance and repair the passivation film by repassivation processes. Pitting can be stopped by repassivation if oxygen concentration in the bath is high enough. Therefore a recommendation was made as corrective action to address this pitting corrosion issue in the most economic way: Continuously sparging air to the stripper bath in order to increase oxygen concentration in the bath. A higher and continuous supply of oxygen is believed to help maintaining the passivating layer on H11 steel surface. After the installation of two air sparging pumps (fish tank air pumps) to the stripper bath, it was reported that the pitting problem was not encountered again. This confirms the mechanism analysis and recommendation. Another practical methods that could help eliminating corrosion factors in the cyanide base stripper bath include the use of a passivating or anodic inhibitor that could enhance the passive film on the H11 steel surface. A further suggestion is to apply some levels of anodic protection through the stainless steel holders. In anodic protection, the metal to be protected is made positive to enhance the formation of a stable oxide film on its surface. The stable oxide film then protects the underlying metal from corrosion. Of course, this method needs extra equipment for careful anodic polarization control and is therefore not a preferred choice.

Pitting due to unanticipated environment changes could occur in different ways. Figure 1.16 illustrates a case of pitting of a milk cooling tank due to unanticipated environment changes as a result of coating degradation. The milk cooler has an inner liner of stainless steel, and an outer casing of mild steel. Chilled brine was used between the steel and stainless steel to cool the milk. The tank experienced steel corrosion in brine and galvanic corrosion between mild steel casing and stainless steel liner. In order to control corroison of the mild steel casing, a layer of organic coating was applied on the internal surface of the mild steel casing. This organic coating layer effectively stopped galvanic corrosion between mild steel casing and stainless steel liner at first; however, with the degradation of the organic coating layer, a small pin-hole was formed in the coating, initiating pitting corrosion. The small pit propagated rapidly as a small corrosion anode,

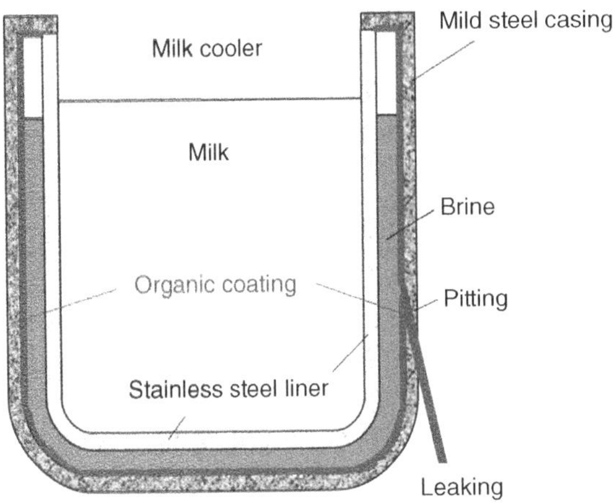

Figure 1.16 Illustrating pitting of cooling tank.

with the large stainless steel liner as the cathode. In this case, what should be painted is actually the stainless steel liner in order to reduce the galvanic effect. It would be even better if both the inner liner and the outer casing of the milk cooler was made of stainless steel to eliminate a galvanic corrosion cell.

The key to understand and address corrosion issues discussed above is the identification of major mechanism changes from general corrosion to passivity breakdown and pitting. In these particular cases, corrosion mechanism changes can relatively easily found through inspection and visual observation. It can be much more difficult to identify corrosion mechanism changes on hidden infrastructures such as underground pipelines and submerged marine structures. More complex localized corrosion cases due to unexpected and dynamic changes in environmental and corrosion mechanisms are discussed in the following chapters of this book.

1.3.3 Multifaceted Corrosion of Underground Pipelines

Buried structures, small as lamp posts [20] and large as pipelines [15, 16, 21–25], are engineering infrastructures exposed to complex environmental conditions and hidden from visual inspection. Multifaceted localized corrosion frequently occurs on these structures due to a number of factors. Firstly, a buried structure such as a pipeline is exposed to a heterogeneous solid/gas/liquid multiphase corrosion medium where liquid water and air occupy the spaces between the soil particles and form discontinuous electrolyte films [26], allowing dynamic and localized corrosion to occur on buried steel infrastructures [27]. Although in principle electrochemical corrosion reactions occurring in soil are the same as those occurring in aqueous solutions, electrode processes associated with soil corrosion can be much more complex and dynamic due to the effects of many factors including soil aeration [28, 29], pH [30], moisture [31–34], ionic species [35], electrical resistivity [36, 37], and microbiological activity [38]. Cole and Marney [27] extensively reviewed earlier work on factors affecting soil corrosion and highlighted work by Romanoff [39] and Fitzgerald [40] who considered aeration, moisture content, soil particle size, pH, bacteria, electrolyte type, and concentration, as well as electrical aspects as major factors affecting soil corrosion. Pipeline coating failure is another issue that can happen due to aging, earth movement, flooding zone, pipe expansion, etc. during pipe in operation. Further complications include localized corrosion at the interfaces of soil and air where multi-environmental conditions exist. Figure 1.17 shows localized pipeline corrosion just below the transition area between soil and air. Detection and identification of such localized corrosion is more difficult, compared to those shown in Figures 1.14–1.16, because of the engineering structure is partly hidden underground.

CUDC is another complexity added to buried pipeline corrosion in soil. When coating is disbonded from the surface of a buried steel pipeline, the disbonded coating layer would stop cathodic protection (CP) currents to reach the inside of the coating disbondment area (usually referred to as cathodic shielding), causing CUDC [41, 42], the worst-case scenario corrosion problems found on buried

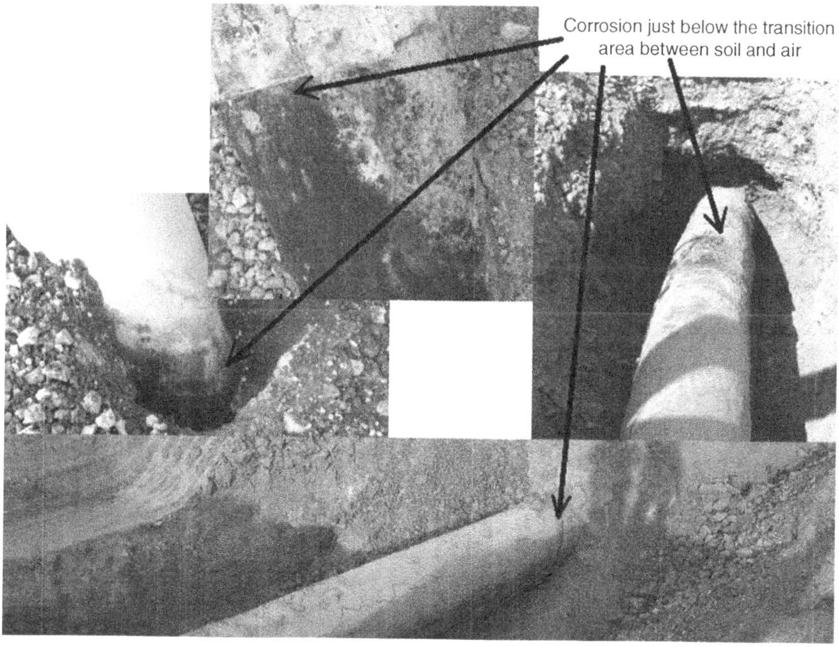

Figure 1.17 Cases of localized corrosion of pipeline at the interface of soil and air.

pipelines. CUDC is not detectable by existing corrosion probes because of difficulties in measuring electrochemical data from the "shielding" disbondment area, therefore knowledge of CUDC is often only from the capture and interpretation of phenomena empirically observed in the industry and inline inspection. Figure 1.18 shows some examples of coating disbondment and CUDC that was further complicated by microbiologically induced corrosion (MIC). As shown in Figure 1.18, severe local attack of a carbon steel underground pipe (wrapped in tape) of a gas supply system occurred. The bandage/coating had become detached at a number of points and water present underneath it probably became anaerobic, allowing sulfate-reducing bacteria to develop, leading to microbiologically induced corrosion. In an anaerobic environment, sulfide can be formed from sulfate by means of sulfate-reducing bacteria.

Although approaches to addressing different corrosion problems in industry can be different, common steps of addressing a complex corrosion problem in industry can include (i) investigate and identify the root cause and mechanisms of a corrosion problem, and determine main factors affecting corrosion processes

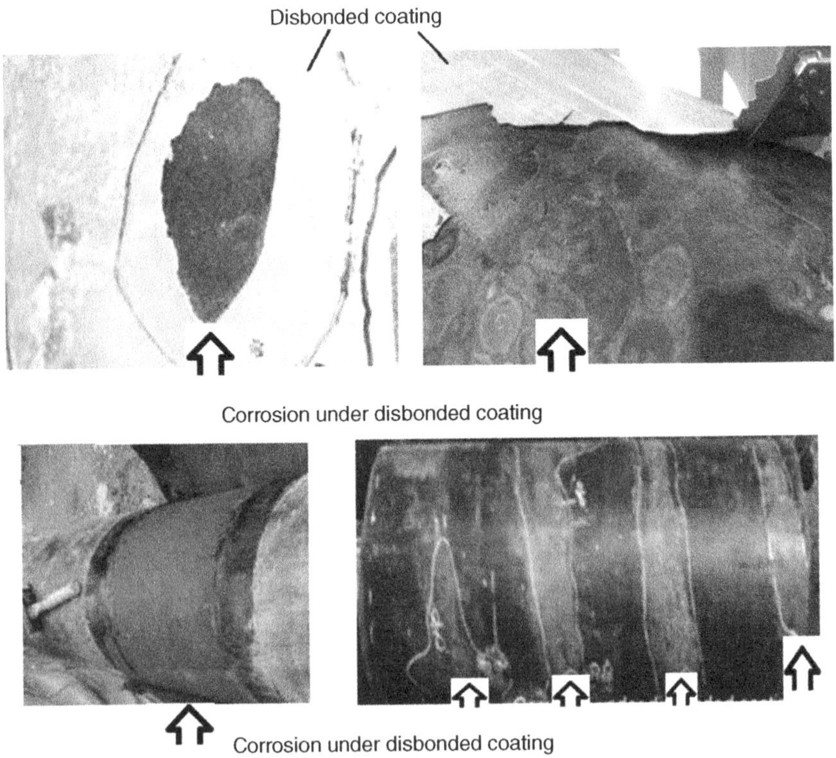

Figure 1.18 Some examples of coating disbondment and CUDC.

and rates; (ii) try to replace the metallic structure with a suitable and practical material that will not corrode or only corrode slowly in the corrosion environment, if not possible, then try to modify the design; (iii) try to modify the corrosion environment, and try to use corrosion inhibitors and protective coatings; (iv) try to apply cathodic protection. For buried pipelines, for instance, specifically corrosion control techniques could include,

i) Materials selection: Mechanical properties including strength (yield strength); toughness (specially fracture toughness); the impact toughness (to prevent dent induced safety problems; depending upon situations) are usually among top three properties that should be considered in pipeline material selection. Many energy pipelines are made of carbon steels, such as X65 or X70, that have an inherently low corrosion resistance in soil. As corrosion poses major challenges for conventional steel pipeline in high-pressure applications, the excellent noncorrosive characteristic of composites has made them possible material choice for pipelines. Since the mechanical properties of some composites such as carbon-fiber-reinforced composites and fiberglass are close to carbon steels' and therefore composites could be suitable substitution of carbon steel. Of course, there are also many other considerations, e.g. costs (however, lower cost should not at the expense of safety); weldability (it is a factor but there are other joining methods available); corrosion (corrosion issues can be addressed through applying corrosion prevention measures).
ii) External coating: While there are various methods for protecting steel against corrosion such as alternate alloys, inhibitors and cathodic protection, for steel in engineering services, however, by far the most commonly used materials for corrosion protection are organic coatings. Protective coatings allow the use of low-cost carbon steel for engineering structure construction in most parts of the world. The most common external coating for buried pipelines is fusion bonded epoxy, chemically curing epoxies as field joint coatings, polyu rethanes, inorganic and organic zinc-rich primers, yellow jacket, and wraps. Epoxies and polyurethanes give the best chemical resistance. While coatings come in many forms, they essentially form a barrier between the steel and the corrosive elements. However, problems can occur with coating cracks, delamination, and defects that need to be protected by cathodic protection. More details are discussed in Chapters 3–5.
iii) Cathodic protection can be utilized in tandem with many different types of coatings, like fusion-bonded epoxy coatings to protect areas where the coating is damaged. This form of corrosion prevention involves making the surface to be protected the cathode of the system. There are two ways by which cathodic protection may be applied; these are by the use of sacrificial anodes

(commonly zinc, magnesium, or aluminum) or via an impressed current system that is more effective method for protecting long buried pipelines. With the sacrificial anode method, a less noble metal anode is physically attached to the steel pipeline to ensure adequate electrical connection. However this method relies on having an electrolyte to conduct electricity and hence are more suitable in aqueous environments or where soil has a high conductivity. Where soil conductivity is low, then an impressed current system must be used. The structure to be protected is made the cathode of the impressed current CP system that is not only more effective than the sacrified anode method but is also more cost effective for larger structures. To mitigate external corrosion of a buried pipeline, CP systems polarize the pipeline catholically, within the from $-850\,mV_{CSE}$ to $-1200\,mV_{CSE}$ potential range. It is widely accepted that corrosion rates within this potential range should decrease to acceptable levels, in practice, it is a great challenge to ensure the complete pipeline is being properly and uniformly polarized. It should be noted that excessive CP in a wet environment can cause coating delaminating from the steel pipe, thus allowing localized corrosion to occur as CUDC. More details are presented in Chapter 4.

In order to ensure the safe and cost-efficient operation of pipelines through the effective application of protective coatings and CP, corrosion needs to be constantly monitored and managed. Information from corrosion inspection and monitoring would enable pipeline owners to prioritize site survey and inspection operations and to develop maintenance strategies to manage pipeline assets. Information from corrosion inspection and monitoring is also used for predicting long-term remnant pipeline life and providing warning of pipeline failure. Unfortunately the collection of corrosion data from buried pipelines and other hidden structures is a major challenge in engineering practice. Due to the lack of corrosion data, the prediction, modelling, management and prevention of engineering structures in complex environmental conditions remain difficulties in corrosion science and engineering.

1.4 Obstacles in Modeling and Managing Complex Localized Corrosion

1.4.1 Challenges in Managing Localized Corrosion

In engineering practice, localized corrosion is a more critical concern than general corrosion because it can take place without showing any outward sign of damage and the failure can be sudden and possibly disastrous. Localized corrosion induced catastrophic infrastructure failures have been widely reported in the literature. For instances localized corrosion induced pipeline explosion in the eastern Chinese city of Qingdao claimed 62 lives and huge economic losses [25]. It was followed by a similar explosion in Taiwan that caused 32 deaths and 321 injuries [43]. A localized corrosion caused gas pipeline explosion at Varanus Island facility in Western Australia was estimated to cost the state's economy between $A1.8 billion and $A6.7 billion [44]. These are just a few publically reported cases among an enormous amount of infrastructure incidents around the world. In the United States alone, there are approximately 240 000 water main breaks every year and corrosion contributed to many of them [45]. These incidents clearly indicate the extreme consequences of infrastructural failure due to localized corrosion-induced damage of high-risk structural components in an unanticipated manner. The lack of visibility and knowledge of complex and localized corrosion and material degradation processes occurring on these structural components is considered to be a major contributor to these infrastructure accidents [46–50].

Buried and submerged steel pipelines are taken as typical examples of industrial and civil infrastructure here because of their significance and complexity in engineering environments. Pipelines could be hundreds of kilometers long exposed to highly complex environmental conditions. According to a report [15], each square kilometer of a major city could host more than 30 km of buried pipes, creating tangled networks of oil and gas lines, water mains, and electrical and telecommunications cables. There are millions of kilometers of pipelines around the world for transporting various types of liquids and gases such as water, crude and refined petroleum, oil, natural gas, biofuels, and hydrogen over long distances, and other fluids including sewage, slurry, beer, hot water, or steam for relatively shorter distances. High pressure oil and gas pipelines are made from steels with inner diameter typically from 100 to 1220 mm. Most pipelines are typically buried at a depth of about 1–2 m underground. A prerequisite to effectively managing complex forms of corrosion on these buried pipelines is sufficient knowledge of, if corrosion is occurring, where it is occurring, and how fast it is occurring on the pipeline. Such knowledge is essential for engineers to alert risks, to predict asset maintenance needs, to select efficient corrosion control technologies, and to monitor the effectiveness of corrosion control methods and materials such as protective coatings, cathodic protection and corrosion inhibitors. Knowledge of the status of pipeline corrosion is also important for the owners of an engineering asset because it would enable owners to prioritize site survey and inspection operations, and to develop maintenance strategies to manage pipeline assets, to predict long-term remnant pipeline life. For instance accurate determination of the location of critical localized corrosion would assist pipeline owners to plan the dig-up inspections of unpiggable pipelines. It is an expensive operation and each dig-up could cost up to tens of thousands of dollars; therefore it is important from asset management perspective to increase the accuracy of identifying dig-up locations in order to save costs. Unfortunately in terms of dig up location, according to industry experts, a "little speculation/guessing" is often needed due to the lack of reliable knowledge of the location of critical localized corrosion. Reasons for corrosion data unavailability include,

1) Invisibility of localized corrosion: Complex forms of localized corrosion often occur on "hidden" infrastructures. Examples include internal pipeline corrosion such as corrosion under deposits and top of the line corrosion, and external pipeline corrosion such as CUDCs. It has been recognized that there is a serious lack of corrosion and materials degradation data from such "invisible" engineering structures, impeding their timely maintenance and protection [15, 16, 51].
2) Dynamically fluctuating environmental conditions and changing corrosion mechanisms: Traditionally the most common approach to collecting pipeline corrosion data is historical field inspection through excavation and/or inline inspection (ILI) using "pigging" tools. Unfortunately, corrosion data collected using these inspection methods leave some critical gaps since field corrosion inspections are commonly infrequent, expensive, and disruptive, and therefore inspection data often do not have sufficient temporal and spatial resolutions required to accurately detect changing localized corrosion processes and mechanisms [51]. Examples include stray current corrosion of buried pipelines and tidal and splash zone corrosion of marine infrastructures.
3) Multi-time and length-scale corrosion: Complex forms of localized corrosion can occur over multiple time and length scales. Corrosion may start from a very small spot on an underground pipeline, and grow to a major localized corrosion problem. This is the case for CUDCs that usually start from a small coating defect where coating disbondment initiates and grow into a large-scale localized corrosion under coating disbondment. Unfortunately, corrosion data collected using conventional inspection methods often do not have sufficient temporal and spatial resolutions required to detect the early and peak stage localized corrosion that can spread over multiple time and length scales and localized corrosion processes associated with several materials degradation and corrosion mechanisms.

To effectively assess and control localized corrosion under complex and dynamically changing environmental conditions, in situ and site-specific corrosion data with sufficient temporal and spatial resolutions are required. Periodic "pigging" and "digging" inspections and field excavation (dig-up) of selected sites and visual inspection (historical excavations) would provide compressive snapshots of general corrosion occurred after a long period of exposure; however they often do not have sufficient temporal and spatial resolutions required for detecting unexpected sudden changes in localized corrosion rates and provide little information about the underlying causes of localized corrosion [51]. Corrosion data collection techniques and inspection methods for timely and accurate monitoring of multi-scale and dynamically changing complex localized corrosion in practical engineering environmental conditions remains a major difficulty in managing localized corrosion in complex industrial environments. There is a need to develop and apply advanced inspection and monitoring techniques in order to facilitate the acquisition of corrosion data for informing and predicting the occurrence of complex forms of localized corrosion on practical engineering structures.

1.4.2 Issues in Modeling and Predicting Complex Localized Corrosion

Corrosion data are often used in conjunction with predictive models to facilitate the prediction and warning of structural health issues. Predictive modeling is often a major component in structural health monitoring and management systems that usually include three major components: inspection and monitoring sensors or probes for data acquisition; a data transmission, storage, mining system for feature extraction; and a health evaluation system for decision-making based on identified data features. Many corrosion predictive models have been developed over the past decades for various industries. In the pipeline industry, for instance, numerous corrosion models and life prediction tools have been designed to inform decisions such as on the levels of cathodic protection and the frequencies of ILI and field excavation inspections. Existing corrosion predictive models are mostly empirical and semi-empirical models that are designed for predicting general corrosion trends based on modeling overall characteristics of general corrosion [52]. They are usually based on the use of so-called ground truth data and the interpretation of existing inspection data.

A major issue in corrosion modeling is that conventional corrosion models are usually based on "retrospective" field inspection and laboratory testing data. Unfortunately, corrosion inspection data often have insufficient temporal resolution required to model corrosion processes affected by dynamic changes in environmental conditions, materials behavior, and corrosion mechanism. Corrosion inspection data often provide little spatial information about corrosion patterns and the underlying localized corrosion mechanisms [51]. Despite of these limitations, historical inspection data are still widely utilized as the only available database for the development and verification of computational probabilistic models in asset management tools. For instance various methods including fuzzy-based models, fuzzy artificial neural network models, Monte Carlo simulation technique, and Euclidean-Support Vector Machines classification approach were used to predict the failure types and rates of oil pipelines using historical pipeline accidents data [51]. These asset management tools are useful in providing an overall and general assessment of the aging of a structure. They could also be useful to predict corrosion at certain locations if historical inspection data have sufficient spatial and temporal resolutions. Unfortunately data with sufficient spatial and temporal resolutions are often unavailable; therefore these models are generally unable to predict the failure of infrastructural systems that are under localized corrosion attack and that are under the effect of dynamically changing environmental conditions.

Most existing models described above could only predict general corrosion rather than localized corrosion. The modeling of localized corrosion in complex engineering environments is much more difficult. A reason is that under industrial environmental conditions, there are too many inter-related variables affecting localized corrosion processes and these variables can change dynamically with time. Materials behavior and corrosion mechanism could also change with time and with environmental conditions. For instance, as shown in Figure 1.15, a usually corrosion resistant low-alloy steel gear face could suffer pitting due to unexpected environmental variations. A highly corrosion-resistant aluminum alloy ship could suffer pitting corrosion if the boat stays in stagnant seawater for a long time due to the lack of sufficient oxygen required for maintaining passivity. It is well-known that passive metals such as stainless steels and aluminum alloys have better resistance to localized corrosion in flowing seawater; therefore the higher the seawater velocity, the more sufficient the oxygen and the less the chance of pitting corrosion to occur on stainless steel and aluminum ships. Conventional corrosion models unable to adequately consider all these interrelated variables affecting localized corrosion processes.

Take buried pipeline for example, Alamilla et al. [53] proposed a statistical mathematical model to make approximate predictions of localized corrosion damage on uncoated metals exposed to soil environments. Although this model was reported to provide predictions that appeared to correlate with experimental and field corrosion data [27, 53], it is only a statistical model for the approximate prediction of maintenance needs of buried structures. The model only considers four variables, soil resistivity, soil pH value, the redox potential and the electric potential, and not includes other soil variables such as soil texture, humidity, bacterial impact, temperature, the chemical concentration, and composition of moisture since these parameters often change over time and vary within the soil profile [53]. For most existing pipeline corrosion models like this one, the difficulty is not only due to the hurdle of including all soil environmental variables in modeling but also due to changes in corrosion mechanisms with time and with location.

It is well recognized that corrosion data especially those related to complex forms of localized corrosion are critical for modeling and forecasting complex multi-scale localized corrosion processes occurring in industry [15, 16, 51]. In an extensive review on modeling of localized corrosion, it is pointed out that existing models lack the resolution to design effective corrosion-resistant systems and that there are gaps in the experimental characterization of the complex corrosion system, particularly at the lower end of the scales [54]. At present there does not appear to have one generally agreed method for predicting localized corrosion on buried gas pipelines, most likely due to the lack of timely and site-specific localized corrosion data.

The need for multiscale and *in-situ* corrosion data is evident if we look into some of the models that were developed for predicting corrosion of critical marine structures [52] and water pipelines [29]. A bi-modal corrosion model was developed for predicting long-term infrastructural corrosion in marine environments [52, 55]. It was verified using long-term field exposure and inspection data that are of relatively low time resolution insufficient to fully explain the underlining corrosion mechanisms responsible for the bi-modal behavior. Similarly a time-dependent multiscale numerical model that was developed to simulate corrosion evolution from microcells to macrocells with time on buried cast iron water pipelines [29] was validated using historical corrosion data from United States National Bureau of Standards cast iron corrosion exposure studies conducted between 1922 and 1940. Such long-term exposure data are "retrospective" information that usually would not have sufficient spatial and temporal resolutions to fully verify dynamic corrosion mechanism changes such as microcell corrosion to macrocell corrosion transition predicted by the model [29].

A possibility of improving these models is to include more environmental variables and corrosion mechanisms in modeling. This, of course, would require more *in situ* and site-specific corrosion data. If such data become available, some of the models described above could be applied to individual section of a pipeline over a period of time of interest. This could facilitate the modeling of localized corrosion at specific sites over a specific time period, providing local and vigorous corrosion prediction. This could also provide direct evidence to fully verify and improve these localized corrosion models for more accurate prediction of localized corrosion events. Modern data analytics and artificial intelligence technologies could also be used to facilitate this. Unfortunately obtaining such *in situ* and site-specific localized corrosion data to enable effective localized corrosion prediction is a major difficulty in corrosion science and engineering, as explained in references [15, 16, 51], because of fundamental technological difficulties in probing dynamic localized corrosion processes [51, 56].

References

1 Frankel, G.S. and Sridhar, N. (2008). Understanding localized corrosion. *Materials Today* 11 (10): 38–44.

2 Evans, U.R. (1981). *An Introduction to Metallic Corrosion*, 3e. London: Edward Arnold Press.

3 Fontana, G. (1987). *Corrosion Engineering*, 3e. McGraw-Hill Book Company.

4 Pourbaix, M. (1973). *Lectures on Electrochemical Corrosion*. New York: Plenum Press.

5 Tomashov, N.D. (1966). *The Theory of Corrosion and Protection of Metals*. New York: The Macmillan Company.

6 Frankel, G.S. (1998). Pitting corrosion of metals – a review of the critical factors. *Journal of the Electrochemical Society* 145 (6): 2186–2198.

7 Macdonald, D.D. (1992). The point defect model for the passive state. *Journal of the Electrochemical Society* 139 (12): 3434–3449.

8 Marcus, P., Maurice, V., and Strehblow, H.H. (2008). Localized corrosion (pitting): a model of passivity breakdown including the role of the oxide layer nanostructure. *Corrosion Science* 50 (9): 2698–2704.

9 Soltis, J. (2015). Passivity breakdown, pit initiation and propagation of pits in metallic materials – review. *Corrosion Science* 90: 5–22.

10 Jones, D.A. (2005). *Prevention and Prevention of Corrosion*, 2e. Hoboken, NJ, USA: Prentice Hall.

11 Newman, R.C. (2001). Understanding the corrosion of stainless steel. *Corrosion* 57 (12): 1030–1041.

12 Ryan, M.P., Williams, D.E., Chater, R.J. et al. (2002). Why stainless steel corrodes. *Nature* 415 (6873): 770–774.

13 Tan, Y.J. (2012). *Heterogeneous Electrode Processes and Localized Corrosion*. Wiley.

14 Trabanelli, G. (1991). Inhibitors – an old remedy for a new challenge. *Corrosion* 47 (6): 410–419.

15 Li, X., Zhang, D., Liu, Z. et al. (2015). Materials science: share corrosion data. *Nature* 527: 441–442.

16 Cheng, F. (2016). Monitor safety of aged fuel pipelines. *Nature* 529: 156.

17 Tan, M.Y.J., Huo, Y., Varela, F. et al. (2021).Corrosion control technology needs for future fuel infrastructure. *Australasian Corrosion Association (ACA) conference: Corrosion & Prevention*, Newcastle, Australia (5–8 June 2022).

18 Tan, Y.J. (1998). Monitoring localized corrosion processes and estimating localized corrosion rates using a wire-beam electrode. *Corrosion* 54: 403.

19 Mansfeld, F. (2006). Classic paper in corrosion science and engineering with a perspective by F. Mansfeld. *Corrosion* 62: 843.

20 The Straits Times (2003). Boy hit on head by lamp post dies. (Wednesday, 7 March).

21 Wikipedia (2008). Western Australian gas crisis. https://en.wikipedia.org/wiki/2008_Western_Australian_gas_crisis (accessed 14 December 2021).

22 Broder, J.M. (2011). BP Is Fined $25 Million for '06 Spills at Pipelines. https://www.nytimes.com/2011/05/04/science/earth/04pipeline.html (accessed 14 December 2021).

23 The Guardian (2015). California oil spill pipeline had been left to rust paper-thin. https://www.theguardian.com/environment/2015/jun/04/us-oil-pipeline-left-to-rust-to-paper-thin-before-145km-pacific-ocean-slick (accessed 14 December 2021).

24 Wikipedia (2014). Kaohsiung gas explosions. https://en.wikipedia.org/wiki/2014_Kaohsiung_gas_explosions (accessed 14 December 2021).

25 Wikipedia (2013). Qingdao oil pipeline explosion. https://en.wikipedia.org/wiki/2013_Qingdao_oil_pipeline_explosion (accessed 14 December 2021)

26 Jiang, J., Wang, J., Wang, W., and Zhang, W. (2009). Modeling influence of gas/liquid/solid three-phase boundary zone on cathodic process of soil corrosion. *Electrochimica Acta* 54: 3623–3629.

27 Cole, I.S. and Marney, D. (2012). The science of pipe corrosion: a review of the literature on the corrosion of ferrous metals in soils. *Corrosion Science* 56: 5–16.

28 Levlin, E. (1996). Aeration cell corrosion of carbon steel in soil: in situ monitoring cell current and potential. *Corrosion Science* 38: 2083–2090.

29 Azoor, R., Deo, R., Birbilis, N., and Kodikara, J. (2018). A coupled electro-chemical-soil model to evaluate the influence of soil aeration on underground metal pipe corrosion. *Corrosion*.

30 Song, G.-L., Cao, C.-N., and Lin, H.-C. (1995). A new technique for correlation of underground corrosion. *Corrosion* 51: 491–497.

31 Murray, J. and Moran, P. (1989). Influence of moisture on corrosion of pipeline steel in soils using in situ impedance spectroscopy. *Corrosion* 45: 34–43.

32 Gupta, S. and Gupta, B. (1979). The critical soil moisture content in the underground corrosion of mild steel. *Corrosion Science* 19: 171–178.

33 Fu, J., Pei, F., Zhu, Z. et al. (2013). Influence of moisture on corrosion behavior of steel ground rods in mildly desertified soil. *Anti Corrosion Methods and Materials* 60: 148–152.

34 Hendi, R., Saifi, H., Belmokre, K. et al. (2018). Effect of black clay soil moisture on the electrochemical behavior of API X70 pipeline steel. *Materials Research Express* 5: 036523.

35 Yan, M., Sun, C., Xu, J. et al. (2014). Role of Fe oxides in corrosion of pipeline steel in a red clay soil. *Corrosion Science* 80: 309–317.

36 Schwerdtfeger, W. (1966). Soil resistivity as related to underground corrosion and cathodic protection. Highway Research Record.

37 Arzola, S., Palomar-Pardave, M., and Genesca, J. (2003). Effect of resistivity on the corrosion mechanism of mild steel in sodium sulfate solutions. *Journal of Applied Electrochemistry* 33: 1233–1237.

38 Mehanna, M., Basséguy, R., Délia, M.-L., and Bergel, A. (2009). Effect of Geobacter sulfurreducens on the microbial corrosion of mild steel, ferritic and austenitic stainless steels. *Corrosion Science* 51: 2596–2604.

39 Romanoff, M. (1957). *Underground Corrosion*, National Bureau of Standards Circular 579, vol. 25. Washington: US Government Printing Office.

40 Fitzgerald, J. (1993). Evaluating soil corrosivity – then and now. *Materials Performance; (United States)* 32.

41 Varela, F., Tan, M., and Forsyth, M. (2015). Understanding the effectiveness of cathodic protection under disbonded coatings. *Electrochimica Acta* 186: 377–390.

42 Varela, F., Tan, M.Y., and Forsyth, M. (2015). Electrochemical method for studying localized corrosion beneath disbonded coatings under cathodic protection. *Journal of the Electrochemical Society* 162: C515–C527.

43 Hsu, J.W. and Liu, F. (2014). Taiwan Gas Blasts Likely Caused By Faulty Pipe. http://online.wsj.com/articles/taiwan-gas-blasts-likely-caused-by-faulty-pipe-1406964902 (accessed 31 August 2017).

44 Gosch, E. (2008). WA gas supply cut 30pc by blast. *The Australian* (7 June).

45 National Institute of Standards and Technology (2009). Technology Innovation Program report. http://www.nist.gov/tip/prev_competitions/upload/ci_wp_031909.pdf (accessed 22 December 2021).

46 Housner, G.W., Bergman, L.A., Caughey, T.K. et al. (1997). Structural control: past, present, and future. *Journal of Engineering Mechanics* 123: 897–971.

47 Li, H.-N., Li, D.-S., and Song, G.-B. (2004). Recent applications of fiber optic sensors to health monitoring in civil engineering. *Engineering Structures* 26: 1647–1657.

48 Chang, P.C., Flatau, A., and Liu, S.C. (2003). Review paper: health monitoring of civil infrastructure. *Structural Health Monitoring* 2: 257–267. https://doi.org/10.1177/1475921703036169.

49 Ko, J.M. and Ni, Y.Q. (2005). Technology developments in structural health monitoring of large-scale bridges. *Engineering Structures* 27: 1715–1725.

50 Brownjohn, J.M.W. (2007). Structural health monitoring of civil infrastructure. *Philosophical Transactions of the Royal Society A* 365: 589–622.

51 Tan, M.Y.J. (2020). A critical overview of monitoring infrastructural health using corrosion probes. *Corrosion Engineering, Science and Technology* 55: 103–117. https://doi.org/10.1080/1478422X.2019.1695390.

52 Melchers, R.E. (2015). Bi-modal trends in the long-term corrosion of copper and high copper alloys. *Corrosion Science* 95: 51–61. 81(2014): 110–116.

53 Alamilla, J.L., Espinosa-Medina, M.A., and Sosa, E. (2009). Modelling steel corrosion damage in soil environment. *Corrosion Science* 51 (11): 2628–2638, ISSN 0010-938X.

54 Gunasegaram, D.R., Venkatraman, M.S., and Cole, I.S. (2014). Towards multiscale modelling of localized corrosion. *International Materials Reviews* 59: 84–114.

55 Melchers, R.E. (2018). A review of trends for corrosion loss and pit depth in longer-term exposures. *Corrosion and Materials Degradation* 1: 42.

56 Tan, Y., Ubhayaratne, K.N.I., Huo, Y. et al. (2019) Predictive maintenance based on smart monitoring and data analytics. *Australasian Corrosion Association (ACA) conference: Corrosion & Prevention*, Melbourne, Australia (24–27 November 2019).

2

Techniques for Localized Corrosion Inspection and Monitoring

Localized corrosion complexity in practical engineering structures especially those exposed to heterogeneous and variable environments makes the prediction and control of industrial structural failure a significant challenge because localized corrosion can take place without showing any outward sign of damage and can cause sudden and disastrous infrastructural failure incidents [1–4]. The lack of visibility of localized corrosion and material degradation processes occurring on "hidden" engineering structures has been considered to be a major contributor to many infrastructure accidents [5–9]. In current industrial practice, risk-based and time-based inspection and fitness-for-service assessments are common methods for the integrity management of engineering structures. This chapter provides an overview of corrosion inspection and monitoring technologies and their limitations, as well as new approaches to *in situ* and site-specific corrosion data acquisition and monitoring. Aged underground pipelines are selected as typical examples of major engineering structures exposed to complex environmental conditions.

2.1 Techniques for Corrosion Detection, Inspection, and Data Acquisition

2.1.1 Issues in Detecting and Predicting Complex Localized Corrosion

Buried and submerged steel pipelines are typical examples of industrial and civil infrastructures that are exposed to complex engineering environmental conditions. There are millions of kilometers of pipelines around the world for transporting various liquids and gases such as water, crude and refined petroleum, oil, natural gas, biofuels and hydrogen over long distances, and other fluids including

Localized Corrosion in Complex Environments, First Edition. Mike Yongjun Tan.

sewage, slurry, beer, hot water, or steam for relatively shorter distances. According to a report [10], each square kilometer of a major city could host more than 30 km of buried pipes, creating tangled networks of oil and gas lines, water mains, and electrical and telecommunications cables. Most pipelines are typically buried at a depth of about 1–2 m underground. Complex forms of localized corrosion often occur on these "hidden" pipelines that are exposed to industrial environments.

A prerequisite to effectively managing complex forms of corrosion on buried pipelines is the availability of sufficient corrosion data that provide information about, if corrosion is occurring, where it is occurring, and how fast it is occurring. Such corrosion data are essential for engineers to alert risks, to predict asset maintenance needs, to select efficient corrosion control technologies, and to monitor the effectiveness of corrosion control methods and materials such as protective coatings, cathodic protection, and corrosion inhibitors. Knowledge of the status of pipeline corrosion is also important for the owners of an engineering asset because it would enable owners to prioritize site survey and inspection operations, and to develop maintenance strategies to manage pipeline assets, to predict long-term remnant pipeline life. For instance accurate determination of the location of critical localized corrosion would assist pipeline owners to plan the dig-up inspections of unpiggable pipelines, which is an expensive operation and it is important from asset management perspective to increase the accuracy of identifying dig-up locations.

Corrosion data are also essential for predictive modeling to facilitate the prediction and warning of corrosion and other structural health issues [11–17]. Predictive modeling is often a major component in structural health monitoring and management systems. Many corrosion predictive models have been developed over the past decades for various industries. In the pipeline industry, for instance, numerous corrosion models and life prediction tools have been designed to inform decisions such as on the levels of cathodic protection and the frequencies of inline inspection and field excavation inspections. Existing pipeline corrosion predictive models are mostly empirical and semi-empirical models that are designed for predicting general corrosion trends based on modeling overall characteristics of general or uniform corrosion [11]. They are usually based on the use of so-called ground truth data and the interpretation of existing inspection data. A major issue in corrosion modeling is that conventional corrosion models are usually based on "retrospective" field inspection and laboratory testing data. Unfortunately, corrosion inspection data often have insufficient temporal resolution required to model corrosion processes affected by dynamic changes in environmental conditions, materials behavior, and corrosion mechanism.

It has been recognized that there is a serious lack of corrosion and materials degradation data and information on major "invisible" engineering structures, impeding their timely maintenance and protection [10, 18, 19]. In current industrial practice, routine inspection is the most widely used method for identifying localized corrosion on engineering structures. Risk-based and time-based inspection and fitness-for-service assessments are common methods for the integrity management of engineering structures. These approaches may not be sufficient for detecting, predicting, and managing complex localized corrosion under dynamic changes in corrosion environments, mechanisms and kinetics. There is a need for more convenient, reliable, and economic asset inspection and management technologies.

2.1.2 Conventional Techniques and Tools for Acquiring Corrosion Data

Visual inspection is obviously the simplest and most direct method for inspecting and examining corrosion damage. For a large engineering structure that is hidden from visual observation such as a long buried pipeline, a large submerged oil platform, a petrochemical facility covered by insulation, a practical issue is to know when and where inspections should be conducted. After that, engineers need to remove the covering or soil over a structure in order to inspect the outer areas of the structure such as a buried pipe for defects. Take buried gas pipeline as an example, visual inspection of exposed or dig-up structures is conventionally used to visually inspect exposed surfaces or internal of a pipeline. However dig-up inspection is an expensive operation that each dig-up could cost up to tens of thousands of dollars; therefore, it is important to increase the accuracy of identifying dig-up locations in order to save costs. Naturally, industries prefer technologies that enable inspection without necessitating the digging up soil or removal of the insulation coverings. Therefore a major need for managing hidden engineering structures such as a pipeline is to visualize the buried, hidden, or internal space of the structure. Various condition assessment technologies and in-line inspection tools have been developed and are used in industry for pipeline inspection and corrosion data acquisition [19–22].

Inline inspection (ILI) is the most valuable and industry standard approach to identifying pipeline anomalies for dig-up inspection and repair. ILI tools (also known as smart pigs) are cylindrical nondestructive testing (NDT) tools that are inserted into the pipe to inspect it while being carried by the fluid flow. There is a wide variety of ILI tools that can be specialized in the detection of metal loss, cracks, geometry deformations, leaks, wax deposition, and other [22–25]. NDT inspection devices are commonly used ILI tools to scan the pipe wall either from inside or outside of the pipe in order to detect localized corrosion damages and other types of damages on internal and external pipe walls. Many different NDT technologies and tools are used in ILI tools for pipeline inspection, including magnetic flux leakage (MFL), ultrasonic testing (UT), radiographic testing, pulsed eddy currents, and electromagnetic acoustic transducers (EMAT) [22, 25]. Visual inspection through dig-ups are usually targeted at anomalies identified by ILI to perform repairs and prevent further corrosion from happening. Figure 2.1 illustrates pigging of a buried gas pipeline that is conventionally protected by a coating and cathodic protection system.

Magnetic flux leakages (MFLs)-based pipeline inspection tools are among the most commonly used techniques for detecting pipeline wall damages and cracks using very strong magnets [22, 25–27]. MFL detects metal losses on the pipe wall by detecting and analyzing changes in magnetic flux leakage using Hall effect sensors [22]. A limitation is that MFL tools do not directly measure damages on pipelines, instead they only estimate damages on pipeline based on measured magnetic flux changes [26, 28]. Ultrasonic transducers (UTs)-based metal thickness loss measurement tools use piezoelectric transducers to generate ultrasonic pulses and calculate internal as well as external pipe metal losses based on the time between the reflexion of the signals produced at each interface [24]. UT tools are considered to be effective, highly accurate, versatile and widely used inspection tools in many industries [23, 29]. UT tools are considerred to have better detection capabilities than MFL tools and produce signals that can be directly related to defects sizes and quantity [30]. However this technology usually needs a homogeneous liquid coupling agent between the transducers and the pipe, which seriously limits its application on gas pipelines. This limitation was overcome by using a train of pigs to isolate and carry a batch of liquid together with the conventional UT tool [23, 27] and by using a coupling agent-free UT tool denominated "elastic wave" [27]. Pulsed eddy current (PEC) uses electromagnetic waves to determine the pipe wall thickness of any material that conducts electricity, which is similar to a traditional MFL tools both physically as well as operationally [31].

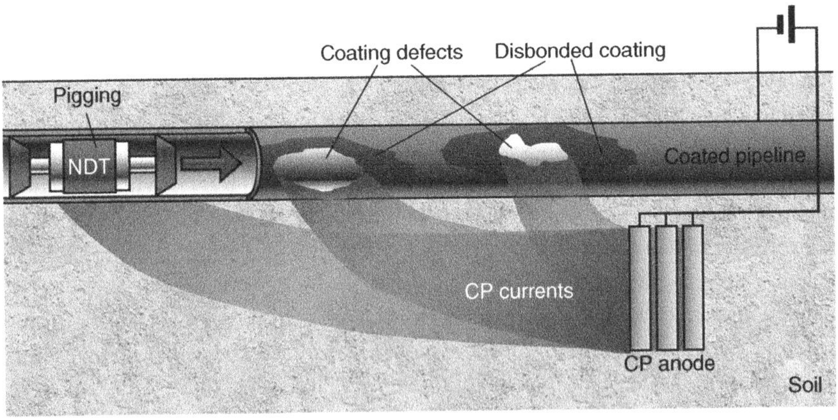

Figure 2.1 Illustrating ILI pigging inspection of a buried gas pipeline.

This technology may not be able to accurately determine localized corrosion such as pits because it only provides an average pipe wall thickness, it is generally only used as a screening technique and not for detecting isolated corrosion damages. Further developments have been made to these NDT technologies, for instance electromagnetic acoustic transducers (EMAT) is a recently developed ultrasonic technology that generates and receives ultrasonic waves without requiring a liquid coupling agent [22]. EMAT is well accepted to be used with other ILI tools for inspecting piggable pipeline, as crack and coating disbondment detection tool for gas pipelines [32–34].

Although NDT tools in the form of pigging tools or probes described above are widely used for inspecting localized corrosion on pipelines, they are not always applicable for some structures. A typical example is unpiggable pipelines that have obvious barriers for applying pigging tools. These pigging barriers include (i) there are physical barriers inside the pipe, therefore a free swimming pigging tool cannot be introduced to or removed from the pipeline; (ii) there is insufficient flow to overcome friction and drive a pig; (iii) there are multi-diameter pipes causing high-velocity excursions making measurements difficult. There are also pipelines that cannot tolerate the interruption of operation by the pigging procedures. Up to the present time, there is no single tool or technology that could meet all requirements for reliably inspecting unpiggable pipelines. Since it is not possible to perform direct visual or NDT inspection of unpiggable pipelines, engineering approaches such as 'direct assessment' (including external corrosion direct assessment (ECDA) and internal corrosion direct assessment (ICDA)) is used as the most realistic approach to assessing the condition of unpiggable pipelines. The accuracy and reliability of ECDA and ICDA relies upon the availability of data that could be acquired from indirect means through the implementation and the combined use of some indirect inspection tools, corrosion monitoring probes, and data analytics techniques. The most commonly used indirect corrosion inspection or monitoring tools include various electrical potential survey techniques, such as Close Interval Potential Surveys (CIPS), Direct Current Voltage Gradient (DCVG), and corrosion probes.

Electrical potential survey techniques are the most commonly used indirect inspection tools for obtaining corrosion and materials degradation-related data from buried unpiggable pipelines. Figure 2.2 shows a comparison of several common potential survey methods and their application for inspecting unpiggable gas pipelines. Although these methods do not measure corrosion rates nor metal losses, they provide a way to locate coating defects and/or assess the effectiveness of the cathodic protection (CP) along a pipeline. Figure 2.2b shows electrical potential monitoring using IR coupons that are steel coupons buried next to the pipe and electrically connected to the pipe in order to assess the operation of a CP system by monitoring the cathodic polarization achieved at the test locations.

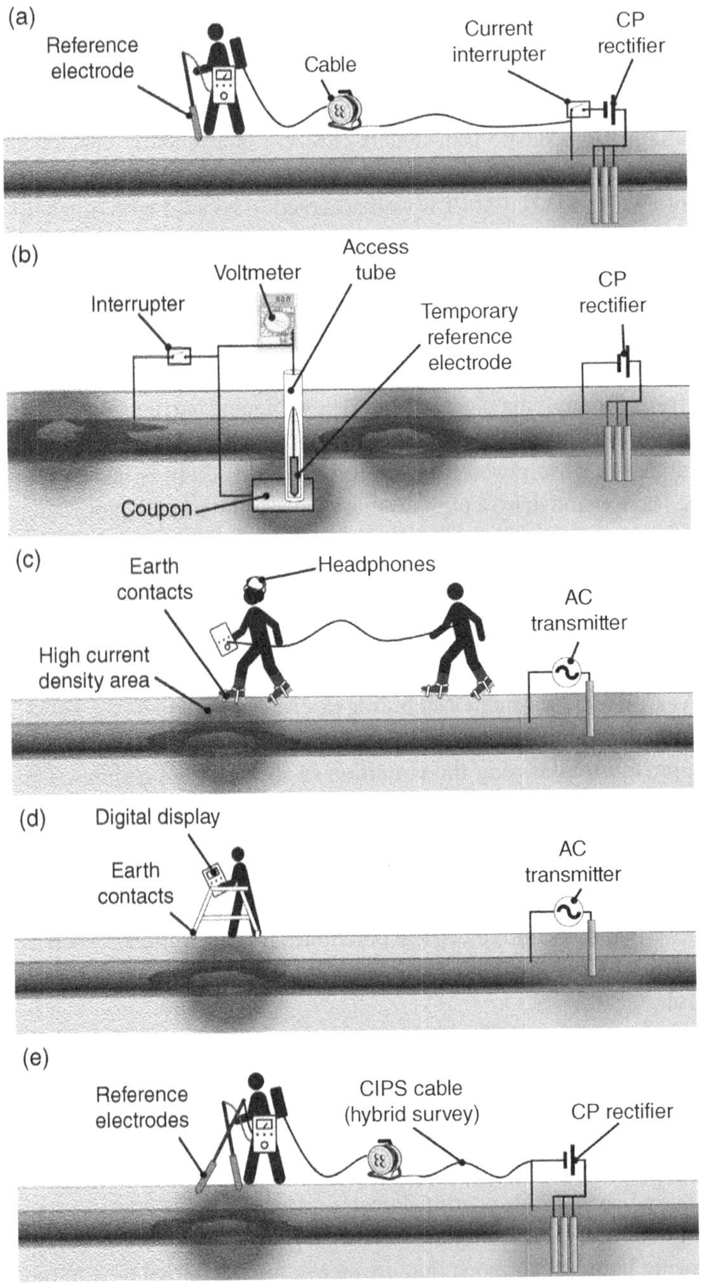

Figure 2.2 Potential survey methods for inspecting unpiggable underground pipeline. (a) CIPS, (b) IR coupons, (c) Pearson, (d) modern ACVG and (e) DCVG. *Source:* Varela et al. [22] / With permission of Taylor & Francis.

They offer a way to minimize IR drops for measuring IR-drop-free potential between the coupons and a reference electrode placed next to the coupon [35, 36]. The CIPS is a family of survey techniques that assess the level of CP through the measurement of the potential difference between a pipeline and reference electrode in contact with the soil at regular intervals (typically 1 m along the pipeline). As illustrated in Figure 2.2a, a trailing cable is used to connect the reference electrode and a high impedance voltmeter to the nearest pipeline testing point. The main purpose is to evaluate the fulfilment of the 100 mV of minimum polarization required by many CP standards like the NACE SP0169-2007 [37–39]. They can also help to locate big defects in the coating, electrical insulation defects, stray current pick up, and discharge areas, as well as undesired electrical contact with other metallic structures because these can cause unusual potential changes [22, 37]. Figure 2.2e illustrates DCVG that is another potential survey method designed to locate defects in the pipeline coating by measuring voltage gradients (IR drop gradients) using two reference electrodes in contact with the soil at a constant distance (typically 1 m) [40, 41]. While this method is a very effective tool to locate coating defects, there is no general relationship between the potentials measured and the defect size because the IR drop gradients measured at each area is not only affected by coating defect sizes, but also affected by many other factors such as soil resistivity. Another major limitation of the method is that it cannot detect disbonded coatings and corrosion under disbonded coatings because of the shielding of current by the coating film [22]. Other relatively less commonly used coating defect survey methods include alternating current voltage gradient (ACVG), as illustrated in Figure 2.2d, and the Pearson survey, as illustrated in Figure 2.2c. More details on these methods can be found in references [22, 42].

Although data from electrical potential survey have a major limitation that they do not indicate corrosion damage on pipelines, these survey data can still be used as useful inputs for direct assessment of unpiggable pipelines. For instance, it is mentioned by engineers that DCVG is a good indicator not to dig-up a section of pipeline because if DCVG data show little coating defects over the section of pipeline, little corrosion is expected on the section of pipeline because good coating condition usually suggests little corrosion risk. It has also been reported that the relative comparison of the DCVG data obtained from different sections of the same pipeline under similar conditions can be useful in establishing a priority order for the maintenance of the pipeline because worse DCVG behaviour likely indicates poorer pipeline conditions. Furthermore, similar DCVG data obtained from different pipelines made of the same materials and exposed to similar conditions could suggest similar conditions for both pipelines. All these information can be useful inputs for direct assessment (ECDA and ICDA) of unpiggable pipelines.

2.1.3 Pipeline Industry Concerns About Integrity Assessment

A survey of the pipeline industry has been conducted to understand current issues in the detection and management of pipeline damages and integrity, as well as technological gaps in detecting localized forms of corrosion [20]. Not surprisingly, ILI pigging survey was widely recognized as the most appreciated approach to detecting pipeline anomalies (dents, gouges, cracking, etc.) and collecting corrosion data from piggable pipelines. Compared to various other corrosion inspection and monitoring technologies such as field data logger, coating survey, dig-up survey, and monitoring sensors, ILI pigging tools have advantages in providing certainty of the pipe wall condition and in assuring operators of the pipeline's safe pressure-containing ability. Dig-up operations are usually targeted at corrosion and other anomalies identified by ILI to verify the ILI results and to perform repairs and prevent further corrosion. The limitations of current ILI technologies are reported to be around the determination accuracy of the geometry of corrosion, in identifying exactly where the dig-up location should be, and in the high costs and low frequency of inspection (e.g. once every 10–15 years for gas pipelines). Another approach to assessing pipeline integrity is hydrostatic testing that is to isolate sections of a pipeline and pressurize them above the maximum allowable operating pressure using fluids such as water and monitor any pressure drop attributable to leaks. This test is useful to assure pipeline fitness for service; however, it provides no information regarding the presence and location of subcritical defects such as localized corrosion. It involves large operational costs; therefore, it is only used as a measure of last resort for existing pipelines [21].

All other asset condition monitoring and inspection tools other than ILI tools, such as CIPS, DCVG, and corrosion probes, are generally only used by the pipeline industry for unpiggable pipelines. The inspection, monitoring, and protection of "unpiggable" pipelines have been identified as prime technological and knowledge gaps and a major challenge in the management of fuel infrastructural networks [20]. In particular, aged pipelines affected by corrosion under disbonded coatings (CUDC) have been identified as prime "high-risk" pipelines responsible for some 90% of gas pipeline corrosion damages. Due to shielding of a pipe by disbonded coatings and heat shrink sleeve (behaving like shielding coating) or by rocks and concrete cases, localized corrosion can occur unnoticed and cannot be detected through normal electrical potential survey or be mitigated by normal CP [20]. Another concern in the industry is about the detection of corrosion and coating damage on "high risk" pipes constructed by horizontal directional drilling (HDD) because these pipes often have suffered coating damages and are subjected to CP shielding. In addition, with the introduction of new fuels especially hydrogen into steel pipelines, issues such as the detection, monitoring, and prevention of hydrogen-induced pipeline integrity issues such as hydrogen embrittlement have been identified as new knowledge gaps and challenge for future fuel infrastructural networks.

Corrosion prediction models are believed to be useful tools for the integrity management of aged pipelines; however, it is still a challenge to cultivate data-driven corrosion prediction models for assisting the prognostic and health management of unpiggable pipelines. The digitalization, standardization, and usage of asset integrity data have been identified as weaknesses in asset data management and analytics. Opportunities exist in developing practical data-driven corrosion prediction models; however, *in situ* and site-specific corrosion data are needed in order to make it a reality because the localized nature of pipeline corrosion. New sensors and inspection tools, advanced information technology and artificial intelligence (AI) technologies, and smart asset repair methods might be required in order to provide solutions to these issues and to provide opportunities for future "predictive maintenance" of aging and complex future fuel infrastructures [20].

2.1.4 Progress in Techniques for Unpiggable Pipelines Corrosion Assessment

The inspection, monitoring, and protection of "unpiggable" pipelines have been identified as critical needs in the management of fuel infrastructural networks [20]. Unfortunately up to the present time, there is no single tool or technology that could meet all requirements for reliably inspecting unpiggable pipelines. An approach currently used by some pipe operators is to use ILI results of piggable pipelines to estimate the condition of unpiggable pipelines with the same coating, similar age, similar surrounding environmental conditions, and comparable DCVG survey data. This approach to judging localized corrosion damage on large pipelines often likes playing "the guessing game" because practical pipelines are all different and their conditions may not be directly comparable, especially at local "high risk" areas with stray currents and disbonded coatings. To improve the accuracy and confidence of pipeline corrosion assessment, there is a need for development of more systematic and reliable condition assessment methods and tools.

A more systematic approach to assessing the condition of unpiggable pipelines is direct assessment, including ECDA for external pipeline corrosion assessment and ICDA for internal pipeline corrosion assessment. For the buried gas pipeline where external pipeline corrosion is the main concern, ECDA is the widely used process for accessing pipeline integrity. ECDA is a risk based assessment process that reviews pipeline integrity-related information from all sources including historical design, construction and maintenance data, CP performance and coating survey data, soil conductivity, moisture and composition data, as well as ILI data from other pipes with similar conditions (if available) [27]. When conducting a direct assessment, four steps to the inspection are followed: a pre-assessment and indirect inspection, and a direct examination and a post-assessment. In the pre-assessment phase of the inspection, all pipeline data, historic and current, are collected in order to determine whether direct assessment is feasible, define direct assessment regions, select indirect inspection tools, and determine if additional integrity data is needed. The second step of the inspection involves the use of unintrusive and aboveground techniques to assess the buried pipelines. In the direct examination phase, the data collected from the pre-assessment and indirect inspection phases are analyzed in order to select sites for excavation and examination of the pipe surface. Finally, in the post-assessment phase, integrity data collected in the previous three phases is analyzed and integrated to assess the effectiveness of the direct assessment process and determine reassessment intervals. Although ECDA is applicable to any pipeline, the method is of particular value for unpiggable pipelines to identify the most corrosion susceptible areas along a pipeline and provide a rational selection criteria for dig-up programs. Based on this information, a probabilistic model could be developed and used to calculate the highest corrosion risk locations. These locations are then excavated, carefully examined using tools such as UT probes and 3D scanners, and if necessary, repaired. Finally, the data collected from the direct inspection activities are feedback to validate the adopted probabilistic model [43, 44]. A detailed description of the ECDA method can be found elsewhere [43, 44].

The accuracy of risk assessment and the efficiency of risk management based on ECDA and ICDA are dependent upon the availability and quality of inspection data. For instance, ECDA integrates data from various methods of inspection such as DCVG, ACVG, Pearson method, potential survey, and CIPS, providing valuable information about the structural integrity of a pipeline and estimating the most external corrosion susceptible areas along the pipeline, the accuracy of direct assessment is obviously dependent upon the reliability of data used for analysis. Some unreliable predictions from ECDA were reported in industry [20]. Currently, the most widely used methods of acquiring corrosion data from unpiggable pipelines include (i) the monitoring of CP conditions by the close interval potential surveys (CIPS) or IR coupons; (ii) the detection of coating defects by direct current voltage gradient (DCVG) surveys; (iii) the detection of metal loss using in-line inspection tools (intelligent pigs); and (iv) historical excavations. It is important to note that only inline inspection and excavations are capable of detecting the pipeline corrosion metal loss. Although these tools produce vital information for prioritizing areas to be repaired, they provide little information regarding the processes and causes of damage such as corrosion. Potential surveys such as CIPS and DCVG can provide valuable information on the effectiveness of the CP system and coating integrity, respectively; however they provide little corrosion information because the coating and CP status are not directly related to corrosion rates or patterns. Currently, there is a lack of technological capability for pipeline engineers to monitor and capture pipeline corrosion processes on a continuous basis and to perform pro-active warning and maintenance on localized corrosion, coating disbondment and degradation, and CP excursion-induced pipeline failures.

The accuracy and reliability of ECDA and ICDA could be increased through the implementation and the combined use of new inspection tools, in-situ corrosion monitoring probes, and data analytics techniques. Over the past decade, significant effort has been made to develop new "pigging-like" ILI tools for inspecting traditionally unpiggable pipelines. Various "pigging-like" tools or devices have been designed and tested over the past decade. These "pigging-like tools" are mostly developed based on innovative application of traditional NDT tools through the making of custom-designed and more flexible devices for detecting corrosion metal loss, cracks, geometry deformations, leaks, and wax deposition in traditional unpiggable pipelines. A typical example of such "pigging-like" ILI tool is an MFL-based tool called ultrahigh resolution MFL caliper that was developed to inspect pipe wall thickness, corrosion metal loss, dents, pitting, welds, internal deposit, bends, deformation, coating defects, and pinhole. MFL tool is also used with other complementary NDT technologies such as eddy currents, EMAT, and UT. Details about the tool can be found in PHMSA report funded by US Department of Transport [45].

Ultrasonics is the most common external corrosion detection technology used in the design of "pigging-like" ILI tools [27]. These include pipe crawlers or tethered tools that are operated in uncircular valves and tight bend pipes. Pipeline crawlers are self-propelled ILI tools capable of inspecting pipes with uncircular valves and tight bends. An example is a robotic platform developed by PHMSA that allows the inspection of presently unpiggable gas transmission pipelines [46]. The platform is supposed to be able to propel itself independently of flow conditions and be able to go round all obstacles encountered in a pipeline, such as bends and plug valves. The platform is equipped with a segmented remote field eddy current sensor, leading to the commercial deployment of the first-ever inspection platform and integrated sensor capable of internal unpiggable gas pipeline inspection [46]. Other innovative design of pigging-like ILI tools include a "Smart Ball" [47], a free-flowing tool used to detect leaks and gas pockets in liquid or high-pressure gas pipelines, and the "INGU Piper" [48, 49] that is a free-swimming, golf-ball-sized tool used to detect leaks, bends, flanges, and diameter changes. The "INGU Piper" [49] currently has sensors for pressure, temperature, position, magnetic fields, and acoustics. It can be inserted into a line without a launcher/receiver setup and can be used while the pipeline is operating under normal service. Although these "pigging-like" tools offer some convenience and interesting features for inspecting unpiggable pipelines, they are generally less accurate than conventional ILI tools and still have limitations for some more complex unpiggable pipelines.

Another relatively new development is aboveground inspection tools for unpiggable pipeline inspection. Over the past decade, various aboveground noncontact techniques have been developed for inspecting unpiggable pipelines, including tomography technology (MTM) [50] and "NoPig" techniques [51]. MTM is reported to be capable of remote detection of stress concentration spots in pipe metal by analyzing its magnetic field. Operator holds device that senses the underground pipe magnetic field and walk along the pipe track. It was claimed that the MTM technique is able to measure buried pipeline magnetic field changing from the above ground device to collect pipe integrity information including dents, corrugations, scores, out of roundness, and differences in wall thickness; metal loss including internal and external corrosion defects of any nature; defects within welded joints; crack-like defects in any orientation; sections with deviations of a level of stress deformed; local corrosion under scaled insulating coating [50]. However, the MTM technique is considered not been sufficiently and independently verified [52] and has not been directly compared with existing ILI and indirect techniques like DCVG. A number of organizations have conducted verification tests on selected aboveground inspection tools, in particular MTM, for unpiggable pipeline inspection. Test results appear to be not conclusive and need "more extensive trials that will further prove the effectiveness of MTM." In a recent PRCI research report [53], it is commented that "the Transkor MTM is a noncontact magnetic tomography survey technology that allows for condition screening of the pipeline while it remains in normal operation. While the technology is commercially available its deployments to-date lack full quantitative validation against a complete set of truth data for metal loss conditions" [53]. In a PRCI research report [54], it is mentioned that an independent evaluation of blind trial has been carried out and found that the Transkor system performed "less well" overall in reliability terms particularly in the concrete-coated pipe section [54]. It was also claimed that the performance of the NoPig technique was significantly higher than existing corrosion detection techniques and it could also detect metal loss [51]. Another emerging aboveground inspection tool, namely Spectrum XLI platform, is claimed to allow operators to complete multiple above-ground surveys in a single pass of the pipeline [55].

Another aboveground technique that was developed based on the use of AC magnetic field to inspect underground pipelines is the above ground inspection tool (AGIT) [56]. AGIT detects changes in AC magnetic field, which is induced around the inspected underground pipeline by a specially generated test current. The test current conducted into the pipeline wall via two electrical contacts on the edges of the inspected range, normally 1 km. AGIT inspection by applying different frequency AC signals from a few Hertz to a few hundred Hertz to the pipe to induce an electromagnetic field on the pipe. The magnetic field differences measured from aboveground sensor is considered to reflect the buried pipe wall integrity. The low-frequency AC currents flow through the whole pipe wall cross-section, whereas the high-frequency AC currents flow mainly through the thin layer under the outer surface of the pipe. The magnetic field on the pipe is frequency dependent. If the pipe wall has defects, the magnetic field at the defect area would have stray field distribution. AGIT has been described to be suitable for most unpiggable pipelines including small diameter pipes for collecting critical information including pipe wall corrosion metal loss, wall thickness change, coating disbondment, and cracking [56].

Various new tools and devices have been invented and are at various stages of development and technology trials. Overall, these technological verification and trails appear still insufficient and more field testing and verification data are needed to fully understand their performance and possible limitations. Some of these have been assessed and summarized in PRCI reports [53, 54]. However unfortunately the advantages, disadvantages, limitations, and principles of these relatively new unpiggable pipeline inspection techniques are not well understood by the pipeline industry. Further development of tools with enhanced inspection capabilities for unpiggable pipelines is an important need in order to increase the accuracy and confidence of inspection and knowledge of corrosion rates on fuel networks [57].

2.2 Corrosion Monitoring Using Sensors and Probes

Another approach to enhancing unpiggable pipeline inspection and management is to engage structural health monitoring (SHM) using sensors and probes. SHM is considered to be a practical method to acquiring *in situ* and site-specific pipeline integrity data required for early and site-specific warning of integrity risks [58]. Initially SHM was mainly used for major bridges, probably due to high awareness of the economic and social effects of aging, deterioration, and extreme events on such important civil infrastructures [59, 60]. Many specialized sensors or probes, such as fiber optic sensors, piezoelectric actuators, vibration sensors, leakage detecting sensors, pressure, flow, and seismic sensors and corrosion detection sensors/probes have been developed and utilized in many infrastructural systems for monitoring a range of structural and environmental parameters such as stress, displacement, acceleration, wind speed, temperature, corrosion, and so on. [60, 61]. These data could be used for enhancing unpiggable pipeline ECDA and ICDA assessment, for estimating corrosion active locations, and for predicting maintenance needs at these sites. The development and application of accurate and robust probes and sensors for acquiring corrosion data have been listed as a critical national need for infrastructure transformational research and a grand challenge for corrosion research [4, 58]. Unfortunately at present successful application of corrosion probes in complex and dynamic engineering environments such as underground pipeline conditions has been limited, and the validity of corrosion monitoring data is often left unexamined. This section presents some fundamental considerations on principles of corrosion probes, issues affecting the application of localized corrosion monitoring probes for practical industrial systems, and possibilities of further technological development.

2.2.1 Overview of Corrosion Sensors and Probes

Corrosion probes and sensors are devices that detect and monitor corrosion processes occurring in a corroding metal structure. They can be used independently or be utilized as sensing components in SHM systems. An attractive feature of corrosion sensors and probes is their capabilities in acquiring *in situ* and site-specific corrosion data required for creating visibility of corrosion and material degradation processes, and facilitating early warning of corrosion-induced structural failure, as well as providing maintenance and protection advices. Corrosion sensors and probes have been used in industry as practical tools for acquiring corrosion data to inform localized corrosion initiation, *in situ* corrosion rates, patterns, and mechanisms with varying degrees of success. Remaining issues in these sensors and probes and difficulties in measuring corrosion in complex industrial environments such as in highly resistive and inhomogeneous media have been discussed previously [19, 62–68].

Corrosion sensors are usually referred to as devices that detect certain corrosion-related changes from a corroding structure; while corrosion probes could be described as devices that detect corrosion-induced changes in the probe surface that simulates a corroding structure surface. Fiber-optic probes can be a corrosion sensor to 'sense' corrosion induced changes, as illustrated in Figure 2.3. Fiber optic sensors use optical fibers to indirectly monitor corrosion issues in pipelines by taking advantage of the influence of temperature and strain levels on their optical properties. Although this measuring principle would only be capable of detecting very significantly advance corrosion, a single fiber could provide sensing and monitoring of more than a hundred kilometers of pipeline [68]. The main disadvantage of these probes besides their low sensitivity is that it requires delicate installation of fragile fibers in proximity to the pipeline wall. The assessment of corrosion is based on an assumption that temperature and stain changes are only due to corrosion, not due to other factors such as weather changes and earth movement. This assumption could not always be valid in practice. Some other types of corrosion sensors such as UT-based corrosion sensors detect corrosion by detecting corrosion-induced metal-losses and surface damages. They have to be exactly installed at the location of localized corrosion, which is not easy in engineering practice.

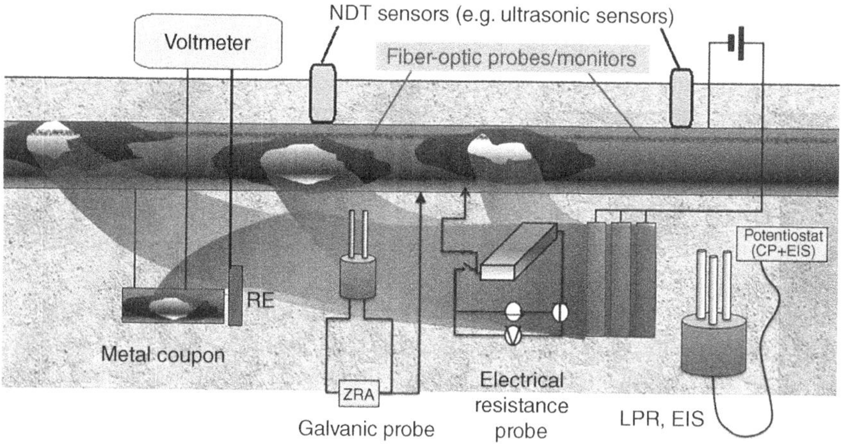

Figure 2.3 Illustrating corrosion probe monitoring of a buried gas pipeline.

Over the past decades, variously designed corrosion probes have been developed to perform corrosion monitoring in various industrial applications. Examples of traditional corrosion monitoring sensors or probes include corrosion coupons, electrical resistance probes, and fiber-optic sensors. Corrosion coupon is one of the most widely applied techniques to obtain corrosion information. The metal coupon's exposed surface can be coated with coating layer containing pre-existent defects or bare. Corrosion coupons can be simply used as weight-loss specimen, they can also be used as electrodes, as illustrated in Figure 2.3, to monitor pipeline corrosion potentials or IR drop-free CP potentials against a reference electrode (RE). After being exposed to a corrosion environment for a period of time, corrosion coupons can be recovered, and corrosion attack can be obtained by visual, microscopic, and weight-loss analysis of the coupon. The main limitation of this technique is that only average corrosion rates can be obtained after the coupons are recovered from corrosion media. This is essentially the same information obtained by examining pipeline surface during dig-up operations. An issue with this technique is that it provides little information on specific corrosion events and mechanisms that may have triggered localized corrosion damage. Another issue is that these coupons and probes generally have difficulties in simulating localized corrosion mechanisms occurring on a pipeline such as corrosion affected by deposits, crevices, disbonded coatings, and galvanic coupling. This is a major issue because the severity of corrosion-induced infrastructural damage is often determined by the corrosion mechanisms, especially localized forms of corrosion. A prerequisite condition for corrosion probes is that they should simulate or reproduce infrastructural corrosion processes and mechanisms on their surfaces, unfortunately this prerequisite condition is often not well appreciated. Although it is well-known in industry that it is important to simulate the actual corrosive environment, relatively little attention has been given to the simulation of localized corrosion mechanisms and processes. Failure in simulating localized corrosion mechanisms and processes can easily be translated into invalid corrosion monitoring results. Other issues with corrosion coupons include the high costs associated with their underground installation and recovery and the slowness in acquiring corrosion data. For these reasons, corrosion coupons are generally used in the pipeline industry mainly for potential monitoring, as illustrated in Figures 2.2b and 2.3.

A corrosion monitoring and data collection tool that has been practically used in the pipeline industry is the electrical resistance (ER) probe. ER probes, as illustrated in Figure 2.3, is often referred to as an "intelligent" weight-loss coupon that monitor the electrical resistance between ends of an elongated "coupon" of constant cross-section exposed to the corrosive environment. As the coupon is corroded, its cross section is reduced, and consequently, its electrical resistance would increase. The increase in electrical resistance can therefore be used to estimate corrosion-caused metal losses. For pipeline application, the ER coupon can be electrically connected to the pipeline to simulate the bare metal exposed in a coating defect under CP [64, 65]. The main disadvantages of these probes are their high sensitivity to temperature variations and relatively low sensitivity to metal loss. Accurate results can only achieved if the corrosion is uniformly distributed over the probe surface because localized corrosion may not lead to major cross-section reduction and electrical resistance increases although thin multifilament probes were developed in recent years that may respond to localized corrosion [66]. For this reason, ER probes are usually only used to provide an indicator of environmental corrosivity. Another corrosion monitoring technique, field signature method (FSM) is considered to be able to detect and monitor cumulative localized corrosion. The FSM technique was developed from the commonly used principle of electrical resistance determination and mapping by impressing a current through several electrodes (sensing pins) on the outer surface of a pipeline and measuring the potential drops between these electrodes. Changes in the geometry in the form of cracks, general corrosion, erosion corrosion, or pitting would disturb the potential field in the metal pipe. The change in the measured potential field pattern is related to a change in pipe wall thickness, and thus the development of corrosion or cracks can be detected. The FSM technique is nonintrusive and is used to measure corrosion damage over a relatively large section of a structure; however, it is generally used for detecting internal pipeline corrosion. The sensing pins are typically separated by a distance of 2–3 times of wall thickness, and therefore FSM has limitations in measurement resolution. Other limitations of the technique include its relatively slow response to corrosion. It was reported to take one year to respond to 1 mpy corrosion rate [67].

Corrosion and materials degradation of infrastructures exposed to practical complex environmental conditions are often affected by dynamically changing local environmental parameters such as local soil and water composition, oxygen level, humidity, salinity, pH, temperature, stray currents, and biological organisms, as well as stray currents, coating disbondment, and cathodic shielding. Although conventional corrosion monitoring devices such as those described above are able to provide some useful information about corrosion and environmental corrosivity, they are generally insufficient for monitoring complex localized corrosion in practical and complex environments. There is a need for *in situ* and site-specific corrosion data to facilitate localized corrosion monitoring, modeling and prediction purposes. "Instantaneous" corrosion sensing and probing techniques including various NDT corrosion sensors and electrochemical corrosion probes are possible solutions to this. At present techniques commonly used for *in situ* and site-specific field corrosion monitoring include ultrasonic sensors, as illustrated in Figure 2.3. The principle of NDT corrosion sensors is simple: they detect cumulative corrosion damages by visual, physical, and electrical means. If measurements are carried out frequent enough using sufficient number of permanent sensing probes, these techniques can actually provide some forms of useful *in situ* and site-specific corrosion data. For instance ultrasonic testing sensors designed based on automatic ultrasonic scanning and recording techniques, as illustrated in Figure 2.3, can be used in the form of arrays and combined with computer techniques to produce maps of the corroded surface. However sensors designed based on physical techniques like ultrasonic sensors have to be placed "completely correctly" on locations where corrosion occurs. This is because these sensors do not interact with the corrosion environment; they can "sense" localized corrosion damages such as pits only if they are installed exactly on pits of sufficient sizes.

Intuitively the use of sensors and probes to monitor corrosion is an uncomplicated task; therefore the application of corrosion sensors and probes is often viewed as an easy task in industry. In reality currently corrosion sensors and probes are not always used successfully in the industry, in fact there are problems in corrosion sensor and probe design, selection and application that often lead to unsuccessful and misleading corrosion testing and monitoring results including those reported in the literature [69–74]. For instance, corrosion coupons and probes are often rarely questioned, as intuitively they seem to perfectly represent the corrosion occurring in an environment, like a normal measurement device like a thermometer. Confusing and misleading corrosion data have been reported, yet little attention has been paid to understand the reasons of test failures. For instance, Srinivasan et al. [69, 70] reported poor correlations between actual corrosion behavior of gas wells and corrosion monitoring data obtained from various techniques including iron counts, caliper surveys, down-hole mounted coupons, radioactive sleeves, electrical corrosion probes, and inspection of pulled tubing. They found that actual pitting rates were between 2 and 15 times the general corrosion rates monitored by these techniques, and in some cases the differences in corrosion rates obtained using different techniques varied by an order of magnitude in value. They reported that corrosion prediction was difficult under high temperature and pressure carbon dioxide corrosion conditions since many oil production wells appeared to suffer little or no corrosion, whereas others in the same area were found to have severe corrosion due to scattered and unpredictable pitting. These results clearly indicate difficulties caused by localized pitting to corrosion monitoring and prediction. Papavinasam et al. [71, 72] also reported significant concerns on some testing and monitoring methodologies for evaluating corrosion inhibitors for oil and gas pipeline applications, confirming complexities in corrosion probe application for practical industry. Another fundamental issue in corrosion monitoring comes from the fact that in the real world corrosion can be dynamic processes with its rates and patterns changing with the time, and thus it is important to identify the specific time periods and environment conditions that of peak or the "worst-case scenario" corrosion rates occur. BP experience at Prudhoe Bay oil field [78] suggests that some forms of localized corrosion such as microbiologically influenced corrosion (MIC) can accelerate rapidly and grow exponentially once initiated, and thus continuous monitoring is important to permit timely response to MIC threat. Tan et al. [75–77] analyzed fundamental limitations in some corrosion testing and monitoring methods, especially in the measurement of localized corrosion, and difficulties, challenges and limitations associated with the industry application of corrosion monitoring techniques and probes.

2.2.2 Electrochemical Probes for Monitoring Localized Corrosion

In order to monitor, predict and manage complex and dynamically changing localized corrosion, "instantaneous" techniques that could continuously measure the prevailing corrosion rates and provide quantitative data for use as a process variable are desirable for integrated and automated corrosion management systems. Electrochemical corrosion monitoring based on the electrochemical nature of metal corrosion is an attractive approach to "instantaneously" acquiring *in situ* and site-specific corrosion data with high spatial and temporal resolutions. Most electrochemical corrosion probes such as those designed based on galvanic effects, electrochemical linear polarization resistance (LPR), and electrochemical impedance spectroscopy (EIS), as illustrated in Figure 2.3, detect corrosion by interacting with the corrosion environment. An advantage of electrochemical corrosion monitoring probes is that they provide not only in situ corrosion data but also sense the "tendency" and mechanism of corrosion. The design and application of electrochemical corrosion probes is underpinned by a basic hypothesis: corrosion probes, despite of their relatively small sizes, can interact with the corrosion environment in the same manner as a larger infrastructural component surface of interest. This is because corrosion and materials failure are not accidental occurrences; they occur as the result of fundamental corrosion thermodynamic spontaneity and kinetic instability of a metal surface in a specific environment. This suggests that corrosion and materials failure occurring on a large structural component surface such as on a buried pipeline should also occur on a smaller steel coupon or probe surface made of the same material and exposed to the same environment. This allows the use of a small-sized corrosion probe to simulate a much larger surface of an engineering structure, and probe/monitor the occurrence of localized corrosion events over the surface.

Many electrochemical methods including corrosion potential measurement, potentiodynamic polarization, LPR, EIS, electrochemical noise analysis (ENA) have been developed over the past decades and widely applied to acquire corrosion data using variously designed electrochemical probes usually designed based on a three electrode system. Corrosion potential recording is probably the most commonly used electrochemical method that monitors corrosion by recording the potential difference between a corroding metal structure such as a pipeline and a stable reference electrode. It is widely used, often in conjunction with Pourbaix diagrams, as an indicator of corrosion thermodynamic status, for instance, to predict active or passive behavior and the effectiveness of cathodic protection. Unfortunately, like any other thermodynamic quantities, the corrosion potential on its own does not provide information on the rate of corrosion. Corrosion rates are usually measured by electrochemical polarization techniques. Electrochemical polarization based techniques, such as the LPR method, are often employed to measure corrosion kinetics based on the Stern and Geary equation that inversely relates the ratio of the applied perturbation potential and the resulting current response, i.e. the polarization resistance, to the uniform corrosion rate at the open circuit potential (OCP). Unfortunately this relationship is not valid for localized corrosion and at the cathodic protection potentials under CP, and consequently this method cannot be used to monitor localized corrosion in underground pipelines.

EIS probes are also made of three identical electrodes, as illustrated in Figure 2.3. To perform the EIS measurement, the working electrode is polarized by a sinusoidal wave of small amplitude ($\pm$20 mV) with frequencies varying in a wide spectrum against the reference electrode by supplying current through the counter electrode. The corrosion system's response changes with the polarization frequency, and in this way, information from electrolyte (high frequencies), charge transference controlled reactions (low frequencies) and diffusion controlled reactions (very low frequencies) can be obtained [79]. Impedance values are employed to measure corrosion kinetics based on the Stern and Geary equation. Despite most EIS probes perform measurements around at OCP and share the same limitation as LPR when monitoring structures under CP, Juchniewicz and Jankowski [80] developed a specific equivalent circuit for interpretation of the spectrum obtained by performing EIS around the structure's CP potential. Although this method has been tested in soils with satisfactory results [80], Jankowski [79] explained that when the electrode is under CP, the cathodic current is raised and the anodic current is reduced, making its accurate determination difficult. More details on these methods can be found in the literature [77, 81, 82].

Galvanic probes, as illustrated in Figure 2.3, are also designed based on the pairs of metals to estimate galvanic corrosion. An example is probes made of steel – stainless steel 304 and steel – copper [83–85]. A linear relationship between corrosion rates estimated by the API Grade A steel – copper probe and those estimated by EIS and linear polarization resistance was found in soils at OCP [83]. Tests under CP in synthetic groundwater were also performed obtaining linear relationships between the probe's and CP current values [84, 85]. However such 'linear' relationships are not a certainty not a common occurence.

A number of fundamental limitations in conventional electrochemical corrosion techniques have been identified, especially in the monitoring of localized forms of corrosion in complex environmental conditions such as in the measurement of localized corrosion of buried steels in highly resistive and inhomogeneous soil media [62, 63, 77]. The accuracy of electrochemical corrosion measurement can be affected by a number of factors such as uncompensated resistances ("ohmic drop") and inappropriate polarization scan rates could cause severe distortions of polarization curves and errors in corrosion rate measurements [81]. It should also be noted that despite of the fact that major effort has been made to develop corrosion probes worldwide and that smart sensors have been identified as a critical national need for infrastructure transformational research in the United States [58], unfortunately at present the reality is that successful application of corrosion monitoring probes in complex infrastructural systems such as underground pipelines has been rather limited. For buried pipelines, the presence of cathodic protection, a highly resistive soil environment and coated surfaces make the accurate electrochemical measurements extremely challenging. This may explain the fact that although corrosion monitoring has been widely applied to many industrial structures such as chemical plants, practical application of existing corrosion monitoring techniques to buried steel pipeline has been limited. For these reasons, the application of sensors and probes for structural health monitoring in complex fuel systems is not always successful. An example is the often unsuccessful application of corrosion sensors for monitoring localized corrosion of infrastructures exposed to complex environmental conditions. Instinctively corrosion probes are considered to be able to acquire *in situ* and site-specific corrosion data; however inadequate attention has been paid on the field application of corrosion monitoring probes, and most of the attention shown in numerous papers in the historical literature has been on the use of corrosion measurement techniques for laboratory corrosion testing and research such as for inhibitor assessment and analysis [86].

2.3 Multi-Electrode Arrays for Probing Complex Corrosion

2.3.1 Localized Corrosion Probes Design Based on Electrode Arrays

It has been shown that the prime challenge in corrosion monitoring is difficulties in simulating and evaluating localized corrosion mechanisms, especially when corrosion mechanisms change with time, affecting the reliability and trustworthiness of corrosion monitoring sensors [87]. Over the past decades, research has been actively carried out to develop sensor and probe-based tools in order to enhance the technological capability for pipeline engineers to monitor and capture pipeline coating deterioration and corrosion processes on a continuous basis. For instance, a pipeline condition monitoring tool has been developed over the past decade to enable site-specific and real-time monitoring of localized corrosion, ineffective CP, coating disbondment, and degradation by placing sensors at strategic and "worst-case scenario" locations of a pipeline [88, 89]. This probe was designed based on an electrochemically integrated multi-electrode array, often referred to as the wire beam electrode (WBE) as illustrated in Figure 2.4 [75], that has been extensively reviewed in references [75–77]. In its basic form, the electrode array consists of many insulated mini-electrodes (namely wires) that is electrochemically integrated by coupling all the wire terminals in the solid phase and by closely packing all the wires in the solid/electrolyte interface in order to effectively simulate a conventional one-piece electrode surface in electrochemical behavior. The results of comprehensive testing have shown that similar corrosion patterns were produced over electrode array and conventional one-piece electrode surfaces when both were exposed to identical corrosion environments and this has been verified theoretically [75, 76]. The electrode array method measures corrosion data without the need for external polarization, avoiding issues associated with the effects of the uncompensated resistance "ohmic drop" and therefore it has shown capabilities for monitoring of localized forms of corrosion in highly resistive and inhomogeneous soil and multiphase media [63, 77]. It simulates and measures various forms of localized corrosion on its electrochemically integrated surface, overcoming difficulties of conventional electrical resistance and electrochemical probes in simulating localized corrosion mechanisms such as

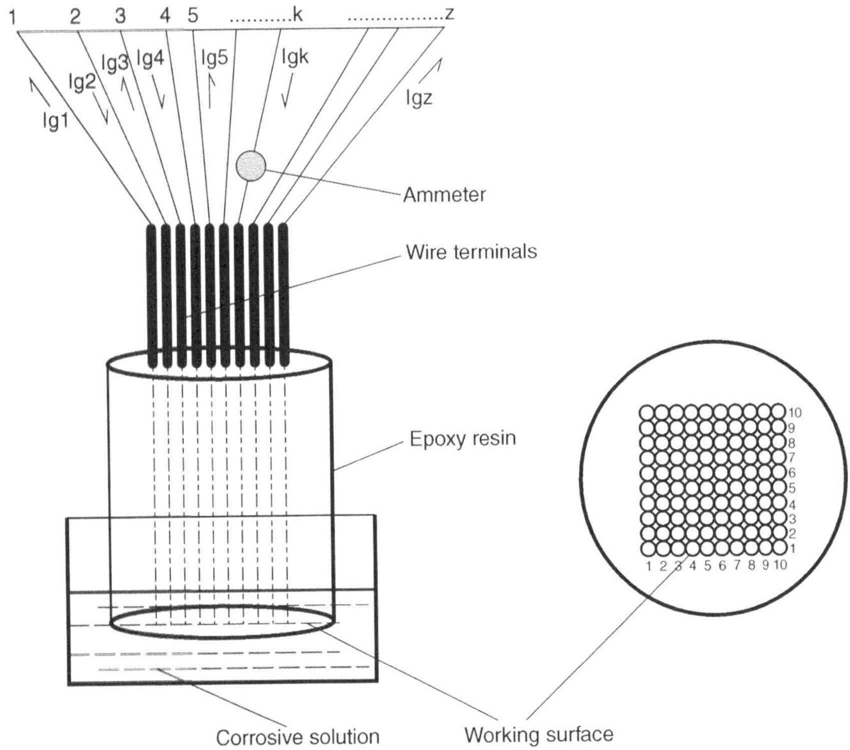

Figure 2.4 Illustrating the concept of electrode array corrosion probe. *Source:* Tan [75] / NACE International.

under-deposit corrosion, crevice corrosion, corrosion under disbonded coatings, and galvanic corrosion. It has shown the temporal resolution required for monitoring dynamic corrosion processes with rates and patterns changing with the time, and corrosion maps can be shown as videos if measurements are taken frequently.

When a large number of closely small packed electrodes are used, high-resolution maps of the electrochemical process can be obtained to study the corrosion process's mechanisms in various environments such as in soil and simulated multiphase conditions [63, 77]. Different versions of the electrode array are reported in the literature such as the coupled multi-electrode array probe that uses small fixed resistances between each electrode and one common terminal (shunt resistors) [90] for measuring the potential drops at each of these resistances. Unfortunately, some of these probes found on the literature showed serious design drawbacks. The distance between electrodes was significantly larger than it should be, reducing the interaction between them and the ability to simulate a continuous metal under corrosion. Also small number of electrodes was used in some of these reported probes, which did not allow the study of interactions between different electrode processes and corrosion patterns. The electrochemically integrated electrode array design uses a programmable switcher to interpose a zero resistance ammeter or a high impedance voltmeter for measuring current or potential distributions among the electrodes, generating maps [75], or videos if measurements are repeated quickly [91].

Over the past two decades, remarkable progresses have been made in the development and implementation of variously designed electrochemically integrated multi-electrode arrays, and this method has now been used globally as a localized corrosion measurement tool. The electrode arrays have been described as a rapid, quantitative method to measure localized corrosion of metals with a capability to monitor quasi *in situ* localized corrosion processes. For instance, Scully and co-workers [92–96] carried out extensive experiments to investigate various forms of localized corrosion and spatial interaction among localized corrosion sites over coupled multi-electrode array surfaces. Using data from electrode array measurements, they modeled the origins of persistent interactions among localized corrosion sites and interactions between early dominating pits and the adjacent electrode surface that were found to develop as regions of enhanced or suppressed pitting susceptibility [94, 95]. Fushimi *et al* [97] studied galvanic corrosion of carbon steel welded with stainless steel with a multi-channel electrode. By detecting the spatial distribution of corrosion currents as a function of immersion time, they found that the weldment acted as a cathode throughout the immersion period, while the other base steel became anodes or cathodes depending on their location, immersion time, and concentration of the electrolyte solution. Liu et al. [98] and Zhang et al. [99] used electrode arrays in the study of various localized corrosion phenomena such as water-line corrosion. Legat [100] found that microelectrode arrays can monitor the time and spatial evolution of steel corrosion in concrete by measuring currents to indicate the temporal anodic and cathodic activities and assess general corrosion rates. Buchheit et al. [101] used coupled electrode arrays to characterize chromate conversion coating formation and breakdown processes. Battocchi et al. [102] employed an Al-Cu array together with a scanning vibrating electrode technique to better interpret the development of localized corrosion. Kallip et al. [103] and Hughes et al. [104–106] researched on the multi-electrode screening technique as an effective method for evaluation of various groups of inhibitors. Overall, remarkable progresses have been made over the past two decades in the development and applications of various forms electrochemically integrated multi-electrode arrays for probing and monitoring localized corrosion that are difficult to measure by conventional electrochemical methods. Figure 2.5 shows examples of various corrosion probe designs based on the electrode array concept.

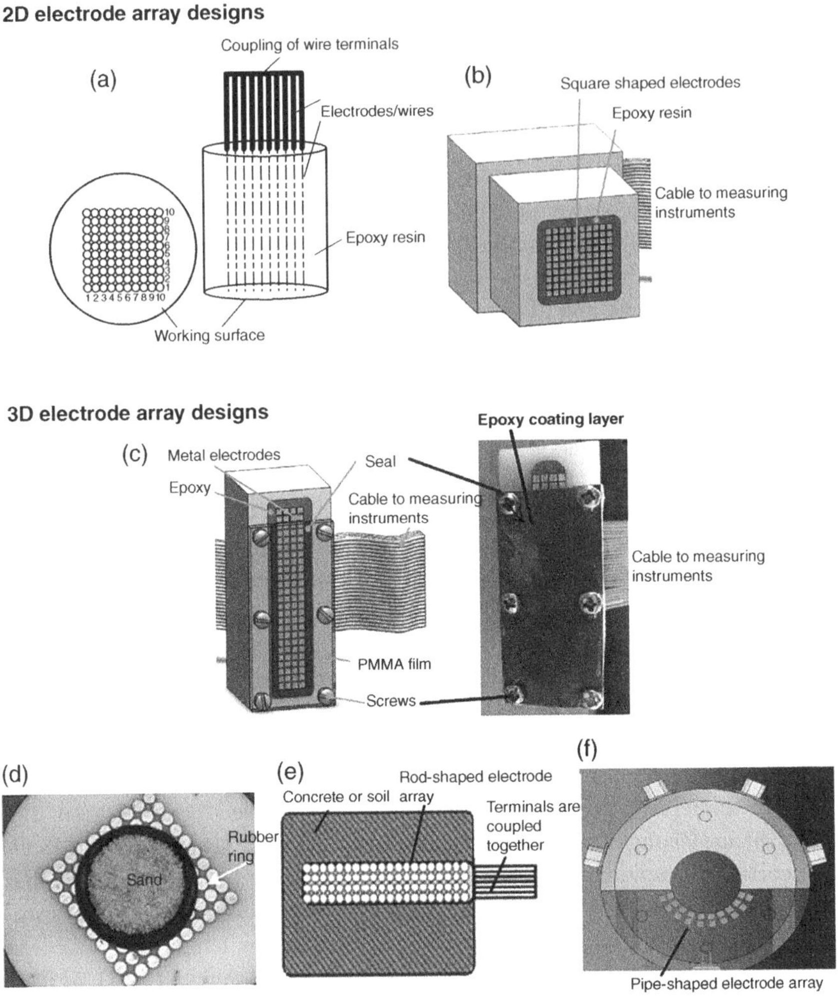

Figure 2.5 Illustration of various a typical corrosion probes based on electrode arrays. *Source:* Adapted from [68]. (a) the original design concept of the WBE; (b) electrode probe designs with different shaped electrodes/wires; (c) the 25x4 electrode array for simulating corrosion under disbonded coating in soil. *Source:* Adapted from Tan, Huo and Varela [89]; (d) electrode array probe for simulating corrosion under deposits. *Source:* Tan et al. [107], ELSEVIER; (e) rod-shaped electrode array probe for simulating steel rebar in concrete structures; (f) pipe-shaped electrode array for studying internal pipeline corrosion and erosion corrosion.

Corrosion data measured from electrode arrays are usually local electrochemical parameters such as local electrode potential, galvanic currents, and impedance from individually addressable electrodes (wires). The analysis of these data is based on the fact that surface of each individual wire in an integrated electrode array can be assumed to be uniform even if the overall surface is nonuniform. This allows the analysis of data from each wire of a nonuniform electrode array using electrochemical theories and methods developed for uniform corrosion processes, i.e. traditional corrosion theories and techniques are extended to study localized corrosion and other nonuniform electrode processes. Based on this concept and the Butler–Volmer equation, new equations describing the electrochemical kinetics of each individual wire have been derived for studying the kinetics of localized corrosion processes [75–77], permitting the mapping of corrosion on an instantaneous and continuous basis. Instantaneous corrosion rate maps were determined from corrosion potential and current distribution maps; and the corrosion rate maps were used to calculate accumulated corrosion depth distributions [77]. In order to achieve the highest possible sensitivity of detecting early signs of localized corrosion, the electrode array has been used in conjunction with electrochemical noise analysis (ENA) since ENA has the potential advantage of providing early signs of localized corrosion initiation [76, 77].

Recently research has been carried out to design electrode array based corrosion probes in order to enable the monitoring of localized corrosion attacks on steel pipelines exposed to complex environmental conditions such as in highly resistive and inhomogeneous soil media, as well as under the effect of dynamically changing environmental conditions [19]. A progress made over the past decade is an effective probe for monitoring buried pipeline corrosion affected by complex environmental factors, such as stray current corrosion, corrosion under disbonded coating, etc.. Stray currents can locally affect the polarization of a structure such as a buried pipeline [108], leading to a major localized corrosion process usually referred to as stray current corrosion. Under a disbonded coating on pipelines, it is believed that the high resistivity of the soil solution trapped within that coating disbondment crevice can shield CP currents, generating unprotected areas where corrosion might be developed [109].

The latest development in pipeline corrosion probe is probably the Pipeline Corrosion Monitoring (PCM) system that was developed for performing *in situ*, site-specific, and sensitive corrosion monitoring of buried pipelines [87–89]. Figure 2.6 illustrates the use of such probe for obtaining corrosion data from buried pipelines with particular focus on unpiggable pipelines. The PCM system was the result of extensive research on the refinement of the electrode array corrosion probe designs, the integration of variously designed corrosion probes, careful consideration on probe installation and application methods, the development of a data logging and analysis, computer visualization software, and IT platform. This PCM system has been successfully tested in the laboratory and in the field using a full-scale prototype built for practical applications. These have included real-life field tests on pipeline sites in Melbourne, Geelong, and Shepparton regions in Australia. Remote monitoring and data transmission via a mobile and satellite networks have been successfully achieved for the PCM system. A data storage and

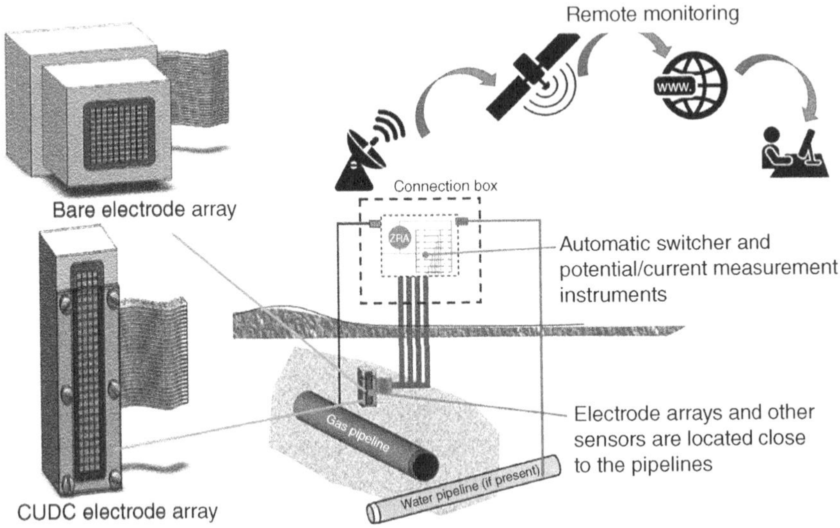

Figure 2.6 Illustrating the use of PCM probe/sensor for obtaining *in situ* and site-specific corrosion data from buried pipelines with particular focus on unpiggable pipelines.

analysis software, as well as a basic web-based IT platform, has been developed to perform data storage, visualization, and corrosion analysis functions. The features of the PCM and associated IT platform include (i) remotely detecting, analyzing, and visualizing data from selected pipeline locations on an interactive map; (ii) displaying basic "warning" of high risk pipeline locations based on local corrosion rates; (iii) a mobile IT platform has been initiated to provide information to engineers. It should be noted that the current PCM system also has its weaknesses and technical gaps: (i) The number of probes deployed on an infrastructure such as a pipeline is limited, therefore the system provides data with excellent temporal resolution but relatively low spatial resolution. It is believed that the full potential of the PCM would be realized if corrosion monitoring data are connected and integrated with other forms of data. (ii) Monitoring probes/sensors are usually designed to simulate and detect the worst-case scenario types of localized corrosion and may not detect all types of corrosion or failure.

Although currently less frequently used in engineering structures, various corrosion probes including electrode array based corrosion monitoring tools hold the promise for obtaining *in situ* localized corrosion information. They have the potential of not only measuring corrosion rates, but more importantly, they provide an insight into the electrochemical processes and mechanisms. This is an invaluable contribution because measuring is the first step to understanding the electrochemical process and eventually, help preventing it.

2.3.2 Various Electrode Array Probe Designs for Examining Crevice Corrosion

Crevice corrosion is a common and insidious form of localized corrosion responsible for many complex forms of corrosion issues in industry. Crevice corrosion occurs when local chemistry develops in confined crevice spaces such as gaps and contact areas between parts, under disbonded coatings, gaskets or seals, inside cracks and seams, spaces under deposits, and sludge piles. Traditionally crevice corrosion is studied and tested in the laboratory using methods such as weight-loss measurement, visual and microscopic examination of corroded specimen, the simulated occluded cell method, microelectrode and freezing methods, and so on. Mathematical modeling is also used extensively to study crevice corrosion. These methods can provide valuable information regarding the process and mechanism of crevice corrosion; however, they do not delineate clearly the crevice corrosion process and they do not give instantaneous crevice corrosion rates. Crevice corrosion is a heterogeneous electrochemical process in nature. When crevice corrosion occurs, there is a distinct separation of the anodic and cathodic regions on the metal surface and different electrochemical reactions occur on the anodic and cathodic areas. To determine the instantaneous rates of crevice corrosion, electrochemical parameters at local areas of a working electrode surface such as local anodic reaction currents have to be determined. However, conventional electrochemical techniques have major difficulties in doing this. For a given crevice type, two factors are believed to be important in the initiation of crevice corrosion: the chemical composition of the electrolyte in the crevice and the potential drop into the crevice. The measurement of both the potential drop and the change in composition of the crevice electrolyte is difficult, which triggered the earliest application of the electrode array for probing crevice corrosion [110].

Corrosion probes designed based on the electrode arrays (The WBEs) has been employed in various crevice corrosion tests. Simple experiments were performed using an initial WBE designed for demonstrating the measurement of electrode potential directly from crevice area in order to understand "oil-stain" on overlapping steel surfaces and its inhibition [110]. Later, local electrochemical parameters measured from simulated crevice areas on electrode arrays were used to calculate instantaneous crevice corrosion kinetics that correlated well with real crevice corrosion appearance [111]. Variously designed WBEs were used to simulate a range of materials and environmental combinations. For instance electrode arrays were used to study crevice corrosion of copper by measuring electrochemical parameters directly from crevice area. It was found that the distribution of corrosion potential in crevice was inhomogeneous. At the early stage of immersion, the corrosion potentials in the crevice shifted toward positive, while with propagation of crevice corrosion began to shift negatively and became homogeneous gradually that is in agreement with the classic theory of crevice corrosion of metal [112]. The WBE was also used to probe the effect of sodium nitrite on the crevice corrosion of mild steel in 5% NaCl and 5% NaCl + 5% $NaNO_2$ solutions by measuring corrosion potentials and the electrochemical impedance spectra inside and outside the crevice. Experimental results showed that in 5% NaCl solution the mild steel inside the crevice was anode and that outside the crevice was cathode. With the addition of 5% $NaNO_2$ solution, the corrosion potentials inside and outside the crevice both shifted toward positive, the difference of corrosion potentials was diminished and the corrosion rate of mild steel was reduced evidently. In 5% NaCl + 5% $NaNO_2$ solution, compact oxide film with a strong adhesion to the surface of the mild steel was formed, and the crevice corrosion of the mild steel was effectively restrained [113].

More complicated electrode arrays were designed in order to better simulate more complex forms of crevice corrosion on industrial structures. For instance an artificial crevice was built by one-dimensional array electrode to investigate corrosion damages to the inner contact surface of screw thread joint between tubing and coupling [114]. An electrode array was designed to assess the severe corrosion of the storage tank bottom plates exposed to a crevice environment filled with sediment water. Results showed that during the initial immersion period of 21 days, the carbon steel suffered from uniform corrosion and its corrosion rate decreased with time. Afterward, the corrosion was transformed to localized corrosion and its corrosion rate increased with time [115]. Another "difficult-to-test" crevice corrosion studied by an integrated multi-electrode array was corrosion under insulation (CUI) [116]. Corrosion of aluminum exposed under insulation materials such as rock wool, glass wool, cotton wool, and tissue paper was successfully simulated and monitored using a WBE. A typical potential noise signature of a major potential jump from an AA1100 WBE was observed, which corresponded to the corrosive species reaching the WBE surface. A good correlation between the galvanic current maps and the corroded surface was also observed [116].

A type of crevice corrosion in oil and gas production pipeline is corrosion under solid deposits such as sand and biofilms. Corrosion inhibitors are used to prevent corrosion under deposits; however the efficiency of corrosion inhibitors is often unknown because it is almost impossible to assess by normal inhibitor testing techniques. Underdeposit corrosion is believed to be controlled by factors including galvanic effects between a large cathode such as the large surface of a pipeline and a small anode under deposits. Failure of inhibitors to penetrate the deposits and the retention of aggressive species such as bacteria in the deposits need to be assessed in order to effectively simulate and evaluate their effects on corrosion rates and patterns. Figure 2.7 shows a case of monitoring underdeposit carbon dioxide corrosion and the performance of corrosion inhibitors using an electrode

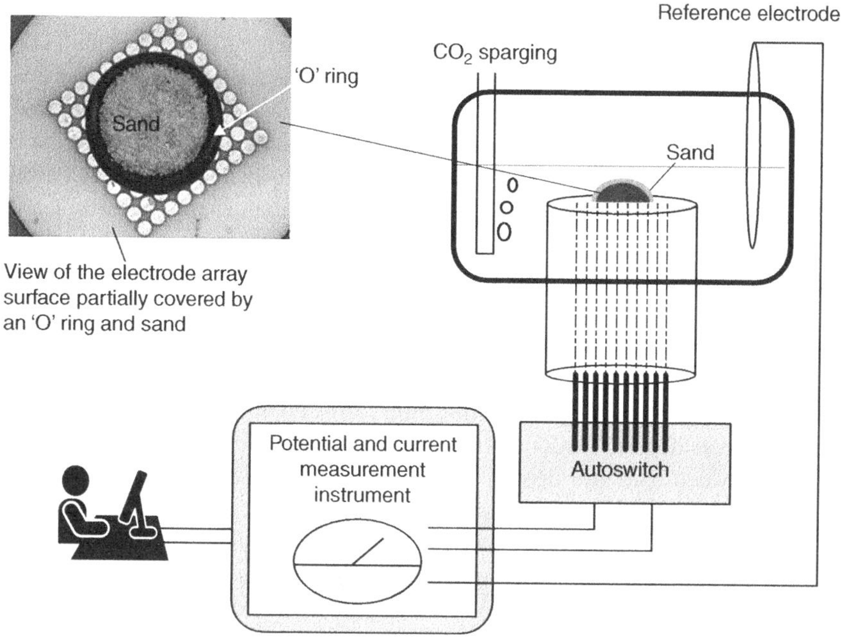

Figure 2.7 Illustration of a typical corrosion under deposit and inhibition test using an electrode array and local galvanic current distributions over its surface that was covered by sand and an inhibitor imidazoline. *Source:* Tan et al. [107] / With permission of Elsevier.

array [107]. As shown in Figure 2.7, the WBE working surface was partially covered with a rubber "O"-shaped ring filled with sand to simulate a localized underdeposit corrosion environment. The partially covered WBE surface was then exposed to synthetic brine under continuous CO_2 sparging to maintain a virtually oxygen-free environment. Underdeposit corrosion processes were monitored by mapping galvanic currents across the multi-electrode array to understand how localized corrosion initiation and propagation under sand and how it changed with the introduction of corrosion inhibitors [107]. When inhibitor imidazoline was added into the brine solution, corrosion anodes and cathodes were found to quickly switch locations, as shown in the current distribution map in Figure 2.7, with corrosion anodes shifted to areas cover by "O" ring and sand, while cathodes located mainly over the four corners where no sand was present. This result suggests that the addition of inhibitor imidazoline initiated the formation of corrosion anode under the sand deposit. In other words, inhibitor imidazoline addition caused underdeposit corrosion although it reduced overall corrosion. This experiment illustrates the complexity of underdeposit corrosion and the effects of inhibitors [107].

Similar experiments were conducted to study underdeposit corrosion of X65 steel using electrode arrays exposed to oxygen containing environments [117, 118]. Results show that the deposit covered area acted as the corrosion anodes due to the oxygen concentration difference between the bare steel area and deposit covered area. The most serious corrosion would occur at the bare steel area, which was close to the sand deposit. This phenomenon was explained to be due to the diffusion of the cations from the deposit covered area to the bare steel area, leading to the decrease of local pH. Once the diffusion path between the bare steel area and the deposit covered area was blocked, the bare steel area would act as a constant cathode, resulting in the formation of a protective calcium carbonate layer on the bare steel surface [117]. In amino trimethylene phosphonic acid and O_2 containing electrolyte solution, localized corrosion was found to occur on steel surfaces covered with silica sand, while little corrosion occurred on steel beneath $CaCO_3$ particle deposit. $CaCO_3$ particle deposit appeared to be affecting galvanic current between bare and deposit covered areas [118]. Corrosion of mild steel under $CaCO_3$ deposit was further investigated by the WBE in the oil-field-produced water [119], and it was shown from potential and galvanic mappings that the wires covered by $CaCO_3$ deposit mainly acted as anode, while the rest bare wire electrodes acted as cathode at the beginning of corrosion, causing apparent corrosion under deposit. However, the deposit-covered electrodes were later transformed from anode to cathode along with the bare electrodes from cathode to anode. This galvanic polarity reversal was explained by the researchers that $FeCO_3$ corrosion products could fill into the deposit layer and form a physical diffusion barrier to block the transportation of aggressive ions from the bulk solution to the metal substrate [119]. In CO_2-containing formation water, the galvanic effect between the mixed deposit-covered steel was assessed by current and potential mapping [120]. It was found that the bare electrode always acted as anode, while the deposit covered electrode always acted as cathode; however, the polarity reversed at the late stage occurred when temperature increased from 25 to 60 °C due to the formation of $FeCO_3$ on the bare electrode [120]. The WBE, in combination of other electrochemical tests, has been considered to become the main stream technology for underdeposit corrosion testing [121]. Copper and aluminum corrosion in highly concentrated sodium chloride solutions with and without solid deposits has also been studied using specially shaped coupons and WBEs [122]. Coupled multi-electrode arrays were implemented to monitor the corrosion processes on copper in a slightly saline environment, which simulated the environmental condition of high-radioactive waste disposal in bentonite [123]. Maximum anodic activity was found to switch between the electrodes in the array that appeared to be consistent with the visual observation that uneven general corrosion without evident pits on copper.

All these reports suggest that the electrode array, if designed and used properly, is capable of monitoring various forms of crevice corrosion including the complex corrosion under deposits. The application of the WBE for probing other forms of crevice corrosion such as corrosion under disbonded coatings and weldment corrosion where the crevice environment also contributes to their initiation and prorogation are discussed in other specific sections of this book.

2.3.3 Further Issues and Challenges in Corrosion Monitoring Probes

Effective control and management of complex forms of localized corrosion is critical for maintaining the integrity of engineering structures such as underground pipelines and offshore infrastructure that are vital for the provision of the world's essential services and the maintenance of its economic activities. An ideal corrosion management and control system should not only ensure the safe operation of these structures but should also significantly extend their life. Corrosion monitoring should be able to contribute to the proactive corrosion management and control system because accurate corrosion monitoring and predictive modeling would enable proactive adjustment of corrosion protection systems in response to changes in corrosion environmental conditions and corrosion mechanisms. Unfortunately there is a significant gap in the ability of engineers to be able to monitor and capture localized corrosion and material deterioration processes reliably and numerically, in order to progress toward proactive corrosion prediction and control. Further technological advancements are needed in designing corrosion probes and in enhancing the reliability of corrosion probes as a structural health monitoring tool. Corrosion probes should become reliable tools for early detection and diagnosis of corrosion, for providing industrial system "health" alarm, for forecasting maintenance requirements, and for generating data for integrated and automated corrosion management system.

Several critical progresses need to be made in order to significantly enhance the reliability and applicability of corrosion probes. Successful corrosion probe application requires the probe surface to simulate not only the corrosion environment but also the critical corrosion mechanisms occurring on the infrastructure surface [19]. It is essential to ensure that the corrosion probes are able to effectively simulate corrosion behavior in actual service environments and reliably evaluate the effects of various factors on corrosion processes, rates, and mechanisms. It is well appreciated that corrosion probes need to simulate the actual service exposure environment; however, relatively less considerations have been given to the effects of environmental parameters on corrosion patterns and mechanisms. This challenge is more acute when corrosion is affected by many inter-related variables such as ununiform temperature and pressure, heterogeneous metallurgy, inhomogeneous soil or solution chemistry and thermo-mechanical conditions, local mechanical stress, coating defects, and cathodic potential and excursions. These effects may have not yet fully quantified and need to be better understood and used in corrosion probe design and application.

In order to fully realize the advantage of corrosion sensors for providing site-specific and *in situ* warning of unexpected localized corrosion-induced structure failures, corrosion probes may need to be placed at strategic and "worst-case scenario" high-risk locations of a structure. For a buried pipeline, for instance, typical high-risk structure sites could be those with high stray current activities, high oxygen concentration and low soil resistivity, changing underground water and rainfall levels, high concentration of corrosive species, and those highly corrosion rate areas identified by pigging, field survey, and historical excavations. Probes embedded at these strategic sites can be used to collect real-time and site-specific data that would contain critical "predictor features" and parameters needed for modeling and predicting localized corrosion, coating disbondment, and degradation. If measurements are carried out frequently enough using sufficient number of corrosion probes located at strategic sites, some forms of useful *in situ* and site-specific corrosion data in the form of videos could be provided to enable effective localized corrosion visualization, monitoring and prediction.

The unique characteristics of electrode-array-based corrosion probes should be further explored. Despite of the progresses made in the development and implementation of electrochemically integrated multi-electrode arrays, challengers and obstacles still remain and prevent the more and wider application of electrode arrays for probing complex corrosion mechanisms and especially for monitoring localized corrosion in industry field conditions. These are believed to include the design of electrode arrays that simulate multiscale and dynamically changing localized corrosion on industry structures, the fabrication of complex-shaped and patterned electrode arrays; the real-time visualization of dynamic localized corrosion processes; data analytics for modeling and interpreting corrosion data. These are discussed in more detail below.

2.3.3.1 Design Electrode Array Probes for Measuring Electrochemical Processes Over Wide Ranges of Length and Time Scales

Corrosion can occur through a series of electrochemical processes over wide ranges of length and time scales without the need of the reactants spatially nearing each other, permitting the localization of electrode chemistry and corrosion products over a corroding electrode surface at various spatial and temporal scales. An example is the corrosion of complex alloys that often consist of a wide range of grains, defects, and intermetallic particles with sizes varying from nano- to micro- or even macro- scales. Interactions between these structures could significantly affect localized corrosion behavior, suggesting the need for technical capabilities of probing and characterizing complex localized corrosion processes occurring over a wide range of length scales. On the other hand, localized corrosion processes are often dynamic, initiating from micro-sized event and propagating into macro-sized corrosion problems. These could be addressed by future three-dimensional and complex-shaped electrode arrays designed and fabricated using microfabrication and additive manufacturing techniques. Innovatively designed and precision-engineered electrode arrays, in conjunction with other electrochemical and data analytical techniques, would enable electrode arrays application to probe these multi-scale corrosion and other electrochemical science issues.

2.3.3.2 Design Electrode Array Probes for Probing Multiple Corrosion Mechanisms Occurring Simultaneously

Another challenge comes from situations that several corrosion mechanisms could occur instantaneously on a corroding industry structure of complex shapes. Cultivating a corrosion probe for such situation would require the identification of most important localized corrosion mechanisms in order to select and design electrode array probes that are able to collect data/parameters related to major mechanisms. An approach that could be taken to simplify corrosion probe design for industry application is to design electrode arrays that simulate and detect the "worst-case scenario" corrosion mechanism such as corrosion under disbonded coatings and stray current corrosion on buried steel pipelines. Another approach is to design electrode arrays that are able to allow the simulation of several mechanisms such as a "sacrificial sensor" that was designed to detect the occurrence and rate of atmospheric corrosion and stress corrosion cracking [124]. Good understanding of the targeted corrosion system and innovative thinking are often needed in order to develop appropriate corrosion probes for probing multiple corrosion mechanisms occur simultaneously over complexly shaped surfaces.

2.3.3.3 Data Acquisition Options Through Combined Use with Various Electrochemical and Surface Analytical Techniques

Suitable measurement techniques that are able to enhance data acquisition from corrosion probes could add extra value to corrosion monitoring and measurement. An example is to combine the electrode arrays with electrochemical noise analysis (ENA) to acquire extra information on localized corrosion "predictor features" [125]. ENA is a technique proposed and utilized for early detection of localized corrosion by detecting "noise signatures" [126]. In a typical experiment, for instance, two characteristic noise patterns were observed from electrode array data: (i) quick potential changes toward more negative direction with no recovery was found to correspond with the initiation and stabilization of the anode inside a crevice; and (ii) cyclic potential oscillation at a constant frequency was found to correspond with the stable anodic dissolution in the occluded cavity site in the WBE current distribution maps. It was further reported that during carbon steel pitting corrosion process, the characteristic "peak" of rapid potential transient, toward less negative direction, followed by recovery was found to correlate with the disappearance of unstable anodes leading to formation and propagation of stable anodes in WBE current distribution maps. This result suggests that the combined WBE-noise signatures method could be applied as a means of early detection and prediction of localized corrosion. On the other hand, it was found that the noise resistance (R_n)-WBE method has advantages over other WBE-based methods when applying to WBE surfaces under general or uniform corrosion [125]. Another approach of obtaining extra corrosion data from electrode arrays is to combine the electrode array with scanning probe techniques [127, 128]. It should be recognized that electrochemical heterogeneity could present at various spatial and time scales, and thus different techniques could be used for probing different heterogeneous electrode processes. For instant scanning probe techniques such as the scanning Kelvin probe force microscopy and scanning electrochemical microscopy could be useful for probe nano-scale electrochemical heterogeneity; the scanning Kelvin probe, the scanning vibrating electrode technique, local electrochemical impedance spectroscopy, and microelectrode array could be used for probing micro-scale electrochemical heterogeneity, while the WBE probe could be more suitable for probing macroscale heterogeneous electrode processes. Since the scale of electrochemical heterogeneity could change with the propagation of heterogeneous electrode processes, these techniques could be applied in various combinations. In order to increase the time resolution, an instrumentation that is able to produce "current flows video" on the WBE surface would be highly desirable. Battocchi et al. combined the electrode array method with electrochemical noise resistance and the scanning vibrating electrode technique (SVET) [127]. From the data obtained, the WBE configuration showed the ability to emulate a continuous surface as the current distributions from the analysis with the SVET has been confirmed [127]. Tan et al. combined the WBE and the scanning reference electrode technique (SRET) to measure electrochemical corrosion parameters simultaneously from both the metallic and electrolytic phases of a corroding metal surface [128]. It was found that the combined WBE-SRET method is able to gain useful information on macro-cell electrochemical corrosion processes that involve macro-scale separation of anodes and cathodes. In such macro-cell corrosion systems, maps measured using WBE and SRET were found to correlate with each other, and both methods were able to detect the locations of anodic sites. However the movement of the scanning probe during SRET measurements was found to affect the SRET detection of cathodic sites [128]. Other combined use of the WBE and electrochemical methods also include the study of coating disbondment over the WBE surface using electrochemical impedance measurement [129, 130]. Extra corrosion and surface metallurgical information could also be obtained through the combined application of electrode arrays and surface analytical techniques such as SEM and even the high-resolution atom probe tomography (APT) [131] for revealing multi-length and time-scale electrochemical corrosion initiation/propagation activities and to correlate them with metallurgical heterogeneities. Numerical simulation and modeling of electrode array with electrochemical measurements could also provide useful information for understanding localized corrosion processes and mechanisms. For instance electrode arrays with different geometrical properties have been numerically investigated to verify a possible extension of the range of variation of the cell constant for electrochemical measurements [132]. All these approaches could be explored further in order to develop techniques with unique capabilities.

2.3.3.4 Electrode Array Manufacturing by Means of Various Fabrication Methods

A critical aspect of the electrode array method is the design and fabrication of the electrode arrays to simulate metal-environment system. We could consider the combined use of various design, materials, and fabrication techniques for developing innovative electrode array probes with desirable and special electrochemical characteristics, multidimensional geometry, sensitivity, compatibility, and long-term chemical and recording stability. To date, electrode arrays have been fabricated using metal or alloy wires, microwires, and machined electrodes. Further development could involve altering the material compositions and geometries of the arrays, as well as the fabrication of flexible multi-electrode arrays, microelectrode arrays, and micro-machined patterned arrays. Flexible arrays provide an advantage over rigid electrode arrays because of the possibility of simulating the geometry of corroding industry structures. The use of additive manufacturing (3D printing) and micromachining techniques provides new options of fabricating the complex geometries and microstructures required for simulating multifaceted corrosion such as those occurring on heterogeneous alloys and in the occluded pitting corrosion cells [133]. Three-dimensional and multi-material microfabrication using focused-ion-beam (*FIB*) chemical-vapor deposition [134] could be useful for fabrication of sub-micro- and micro-sized electrode arrays for studying localized corrosion of additively manufactured alloys. Another challenge in fabricating electrode arrays is finding appropriate coating materials that are compatible to corrosion environmental conditions such as high temperature and biological conditions. Although coupled multi-electrode arrays have been made for high temperatures (e.g. up to 150 °C in corrosive environments), the degradation of the insulating mounting material (e.g. epoxy) remain a challenge. Chiang *et al.* studied the effectiveness of a diamond-like carbon film to protect the Alloy 22 (UNS N06022) electrodes in an acidic sodium chloride-sodium nitrate-potassium nitrate ($NaCl$-$NaNO_3$-KNO_3) salt mixture at 150 °C. Post-test SEM analysis and cyclic polarization tests revealed that the coating was adherent and prevented crevice formation between the Alloy 22 electrode and the epoxy-mounting material [135, 136].

2.3.3.5 Design Electrode Array Probes for Industry Application Conditions

The most significant challenge is probably the design and selection of suitable electrode array probes that accurately simulate and measure localized corrosion with complicated and changing corrosion mechanisms. This challenge is more acute when corrosion is affected by many inter-related variables such as ununiform temperature and pressure, heterogeneous metallurgy, inhomogeneous soil or solution chemistry and thermo-mechanical conditions, local mechanical stress, coating defects, and cathodic potential and excursions. On the other hand corrosion mechanisms can be significantly affected not only by environmental parameters such as specimen surface conditions, wear, abrasion and time of exposure and others, but also by materials involved. It is important to note that different metals and materials could respond differently to changes in environmental conditions, for instance passive metals such as stainless steel and active metals such as mild steel can respond differently to aeration. In the case of stainless steels, the corrosion controlling factor is usually the passivity of the metal surface rather than oxygen transportation. All these factors could affect the design of corrosion probes for a specific industry application. Although various forms of electrode array probes have been designed and used in various research applications, as described above, more attention should still be given to the simulation of localized corrosion mechanisms occurring in industry practice, especially the possibility of corrosion mechanism change, corrosion environment change, and possible transitions from general to localized corrosion.

2.3.3.6 Development of Data Systems and IT Platforms

Although corrosion probes can provide useful *in situ* data from selected locations of an asset, there is a need to integrate data from limited monitoring sites into the whole database by suitable models in order to provide fuller coverage of a huge structure (e.g. a 1000 km underground pipeline). An information platform would enable the integration of various data inputs and allow industry to cost effectively gather information on the in-service integrity of assets/infrastructure, gain higher levels of confidence in the condition of the asset, timely maintenance, safety, and continuous availability/operation of the asset. A solution is probably an IT and data analytical platform for infrastructure health monitoring, failure prediction, and life extension. The idea is to develop a web-based information platform that can linkup multiple industries and multi-disciplinary areas of research. This platform "visualizes" what's happening underground by taking and linking up different sources of corrosion and materials degradation data available. It can include corrosion probes as an *in situ* corrosion monitoring information source. This platform would allow continued data input from various information sources such as remote corrosion monitoring, ILI, and other industry inspection; and the analysis and application of data for various purposes. For instance, risk assessment models can be employed to prioritize the assets for maintenance and renewal.

Overall it is shown in literature that corrosion monitoring using sensors and probes is a practical approach to *in situ* and site-specific localized corrosion simulation and monitoring for corrosion prediction and management purposes. Findings and conclusions from buried pipelines are believed to be useful references to many other engineering structures although the actual design of the inspection tools and monitoring probes may need to be modified based on the nature of each engineering structure.

References

1 Wikipedia (2013). Qingdao oil pipeline explosion. http://en.wikipedia.org/wiki/2013_Qingdao_oil_pipeline_explosion (accessed 23 August 2022).

2 Hsu, J.W. and Liu, F. (2014). Taiwan Gas Blasts Likely Caused By Faulty Pipe. http://online.wsj.com/articles/taiwan-gas-blasts-likely-caused-by-faulty-pipe-1406964902 (accessed 31 August 2017).

3 Gosch, E. (2008). WA gas supply cut 30pc by blast. *The Australian* (7 June).

4 National Institute of Standards and Technology (2009). Technology Innovation Program report. http://www.nist.gov/tip/prev_competitions/upload/ci_wp_031909.pdf (accessed 22 December 2021).

5 Housner, G.W., Bergman, L.A., Caughey, T.K. et al. (1997). Structural control: past, present, and future. *Journal of Engineering Mechanics* 123: 897–971.

6 Li, H.-N., Li, D.-S., and Song, G.-B. (2004). Recent applications of fiber optic sensors to health monitoring in civil engineering. *Engineering Structures* 26: 1647–1657.

7 Chang, P.C., Flatau, A., and Liu, S.C. (2003). Review paper: health monitoring of civil infrastructure. *Structural Health Monitoring* 2: 257–267. https://doi.org/10.1177/1475921703036169.

8 Ko, J.M. and Ni, Y.Q. (2005). Technology developments in structural health monitoring of large-scale bridges. *Engineering Structures* 27: 1715–1725.

9 Brownjohn, J.M.W. (2007). Structural health monitoring of civil infrastructure. *Philosophical Transactions of the Royal Society A* 365: 589–622.

10 Li, X., Zhang, D., Liu, Z. et al. (2015). Materials science: share corrosion data. *Nature* 527: 441–442.

11 Melchers, R.E. (2015). Bi-modal trends in the long-term corrosion of copper and high copper alloys. *Corrosion Science* 95: 51–61. 81(2014): 110–116.

12 Alamilla, J.L., Espinosa-Medina, M.A., and Sosa, E. (2009). Modelling steel corrosion damage in soil environment. *Corrosion Science* 51 (11): 2628–2638, ISSN 0010-938X.

13 Marney, D. and Cole, I. (2012). The science of pipe corrosion: a review of the literature on the corrosion of ferrous metals in soils. *Corrosion Science* 56: 5–16.

14 Koch, G.H., Brongers, M.P.H., Thompson, N.G. et al. (2001). Corrosion cost and preventive strategies in the United State. National Technical Information Service Report No. FHWA-RD-01-156.

15 Gunasegaram, D.R., Venkatraman, M.S., and Cole, I.S. (2014). Towards multiscale modelling of localized corrosion. *International Materials Reviews* 59: 84–114.

16 Azoor, R.M., Deo, R.N., Birbilis, N., and Kodikara, J.K. (2018). Coupled electro-chemical-soil model to evaluate the influence of soil aeration on underground metal pipe corrosion. *Corrosion* 74: 1177–1191.

17 Melchers, R.E. (2018). A review of trends for corrosion loss and pit depth in longer-term exposures. *Corrosion and Materials Degradation* 1: 42.

18 Cheng, F. (2016). Monitor safety of aged fuel pipelines. *Nature* 529: 156.

19 Tan, M.Y.J. (2020). A critical overview of monitoring infrastructural health using corrosion probes. *Corrosion Engineering Science and Technology* 55: 103–117. https://doi.org/10.1080/1478422X.2019.1695390.

20 Tan, Y., Ubhayaratne, K.N.I., Huo, Y. et al. (2019). Predictive maintenance based on smart monitoring and data analytics. *Australasian Corrosion Association (ACA) Conference: Corrosion & Prevention*, Melbourne, Australia (24–27 November 2019).

21 Kishawy, H.A. and Gabbar, H.A. (2010). Review of pipeline integrity management practices. *International Journal of Pressure Vessels and Piping* 87 (7): 373–380.

22 Varela, F., Tan, M.Y., and Forsyth, M. (2015). An overview of major methods for inspecting and monitoring external corrosion of on-shore transportation pipelines. *Corrosion Engineering Science and Technology* 50: 226–235.

23 Cordell, J. and Vanzant, H. (2003). In-line inspection pigging. In: *Pipeline Pigging Handbook*, 3e, 1–15. Houston, TX, USA: Clarion Technical Publishers.

24 Brockhaus, S., Ginten, M., Klein, S. et al. (2014). In-line inspection (ILI) methods for detecting corrosion in underground pipelines. In: *Underground Pipeline Corrosion* (ed. M.E. Orazem), 255–285. Cambridge: Woodhead Publishing.

25 Bai, Y. and Bai, Q. (2014). Pipeline inspection and subsea repair. In: *Subsea Pipeline Integrity and Risk Management*, 73–99. Boston: Gulf Professional Publishing.

26 Bubenik, T. (2014). Electromagnetic methods for detecting corrosion in underground pipelines: magnetic flux leakage (MFL). In: *Underground Pipeline Corrosion* (ed. M.E. Orazem), 215–226. Cambridge: Woodhead Publishing.

27 J. Tiratsoo: 'Pipeline Pigging and Integrity Technology', Chap. 4, 'Intelligent pigging', 375–389; 2003, Beaconsfield, UK, Scientific Surveys Ltd.

28 Miller, S. and Clouston, S. (2013). Optimizing magnetic-flux leakage inspection sizing model performance using high-resolution nondestructive examination data. In: *Pipeline Pigging and Integrity Technology*, 4e (ed. J. Tiratsoo), 299–308. Beaconsfield, UK: Tiratsoo Technical.

29 Hrncir, T., Turner, S., Polaski, S.J. et al. (2013). A case study of the crack sizing performance of the GE ultrasonic phased-array inspection tool on the Centennial pipeline. In: *Pipeline Pigging and Integrity Technology*, 4e (ed. J. Tiratsoo), 317–328. Beaconsfield, UK: Tiratsoo Technical.

30 Barbian, A. and Beller, M. 2012 In-line inspection of high pressure transmission pipelines: state-of-the-art and future trends. *Proceedings of 18th World Conference on Nondestructive Testing*, Durban, South Africa (April 2012). South African Institute for Non-Destructive Testing (SAINT) Paper 9 (Invited Overview Talks).

31 Kania, R. and Coleman, G.A. (2013). Self-excited eddy currents for the detection of SCC. In: *Pipeline Pigging and Integrity Technology*, 4e (ed. J. Tiratsoo), 487–496. Beaconsfield, UK: Tiratsoo Technical.

32 Gao, M., Tandon, S., and Krishnamurthy, R. (2013). Evaluation of EMAT tool performance by monitoring industry experience. In: *Pipeline Pigging and Integrity Technology*, 4e (ed. J. Tiratsoo), 137–160. Beaconsfield, UK: Tiratsoo Technical.

33 Marr, J., Kania, R., Rosca, G. et al. (2013). Combining EMAT ILI and multiple data sets for crack detection in natural gas pipelines to reduce validation costs. In: *Pipeline Pigging and Integrity Technology*, 4e (ed. J. Tiratsoo), 161–168. Beaconsfield, UK: Tiratsoo Technical.

34 Al-Qahtani, H., Beuker, T., and Damaschke, J. (2013). ILI with high-resolution EMAT technology: crack detection and coating disbondment. In: *Pipeline Pigging and Integrity Technology*, 4e (ed. J. Tiratsoo), 115–122. Beaconsfield, UK: Tiratsoo Technical.

35 C. D. Stears, R. M. Degerstedt, O. C. Moghissi and L. Bone III: 'Field program on the use of coupons to monitor cathodic protection of an underground pipeline', Proc. Corrosion 97, New Orleans, Louisiana, March, 1997, NACE international, Paper 564, 1–12.

36 NACE, The use of coupons for cathodic protection monitoring applications, RP0104:2004, NACE, Houston, TX, USA, 2004.

37 Hevle, D. and Kowalski, A. (2006). Close-interval survey techniques. In: *ASM Handbook*, 13C, Corrosion: Environments and Industries, Rev.e (ed. S.D. Cramer and S. Covino Jr.), 84–88. Ohio, USA: ASM International.

38 ISO 15589-1:2015 Petroleum, petrochemical and natural gas industries — Cathodic protection of pipeline systems — Part 1: On-land pipelines, ISO, Geneva, Switzerland, 2015

39 NACE, Control of External Corrosion on Underground or Submerged Metallic Piping Systems, SP0169-2007, NACE, Houston, TX, USA, 2007
40 Masilela, Z. and Pereira, J. (1998). Using the direct current voltage gradient technology as a quality control tool during construction of new pipelines. *Engineering Failure Analysis* **5** (2): 99–104.
41 Shawki, S. (2011). Above ground coating integrity assessment: experience with SUMED pipelines. In: *Integrity of Pipelines Transporting Hydrocarbons* (ed. G. Bolzon, T. Boukharouba, G. Gabetta, et al.), 311–322. Biskra, Algeria: Springer.
42 Eyre, D. (2014). The Pearson survey method for detecting corrosion in underground pipelines. In: *Underground Pipeline Corrosion* (ed. M.E. Orazem), 247–254. Cambridge: Woodhead Publishing.
43 Pikas, J. (2006). External corrosion direct assessment integrated with integrity management. In: *ASM Handbook*', 13C (ed. S.D. Cramer and B.S. Covino Jr.), 79–83. Ohio, USA: ASM International.
44 NACE, 'Pipeline external corrosion direct assessment methodology', RP-0502:2002, NACE, Houston, TX, 2002.
45 Department of Transport USA 2011. Application of Remote-Field Eddy Current Testing to Inspection of Unpiggable Pipelines, https://primis.phmsa.dot.gov/rd/articles/CorrosionMitigation10.pdf (assceesed 21 December 2021).
46 PHMSA, Design Construction and Demonstration of a Robotic Platform Capable of Inspecting Unpiggable Gas Pipelines, DTRS56-05-T-0002, US Patent under DOT Contract 7,459,999B2 https://primis.phmsa.dot.gov/rd/articles/CorrosionMitigation2.pdf (accessed 21 December 2021).
47 Puneet Agarwal, Elizabeth Bowlin, Integrity Assessment of DTI Pipelines Using High Resolution NDE in Select Areas, Year 3, PRCI Contractor Project Number: PR-201-153718, March 20, 2018.
48 NDE 3-1 Project Team, Evaluation of Large Standoff Magnetometry Techniques for Above the Pipe Integrity Assessment Project NDE-3-1, Phase 2, PRCI Contractor Project Number: Contract PR-398-133725, May 2017
49 INGU Pipers, Get uninterrupted condition-critical data to optimise your pipeline integrity programs, https://www.aquip.com.au/well-and-pipeline-monitoring/inline-inspection/pipers/. Accesses January 2022
50 Transkor Group Inc. Magnetic Tomography Method (MTM) safety programs for underground and underwater pipelines. 2013. http://transkorgroup.com/safetyprograms, assceesed 21/2/2021
51 Guennadi, S. and Krivoi, N.P. Inspection Services GmbH, NoPig: a new inspection technique for non-piggable pipelines. *Pipeline Technology Conference 2007*. Hannover Messe, Hannover, Germany (16 April 2007–17 April 2007). https://www.pipeline-conference.com/sites/default/files/papers/6.5%20Krivoi.pdf (accessed 22 December 2021).

52 Li, Z., Jarvis, R., Nagy, P.B. et al. (2017). Experimental and simulation methods to study the Magnetic Tomography Method (MTM) for pipe defect detection. *NDT and E International* 92: 59–66.

53 Puneet Agarwal, Elizabeth Bowlin, Integrity Assessment of DTI Pipelines Using High Resolution NDE in Select Areas, Year 3, PRCI Contractor Project Number: PR-201-153718, March 20, 2018.

54 NDE 3-1 Project Team, Evaluation of Large Standoff Magnetometry Techniques for Above the Pipe Integrity Assessment Project NDE-3-1, Phase 2, PRCI Contract PR-398-133725, May 2017

55 PureHM, Spectrum XLI system, https://www.purehm.net/technologies/spectrum-xli/, assceesed 21/12/2021

56 Kriboi, G.S. and Glinka, M. (2018). Improvements of an inspection technique for non-piggable pipeline diagnostics from above ground. *Pipeline Technology Conference*. Estrel, Berlin, Germany (12 March 2018–14 March 2018).

57 Tan, M.Y.J., Huo, Y., Varela, F. et al. (2021), Corrosion control technology needs for future fuel infrastructure. *Australasian Corrosion Association (ACA) conference: Corrosion & Prevention*. Newcastle, Australia (5–8 June 2022).

58 National Research Council. (2011). Committee on Research Opportunities in Corrosion Science and Engineering, Research Opportunities in Corrosion Science and Engineering, National Research Council. Washington, DC: The National Academies Press.

59 Chang, P.C., Flatau, A., and Liu, S.C. (2003). Review paper: health monitoring of civil infrastructure. *Structural Health Monitoring* 2: 257–267. https://doi.org/10.1177/1475921703036169.

60 Ko, J.M. and Ni, Y.Q. (2005). Technology developments in structural health monitoring of large-scale bridges. *Engineering Structures* 27: 1715–1725.

61 Brownjohn, J.M.W. (2007). Structural health monitoring of civil infrastructure. *Philosophical Transactions of the Royal Society A* 365: 589–622.

62 Tan, M.Y.J. (2017). Principles and issues in structural health monitoring using corrosion sensors. Corrosion/2017, paper no. 9240 (NACE).

63 Tan, Y.J. (2011). Experimental methods designed for measuring corrosion in highly resistive and inhomogeneous media. *Corrosion Science* 53: 1145.

64 Li, S.Y., Jung, S., Park, K.W. et al. (2007). Kinetic study on corrosion of steel in soil environments using electrical resistance sensor technique. *Materials Chemistry and Physics* 103 (1): 9–13.

65 Kioupis, N. and Maroulis, K. 2006. AC-corrosion detection on electrical resistance probes connected to a natural gas transmission pipeline. *Proceedings of 8th International Conference Pipeline Reabilitation and Maintenance*, Instanbul, Turkey (September 2006). IBC Technical.

66 Li, S.Y., Kim, Y.G., Jung, S. et al. (2007). Application of steel thin film electrical resistance sensor for in situ corrosion monitoring. *Sensors and Actuators B: Chemical* 120 (2): 368–377.

67 Joosten, M.W., Fischer, K.P., Strommen, R., and Lunden, K.C. (1995). Internal corrosion monitoring of subsea oil and gas production equipment. *Materials Performance* 34 (4): 44–48.

68 Yan, S.Z. and Chyan, L.S. (2010). Performance enhancement of BOTDR fiber optic sensor for oil and gas pipeline monitoring. *Optical Fiber Technology* 16 (2): 100–109.

69 Kane, R.D. and Srinivasan, S. 1996. Experience survey on corrosion monitoring and mitigation techniques for sweet well production. IntelCorr/96 Technical Sessions, http://www.corrosionsource.com/events/intercorr/techsess/papers/session6 (visited June 2009).

70 Srinivasan, S. and Kane, R.D. (1996). 'Prediction of Corrosivity of CO_2/H_2S Production Environments', Corrosion/96, Paper No. 11, Denver, CO.

71 Papavinasam, S., Revie, R.W., Attard, M. et al. (2003). Comparison of laboratory methodologies to evaluate corrosion inhibitors for oil and gas structures. *Corrosion-NACE* 59: 897–912.

72 Papavinasam, S., Revie, R.W., Attard, M. et al. (2003). Comparison of techniques for monitoring corrosion inhibitors in oil and gas structures. *Corrosion-NACE* 59: 1096–1111.

73 Corrview (2004). The benefits and limitations of corrosion coupons. https://pdf4pro.com › cdn, (visited 22 December 2021).

74 Kane, R. (2007). A new approach to corrosion monitoring, Chemical Engineering, pp. 34–41. www.che.com (accessed 23 August 2022).

75 Tan, Y.J. (1998). Monitoring localized corrosion processes and estimating localized corrosion rates by means of a wire beam electrode. *Corrosion-NACE (USA)* 54 (5): 403–413.

76 Tan, Y. (2009). Sensing electrode inhomogeneity and electrochemical heterogeneity using an electrochemically integrated multi-electrode array. *Journal of the Electrochemical Society* 156: C195–C208.

77 Tan, Y. (2013). *Heterogeneous Electrode Processes and Localized Corrosion*, 1–246. Hoboken, NJ: Wiley.

78 Bailey, A. (2006). Petroleum News, Vol. 11, No. 20 Week of May 14.

79 Jankowski, J. (2002). Electrochemical methods for corrosion rate determination under cathodic polarisation conditions – a review part 2 – AC methods. *Corrosion Reviews* **20** (3): 179–200.

80 Juchniewicz, R. and Jankowski, J. (1993). Application of impedence spectroscopy to the assessment of cathodic protection effectiveness. In: *Progress in Underdtanding and Prevention of Corrosion*, vol. 2 (ed. J.M. Costa and A.D. Mercer), 1401–1408. London, UK: EFC.

81 Mansfeld, F. (1988). Don't be afraid of electrochemical techniques – but use them with care! 1988 Whitney Award Lecture. *Corrosion* 44 (12): 856–868. https://doi.org/10.5006/1.3584957.

82 Frankel, G.S. (2008). Electrochemical techniques in corrosion: status, limitations, and needs. *Journal of ASTM International* 5: 101241. https://doi.org/10.1520/JAI101241.

83 Choi, Y.S. and Kim, J.G. (2005). A galvanic sensor for monitoring the corrosion damage of buried pipelines: Part 1 – electrochemical tests to determine the correlation of probe current to actual corrosion damage in synthetic groundwater. *Corrosion* **61** (3): 293–301.

84 Choi, Y.S., Kim, J.G., and Yang, S.J. (2006). A galvanic sensor for monitoring the corrosion damage of buried pipelines: Part 2 – correlation of sensor output to actual corrosion damage of pipeline in soil and tap water environments. *Corrosion* **62** (6): 522–532.

85 Choi, Y.S., Kim, J.G., and Koo, J.Y. (2007). A galvanic sensor for monitoring the corrosion damage of buried pipelines: Part 3 – correlation of probe current to cathodic protection and stray current. *Corrosion* **63** (10): 951–957.

86 Tan, Y., Huang, Y., and Mansfeld, F. (2014). Testing and analysis techniques in rare earth metal corrosion inhibitor research. In: *Rare Earth Based Corrosion Inhibitors* (ed. M. Forsyth and B. Hinton), 38–83 (46 pages). Cambridge, England: Woodhead Publishing.

87 MYJ Tan, F Varela, Y Huo, F Mahdavi and K. Wang, "Acquisition, visualisation and interpretation of pipeline corrosion monitoring data", EUROCORR 2017 - The Annual Congress of the European Federation of Corrosion, 20th International Corrosion Congress and Process Safety Congress 2017. Czech Association of Corrosion Engineers. Prague, Czech Republic, 03 Sep 2017 - 07 Sep 2017.

88 Tan, Y., Huo, Y., Varela, F., and Wang, K. (2021). A new approach to probing localized corrosion of pipelines. *Materials Performance* 60 (2): 5.

89 Tan, Y., Huo, Y., and Varela, F. (2021). Field and laboratory assessment of electrochemical probes for visualizing localized corrosion under buried pipeline conditions. *Journal of Pipeline Science and Engineering* 1 (1): 88–99.

90 Yang, L., Sridhar, N., Pensado, O., and Dunn, D.S. (2002). An in-situ galvanically coupled multielectrode array sensor for localized corrosion. *Corrosion* **58** (12): 1004–1014.

91 Varela, F., Tan, M.Y.J., and Forsyth, M. (2015). An electrochemical method for measuring localized corrosion under cathodic protection. *ECS Electrochemistry Letters* 4 (1): C1–C4. Video in Supplemental Material: https://iopscience.iop.org/article/10.1149/2.0091501eel.

92 Lunt, T.T., Brusamarello, V., Scully, J.R. et al. (2000). Interactions among pitting corrosion sites investigated with electrode arrays. *Electrochemical and Solid State Letters* 3: 271–274.

93 Lunt, T.T., Scully, J.R., Brusamarello, B. et al. (2002). Spatial interactions among localized corrosion sites. *Journal of the Electrochemical Society* 149: B163–B173.

94 Budiansky, N.D., Hudson, J.L., and Scully, J.R. (2004). Origins of persistent interaction among localized corrosion sites on stainless steel. *Journal of the Electrochemical Society* 151: B233–B243.

95 Budiansky, N.D., Bocher, F., Cong, H. et al. (2007). Use of coupled multi-electrode arrays to advance the understanding of selected corrosion phenomena. *Corrosion* 63: 537–554.

96 Bocher, F., Presuel-Moreno, F., and Scully, J.R. (2008). Investigation of crevice corrosion of AISI 316 stainless steel compared to Ni-Cr-Mo alloys using coupled multielectrode arrays. *Journal of the Electrochemical Society* 155: C256–C268.

97 Fushimi, K., Naganuma, A., Azumi, K., and Kawahara, Y. (2008). Current distribution during galvanic corrosion of carbon steel welded with type-309 stainless steel in NaCl solution. *Corrosion Science* 50: 903–911.

98 Liu, Z., Wang, W., Wang, J. et al. (2014). Study of corrosion behavior of carbon steel under seawater film using the wire beam electrode method. *Corrosion Science* 80: 523–527.

99 Zhang, M., Ge, H.-H., Wang, X.-J. et al. (2015). Research into the corrosion behavior of carbon steel in simulated reverse osmosis product water and seawater using a wire beam electrode. *Anti-Corrosion Methods and Materials* 62 (3): 176–181.

100 Legat, A. (2007). Monitoring of steel corrosion in concrete by electrode arrays and electrical resistance probes. *Electrochimica Acta* 52: 7590–7598.

101 W. Zhang, B. Hurley and R. G. Buchheit, Characterization of chromate conversion coating formation and breakdown using electrode arrays, Journal of the Electrochemical Society, 149 (2002), B357–B365

102 Battocchi, D., He, J., Bierwagen, G.P., and Tallman, D.E. (2005). Emulation and study of the corrosion behavior of Al alloy 2024-T3 using a wire beam electrode (WBE) in conjunction with scanning vibrating electrode technique (SVET). *Corrosion Science* 47: 1165–1176.

103 Kallip, S., Bastos, A.C., Zheludkevich, M.L., and Ferreira, M.G.S. (2010). A multi-electrode cell for high-throughput SVET screening of corrosion inhibitors. *Corrosion Science* 52: 3146–3149.

104 García, S.J., Muster, T.H., Özkanat, Ö. et al. (2010). The influence of pH on corrosion inhibitor selection for 2024-T3 aluminium alloy assessed by high-throughput multielectrode and potentiodynamic testing. *Electrochimica Acta* 55: 2457–2465.

105 Garcia, S.J., Muster, T.H., Hughes, A.E. et al. (2010). Validation of a fast scanning technique for corrosion inhibitor selection: influence of cross-contamination on AA2024-T3. *Surface and Interface Analysis* 42: 205–210.

106 Muster, T.H., Hughes, A.E., Furman, S.A. et al. (2009). A rapid screening multi-electrode method for the evaluation of corrosion inhibitors. *Electrochimica Acta* 54: 3402–3411.

107 Tan, Y., Fwu, Y., and Bhardwaj, K. (2011). Electrochemical evaluation of under-deposit corrosion and its inhibition using the wire beam electrode method. *Corrosion Science* 53 (4): 1254–1261.

108 Huo, Y. and Tan, M.Y. (2018). Localized corrosion of cathodically protected pipeline steel under the effects of cyclic potential transients. *Corrosion Engineering, Science and Technology* 53: 348–354.

109 Varela, F., Tan, M.Y.J., and Forsyth, M. (2015). Understanding the effectiveness of cathodic protection under disbonded coatings. *Electrochimica Acta* 186: 377–390.

110 Tan, Y.J. (1994). A new method for crevice corrosion studies and its use in the investigation of oil-stain. *Corrosion* 50: 266–269.

111 Tan, Y.J. (1999). A novel crevice corrosion experiment using a wire beam electrode. *Journal of Corrosion Science and Engineering* 1: 1–5. paper 11. https://www.jcse.org/.

112 Zhong, Q.D. (1999). Study on crevice corrosion of copper using wire beam electrode. *Journal of the Chinese Society of Corrosion and Protection* 19 (3): 189–192.

113 Li, G.X., Wang, D., Zhu, R.L. et al. (2005). Effect of sodium nitrite on crevice corrosion of mild steel using wire beam electrode. *Hunan Daxue Xuebao/ Journal of Hunan University Natural Sciences* 32 (6): 91–93.

114 Ren, C., Chen, J., Pu, L. et al. (2015). Electrochemical investigation on crevice corrosion of P110 steel by using one-dimensional array electrode. *International Journal of Electrochemical Science* 10 (10): 8210–8223.

115 Cheng, Q., Tao, B., Liu, S. et al. (2017). Corrosion behaviour of Q235B carbon steel in sediment water from crude oil. *Journal of the Chinese Society of Corrosion and Protection* 37 (2): 126–134.

116 Aung, N.N., Wai, W.K., and Tan, Y.J. (2006). A novel electrochemical method for monitoring corrosion under insulation. *Anti-Corrosion Methods and Materials* 53 (3): 175–179.

117 He, L., Xu, Y., Wang, X., and Huang, Y. (2019). Understanding the propagation of nonuniform corrosion on a steel surface covered by marine sand. *Corrosion* 75 (12): 1487–1501.

118 Xu, Y., Huang, Y., He, L. et al. (2017). Experimental study on under-deposit corrosion and its inhibition using electrochemical methods and electronic coupon technique. *Anti-Corrosion Methods and Materials* 64 (2): 148–161.

119 Wu, Y.L., Zhang, D.P., Cai, G.Y. et al. (2020). Effects of temperature on polarity reversal of under deposit corrosion of mild steel in oilfield produced water. *Corrosion Engineering, Science and Technology* 55 (8): 708–720.

120 Zhang, G.A., Yu, N., Yang, L.Y., and Guo, X.P. (2014). Galvanic corrosion behavior of deposit-covered and uncovered carbon steel. *Corrosion Science* 86: 202–212.

121 Chen, J., Ju, H., Sun, C. et al. (2017). Application of electrochemical testing technology for corrosion under-deposits. *Journal of the Chinese Society of Corrosion and Protection* 37 (3): 207–215.

122 Huo, Y., Tan, M.Y., and Shu, L. (2013). Effects of high salt concentration and residue on copper and aluminum corrosion. *Chemical Research in Chinese Universities* 29 (3): 538–544.

123 Kosec, T., Hren, M., and Legat, A. (2017). Monitoring copper corrosion in bentonite by means of a coupled multi-electrode array. *Corrosion Engineering, Science and Technology* 52: 70–77.

124 Hurley, M.F., Cunningham, A., Lysne, D. et al. (2018). A condition monitor for atmospheric induced stress corrosion cracking. *NACE - International Corrosion Conference Series*, Corrosion, Phoenix, AZ, USA (15–19 April 2018).

125 Tan, Y. (2009). Sensing localised corrosion by means of electrochemical noise detection and analysis. *Sensors and Actuators, B: Chemical* 139 (2): 688–698.

126 Hladky, K. and Dawson, J.L. (1981). The measurement of localized corrosion using electrochemical noise. *Corrosion Science* 21: 317–322.

127 Battocchi, D., He, J., Bierwagen, G.P., and Tallman, D.E. (2005). Emulation and study of the corrosion behavior of Al alloy 2024-T3 using a wire beam electrode (WBE) in conjunction with scanning vibrating electrode technique (SVET). *Corrosion Science* 47: 1165–1176.

128 Tan, Y.-J., Liu, T., and Aung, N.N. (2006). Novel corrosion experiments using the wire beam electrode: (III) measuring electrochemical corrosion parameters from both the metallic and electrolytic phases. *Corrosion Science* 48: 53–66.

129 Mahdavi, F., Tan, M.Y.J., and Forsyth, M. (2016). Communication – an approach to measuring local electrochemical impedance for monitoring cathodic disbondment of coatings. *Journal of the Electrochemical Society* 163: C228–C231.

130 Thu, Q.L., Bonnet, G., Compere, C. et al. (2005). Modified wire beam electrode: a useful tool to evaluate compatibility between organic coatings and cathodic protection. *Progress in Organic Coatings* 52: 118–125.

131 Parvizi, R., Hughes, A.E., and Tan, M.Y.J. (2021). New approach to probing localised corrosion processes over wide length and time scales using integrated multi-scale electrode arrays. *Corrosion Science* 181: 109238.

132 Manzin, A., Boveri, C., Capra, P.P. et al. (2009). Experimental and numerical characterization of an electrode-matrix cell for electrochemical measurements. *Sensors and Actuators, B: Chemical* 138: 326–335.

133 Daryadel, S., Behroozfar, A., Morsali, S.R. et al. (2018). Additive manufacturing of metals at micro/nanoscale by localized pulsed electrodeposition: Nanotwinned copper nanowires. *ASME 2018 13th International Manufacturing*

Science and Engineering Conference, MSEC. College Station, TX, USA (18 June 2018)

134 Hoshino, T., Kawamori, M., Suzuki, T. et al. (2004). Three-dimensional and multimaterial microfabrication using focused-ion-beam chemical-vapor deposition and its application to processing nerve electrodes. *Journal of Vacuum Science and Technology B: Microelectronics and Nanometer Structures* 22: 3158–3162.

135 Chiang, K.T. and Yang, L. (2008). Development of crevice-free electrodes for multielectrode array sensors for applications at high temperatures. *Corrosion, NACE* 64: 805–812.

136 Chiang, K.T. and Yang, L. (2010). High-temperature electrochemical sensor for online corrosion monitoring. *Corrosion* 66: 0950021-0950028.

3

Localized Corrosion in Changing Environments

In practical engineering systems, environmental conditions can be highly nonuniform and dynamically changing. Examples include steel pipelines buried in inhomogeneous soil under the effects of seasonal rainfall, cathodic protection disruption, and cyclic stray currents. A nonuniform and changing environment creates major challenges to the understanding, modeling, prediction, and control of corrosion. Despite extensive investigation over the past several decades, localized corrosion in complex and changing environmental conditions remains a common cause of pre-mature and unexpected infrastructure failure. A fundamental factor contributing to the lack of predictability of such infrastructure failure is the variability of practical engineering environments, which poses significant challenges to the detection, prediction, and control of corrosion. This chapter firstly identifies major obstacles in detecting and managing localized corrosion in dynamic changing environmental conditions. Methods and tools including the WBE-video technique are presented for facilitating the real-time visualization of dynamic localized corrosion and *in situ* corrosion data acquisition. Electrode-array-based corrosion probes have been designed and used to acquire data for detecting, modeling, predicting, and managing such complex and changing localized corrosion. Corrosion behaviors of buried steel pipelines with damaged coatings, disrupted cathodic protection, and stray currents are selected to discuss as typical cases. It is shown that localized corrosion on buried pipelines can be affected by factors such as coating defect sizes, changing soil conditions, cathodic protection disruptions, and stray currents. Knowledge reported here is mainly from buried steel pipelines; however, it is believed to be useful references to many other engineering structures although the actual corrosion behavior may vary with the environmental conditions and engineering materials.

3.1 Probing Localized Corrosion in Nonuniform and Changing Environments

3.1.1 Challenges of Localized Corrosion in Diverse and Changing Environments

Localized corrosion due to unexpected environment changes is a common occurrence. Figure 3.1 shows an example of pre-mature infrastructure failure that a concrete building collapsed due to accelerated localized corrosion of a small section of a steel reinforcing bar [1]. Localized corrosion of the rebar was due to seepage of corrosive water to the concrete structure, causing an unexpectedly local corrosive environment and rapid attack on a small portion of the steel reinforcing bar buried in the otherwise passive concrete environment. Similar cases of

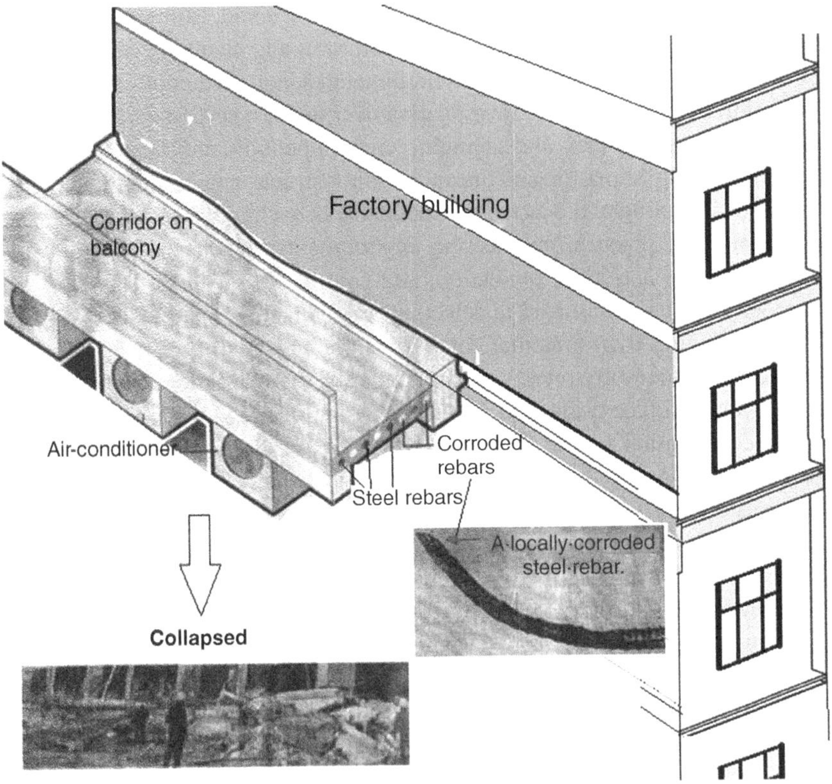

Figure 3.1 A collapsed building due to localized corrosion of a small section of a steel reinforcing bar buried in concrete factory building.

localized corrosion-induced structural failures also occurred on underground and aboveground pipelines, such as those shown in Figure 1.1, where steel pipelines were exposed to nonuniform and sometimes cyclically changing solid/gas/liquid multiphase media [2], allowing dynamic and localized corrosion to occur unnoticed on steel surfaces [3]. More cases of localized corrosion-induced failure in dynamically changing industrial environments can be found in Section 1.3.

Cathodic protection (CP) and coatings are the most common and effective techniques for protecting buried and submerged steel structures such as pipelines from corrosion. Unfortunately the protectiveness of CP could be impaired under some complex pipeline conditions such as under the effects of stray currents, CP disruptions, and shielding coatings. For instance, it is well-known that CP may lose its effectiveness due to potential excursions caused by various forms of electrical interference signals, commonly referred to as stray currents [4, 5]. Potential excursions, in particular anodic transients, are real occurrences that have been shown in the literature to be a major factor contributing to the CP failure of buried pipelines. Despite major efforts made during the past decades, there are still unanswered questions regarding the effects of potential excursions on CP efficiency and buried steel pipeline corrosion. For instance, there were issues about how "big" these excursions (either in amplitude, duration, or frequency) need to be in order to trigger major pipeline corrosion problems. This uncertainty caused significant difficulties in selecting suitable parameters in relevant industry CP standards.

Technological difficulties in effectively quantifying the effects of dynamically changing potential excursions on buried steel corrosion are considered to be the prime reason for the lack of conclusive experimental evidence needed to make significant progress in understanding and preventing corrosion due to stray currents [4, 5]. Conventional corrosion testing and measurement methods have difficulties in dealing with rapidly changing environments under stray currents, since generally they are designed for simplified stable industrial environments. This leaves a major gap in measuring, predicting, managing and controlling unexpected localized corrosion in nonuniform and changing industrial environmental conditions.

Take soil corrosion as an example, although electrochemical corrosion reactions occurring in soil are the same as those occurring in aqueous solutions, electrode processes associated with soil corrosion can be much more complex and dynamic due to the effects of many factors including uneven and varying soil moisture contents [2, 6–9], aeration [10, 11], electrical resistivity [12, 13], pH [14], ionic species [15], and microbiological activity [16]. Soil itself is an inhomogeneous media with different texture, particle size, and structure [3], and these soil properties often change over time. The presence of fluctuating moisture levels, temperatures, and oxygen contents during the wet/dry weather cycles makes soil corrosion rather complex to simulate and predict. Probably due to the complex interplay of various soil properties and influence factors, the actual effect of each factor on corrosion sometimes remains rather uncertain. Discrepancies in soil corrosion testing results are often reported in the literature, suggesting the complexities in soil corrosion. Cole and Marney [3] extensively reviewed earlier work on factors affecting soil corrosion. They highlighted work by Romanoff [17] and Fitzgerald [18] who considered aeration, moisture content, soil particle size, pH, bacteria, electrolyte type and concentration as major factors affecting soil corrosion. Romanoff indicated that soil resistivity appeared to correlate well with soil corrosion because it is related to the concentration of dissolved salts in soil. This view was backed by Scully and Bundy [19] who supported the inverse dependence of soil corrosion on soil resistivity based on corrosion rate data obtained from linear polarization resistance (LPR) and electrochemical impedance spectroscopy (EIS) measurements. However, Murray and Moran [6] reported different findings that the corrosion current density depended primarily on soil moisture content rather than soil resistivity. On the other hand, Nie et al. [20] postulated that the mass transfer of dissolved oxygen played an essential role in the kinetics of the soil corrosion process.These controversaries confirm complicities in soil corrosion and its controlling factors.

Soil environment can be further complicated by the application of CP. Steel structures buried underground are commonly protected by CP, which would induce chemical environment on steel surface to change locally and dynamically. When CP is applied on a pipeline, environmental conditions on bare pipeline surface especially aeration [10, 11], electrical resistivity [12, 13], and especially pH [14] would also change locally because of CP reactions. However it is usually unknown what exact changes CP would induce on a buried pipeline and if the changes will be uniformly or locally applied on it. There are possibilities that CP could cause localized changes to the environment around the pipeline, leading to changes in local corrosion kinetics and mechanisms. Historically, it was found that CP cause local changes in pipeline corrosion [21–24] and local pH [25, 26] in sandy soil until reaching a stable value after extended exposure period (e.g. 78–90 days). Soil corrosion of pipelines can be further complicated by other changing factors in particular coating disbondment and cathodic shielding [22], disruption of the CP systems [27], or interfered by stray currents [21, 28]. These suggest a rather nonuniform and changing soil environment over a practical CP-protected buried pipeline.

Under the effects of so many different variables and influencing factors, accurate prediction and modeling of localized soil corrosion becomes highly difficult. In a dynamically changing environment, corrosion management and protection measures may also need to be adjusted based on the dynamically changing nature of the corrosion condition, activities, and mechanisms [4]. Over the past decades, extensive research has been carried out to clear up discrepancies in soil corrosion through understanding and quantifying the effects of soil variables on corrosion by acquiring *in situ* and site-specific corrosion data from monitoring dynamically changing soil corrosion processes as a function of experimental variables. A prerequisite condition to accurately predict and effectively control buried pipeline corrosion is good knowledge of the actual corrosion behavior [4, 5]. Unfortunately the initiation, location, and rates of localized corrosion are considered to be extremely challenging to detect or monitor in practical engineering infrastructure [4, 5]. Currently the most commonly employed method of assessing the effects of localized corrosion on the integrity of infrastructure is visual inspection and inline inspection (ILI). Periodic visual and ILI inspections are useful for the assessment of various forms of infrastructure damage; however personnel-intensive inspections are not only expensive, disruptive, and infrequent, more importantly inspection data often have insufficient temporal and spatial resolutions required to accurately detect dynamically changing localized corrosion processes and mechanisms [12].

In order to enhance inspection capabilities, smarter inspection and sensing technologies have been developed and employed over the past decades. These include sensing paints, dye penetrant testing, ultrasonic techniques, ground-penetrating radars, and sound-wave propagation methods [29, 30]. Various forms of smart sensors have also been developed to detect infrastructural damage by monitoring a range of physical or chemical parameters, including fiber-optic sensors that are surface mounted on existing infrastructures or embedded in newly constructed infrastructures to obtain information about strain, temperature, delamination, cracks, corrosion, and concentration of chloride ions in structures [29–33]. Other types of smart sensors such as piezoelectric sensors have also shown strong potentials for structural health monitoring [34]. However it should be noted that existing sensors have a common limitation: They detect corrosion damages only after significant extent of general corrosion occurred. They are not sensitive enough to localized corrosion because localized corrosion, by its nature, occurs in small areas and may not cause any significant extent of general metal dissolution. It has been noted that material damage is very difficult to be identified using these sensors, which could be due to the damage scale is not sufficiently large [35]. At present localized corrosion monitoring remains a critical weakness; therefore, it is critical to develop techniques that are able to facilitate the real-time visualization and corrosion data acquisition of complex localized corrosion in dynamic changing conditions. If such techniques become available for providing *in situ* and site-specific data, better modeling and prediction tools would become possible for significantly improving our capability of managing infrastructures affected by complex and dynamically changing localized corrosion.

3.1.2 Probing Localized Corrosion in Diverse and Changing Environments

A critical need for probing and monitoring localized corrosion in nonuniform and changing environments is the capability to acquire data with high enough temporal and spatial resolutions. A strategy is to develop corrosion probes that are able to simulate and visualize localized corrosion under dynamically changing environments through rapid measurements and display of measurement data as a "video." The design of a corrosion probe for such complex environments could focus on the worst-case scenario corrosion condition. Based on the electrochemical nature of metal corrosion, electrochemical monitoring would be an attractive option for acquiring *in situ* and site-specific corrosion data with high spatial and temporal resolutions. However to perform electrochemical monitoring in highly inhomogeneous, resistive, and varying media such as soil and concrete, several challenging issues need to be addressed. The first issue is practical difficulties associated with the setup and maintenance of electrochemical testing cells for monitoring corrosion in highly resistive and inhomogeneous media. For conventional three-electrode electrochemical measurements, even if electrical continuity between working, reference, and auxiliary electrodes is carefully and successfully maintained, significant *IR* potential drops and nonuniform polarization current distribution problems can still cause major uncertainties in data analysis. It has been recognized for a long time that the nonuniform ohmic potential drop would lead to major errors in the determination of kinetic parameters [36, 37]. Another issue is theoretical limitations associated with conventional electrochemical methods in measuring localized corrosion. In high resistance and inhomogeneous media, localized corrosion is often the key process triggering corrosion failure; conventional electrochemical methods have common limitations in measuring localized electrochemical thermodynamics and kinetics because they are developed based on a uniform corrosion mechanism. Relatively new methods such as scanning probe techniques have been developed into useful research tools for localized corrosion studies; however, scanning probes are unable to scan corrosion occurring in soil, inside a crevice, or under solid deposits. These issues have been extensively discussed in Reference [38].

Obviously successful corrosion testing in highly resistive and inhomogeneous media requires technological innovations. One approach to overcoming some of these practical difficulties is the innovative design of miniaturized electrochemical probes that can be embedded in, or emplaced on, various structures. Miniature probes offer the possibility of overcoming practical difficulties associated with the installation of corrosion probes in difficult-to-access areas such as configurational and structural overlaps on an engineering structure such as aircraft. Miniature probes also make it convenient to setup and maintain electrochemical testing cells. For instance, Glass et al. [39] designed a miniaturized corrosion probe system incorporating individual sensing elements and a miniaturized electrochemical linear polarization resistance (LPR) testing cell for acquiring environmental corrosivity information. Such miniature probe is believed to provide electrochemical data that are useful for early warning of environmental corrosivity. However in principle miniaturized electrochemical probes do not solve issues with the *IR* potential drops and nonuniform polarization current distribution in highly resistive and inhomogeneous media. They still have restrictions in applying to highly resistive media such as soil and concrete and have difficulties in measuring localized corrosion. Another approach is the development of novel electrochemical corrosion sensors that are able to avoid *IR* potential drops and nonuniform polarization current problems and that are able to detect localized corrosion. Macrocell sensors [40], the so-called anode-ladder and expansion-ring, have been developed to monitor the corrosion risk of concrete structures. The anode ladder system consists of several single black steel anodes positioned 50 mm from the next one to prevent interactions between the anodes [40]. In an experiment, the macrocell expansion-ring sensor consisted of six measuring rings ("anodes") separated by sealing rings as parts of the main sensor and a "cathode-bar," which are installed in small holes, to be drilled into the concrete structure. The basic measuring principle is to place several electrodes into a concrete structure at different depths and to measure the macrocell current, potentials, and the electrical resistance of the concrete around the sensors. This sensor system detects the depth profile of the critical chloride content that may initiate depassivation and corrosion, and thus the time-to-corrosion could be estimated, enabling the owners of buildings to initiate preventive protection measures before cracks and spalling occur. In a typical experiment, it was found that the critical chloride content reached a depth of 5 mm into a concrete specimen in about 80 days after concrete placement, which caused a significant increase of the macrocell current [40]. It was also shown that high-performance concrete without addition of chlorides recorded very low macrocell currents, lower than 15 μA, while considerable macrocell currents in the range of 50 μA were measured when 1–5% chlorides cement had been added to the concrete. However it was also found that macrocell currents did not increase with chloride content, and this result suggested that the sensor only gave an indication of galvanic activities due to the ingress of chlorides. It did not give information on corrosion rates of steel in a concrete structure [40].

Over the past decades variously designed corrosion probes based on the electrochemically integrated multi-electrode arrays, often referred to as the wire beam electrode (WBE), have been developed to facilitate the monitoring of difficult-to-measure forms of localized corrosion [38, 41, 42]. As discussed in Chapter 2, the WBE has unique characteristics that enable the effective simulation of corrosion of a continuous metal surface in highly resistive and inhomogeneous media. The

WBE concept has facilitated the design of novel electrode array corrosion probes that overcome limitations of conventional corrosion probes in probing corrosion in highly resistive and inhomogeneous media by addressing issues with potential "ohmic drops" and nonuniform polarization current distribution [41–43]. Variously designed electrode array probes have been used in localized corrosion tests and experiments, as illustrated in Figure 3.2, that simulate localized corrosion in various engineering structures. All corrosion probe designs shown in Figure 3.2 are based on close-packed electrode arrays that enabled unprecedented spatial and temporal information on the behavior and interactions of local corrosion anodes and cathodes. It should also be noted however that some of the far spaced multi-electrode arrays described in the literature would not allow sufficient chemical and electrochemical interaction between electrodes (wires) in an array; therefore they would not effectively simulate a conventional electrode surface in localized corrosion behavior. This is particular important for highly resistive media such as concrete and soil.

Figure 3.2a shows an experimental setup for studying steel corrosion in soil and its inhibition [41]. In a typical experiment, measurements of galvanic current distribution maps were carried out during the exposure of the WBE to dry, damp, and chlorinated sand environments [41]. Characteristic changes in galvanic current distribution maps were observed during the initiation and propagation of localized corrosion processes. Specifically, during corrosion initiation in damped sand, new corrosion anodes were found to initiate and corrosion appeared to be in a general form. When the WBE was exposed to chlorinated sand, massive disappearance of anodic sites was found to occur after the initial corrosion anode initiation period, resulting in accelerated anodic dissolution of a small number of remaining anodic sites, i.e. localized corrosion. Addition of corrosion inhibitor potassium dichromate ($K_2Cr_2O_7$) to the sand environment was found to significantly reduce galvanic current only after an initial increase in galvanic current, suggesting that $K_2Cr_2O_7$ behaved as an anodic inhibitor that prompted initial corrosion to form protective oxide film on electrode surfaces. This experiment demonstrated the ability of the WBE probe in measuring galvanic current distribution maps from a steel piece when the soil environment was altered by water and corrosion inhibitor $K_2Cr_2O_7$ addition [41]. This also suggests the applicability of the setup shown in Figure 3.2a for simulating and measuring changing corrosion of steel and its inhibition in soil.

Figure 3.2b shows a schematic experiment setup that the WBE is used for monitoring the penetration of corrosive species under simulated corrosion under insulation conditions. Corrosion under insulation is a problem frequently occurring to the external metal pipes when they are encapsulated in highly resistive and uneven thermal insulation. Corrosion under insulation often permits corrosion at

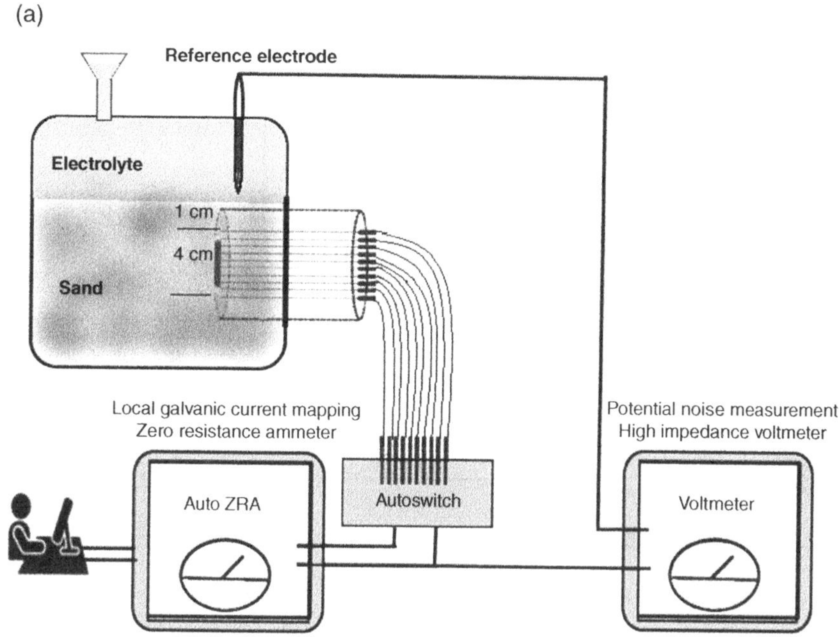

Figure 3.2 Variously designed electrode array probes and measurement systems selected have been used for different engineering structures. (a) Illustration of a typical soil corrosion and inhibition test using an electrode array and a local corrosion map and photo of the corroded electrode array surface. *Source:* Aung and Tan [41], ELSEVIER. (b) Experimental setup for mapping galvanic current and detecting potential noise over a multi-electrode array (WBE) buried under insulation materials. *Source:* Adapted from Aung et al. [43], Emerald Group Publishing Limited. (c) Schematic design for simultaneously recording current flows and electrochemical noise from a rod-shaped WBE embedded in a reinforced concrete structure simulating a sewer or wastewater pipe. (d) An experimental method using the WBE sensor for probing localized corrosion initiation behavior of steel in soil with different moisture contents under the scenario that CP was disrupted and affected by stray currents. *Source:* [21, 28].

local areas of a structure to proceed in an insidious manner and is thus a major concern in petroleum, chemical, food processing, and many other types of industrial structures. Currently the detection of corrosion under insulation relies on methods such as radiography, ultrasonic testing, and insulation removal inspection. Traditional electrochemical techniques such as linear polarization measurements are generally difficult to perform under insulation conditions because it is too difficult to setup a three-electrode corrosion testing cell over a metal surface that is covered by an insulation of high electrical resistance. As shown in Figure 3.2b, corrosion of an aluminum WBE exposed under insulation materials

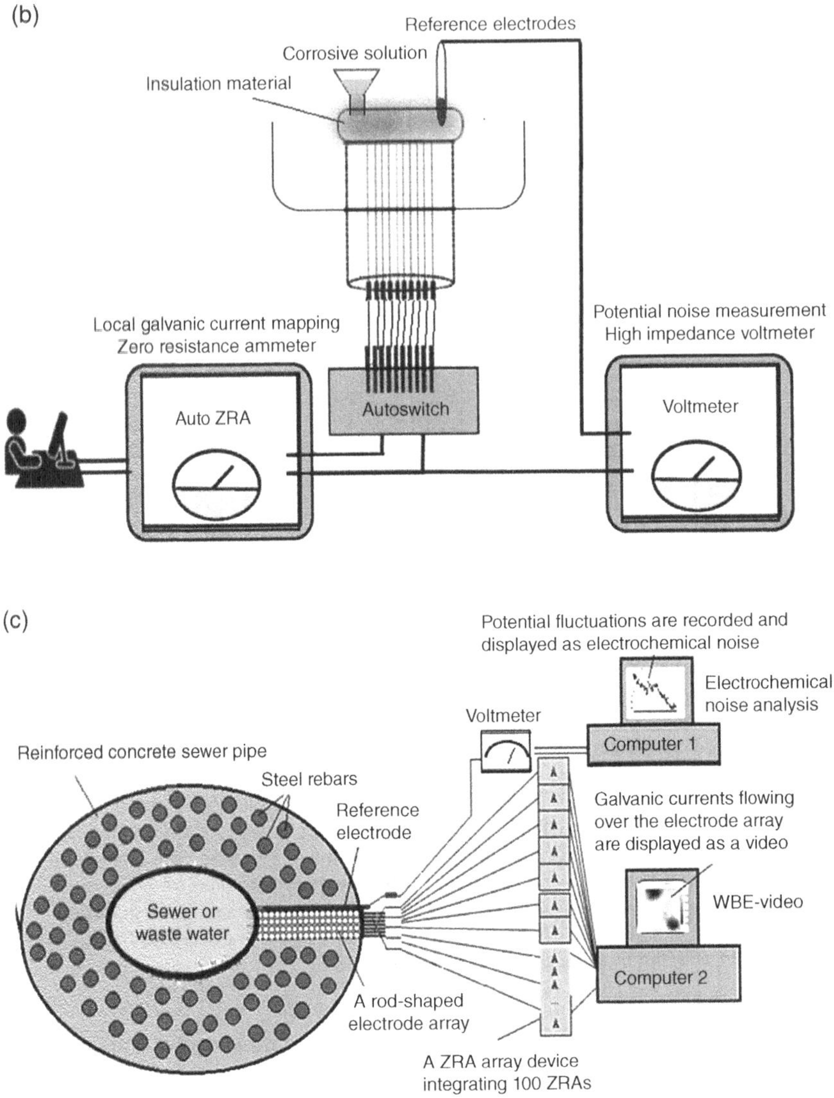

Figure 3.2 (Continued)

such as rock wool, glass wool, cotton wool, and tissue paper was simulated and monitored based on the observation of typical potential noise signature [43]. Corrosion potential noise and galvanic current distribution over the WBE surface were measured continuously to determine the time required for corrosive solution

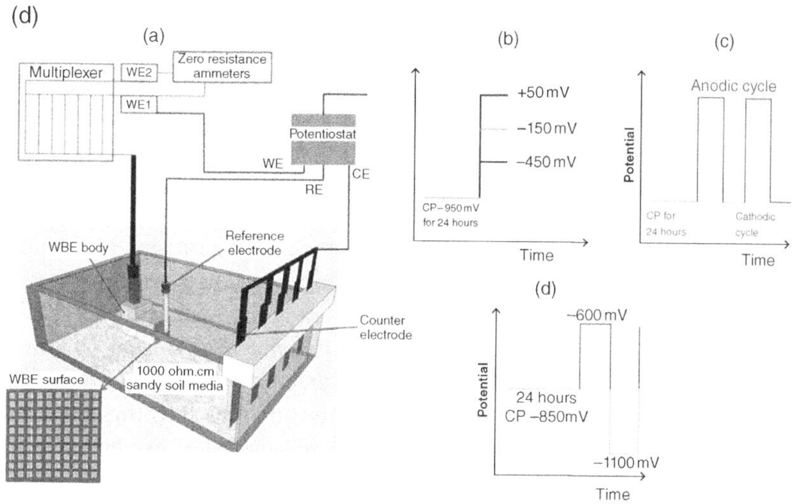

Figure 3.2 (Continued)

to reach the aluminum WBE surface and the subsequent initiation of corrosion. Good correlation between the WBE galvanic current maps and the corroded surface was also observed [43]. This experiment also suggests the applicability of the setup shown in Figure 3.2b for simulating and measuring changing corrosion under various covering and wrapping materials.

Figure 3.2c shows an electrode array designed for measuring and probing corrosion of steel rebar in concrete. This rod-shaped WBE probe is a three dimensional (3D) design for embedding in concrete structures. Closely packed steel wires of mm diameters are used to make the electrode array that allows high spatial resolution simulation and measurement of galvanic currents flowing across wires located over different areas of the WBE rod surface. This probe enables continuous, *in situ*, and real-time monitoring of the initiation and propagation of localized corrosion in reinforced concrete infrastructures, by strategically embedding it in the structures. In an earlier work by Legat [44], both electrode arrays and electrical resistance probes were placed in concrete specimens exposed to periodic wetting and drying cycles for monitoring changing corrosion in concrete structures. It was found that electrode arrays can monitor the time and spatial evolution of steel corrosion in concrete, while the electrical resistance probe was able to measure the cumulative corrosion damage. The measured currents from electrode arrays indicated the temporal anodic and cathodic activities of individual electrodes. Two characteristic fluctuations of the measured currents from the electrode arrays were found: slower fluctuations and shorter transients. It was believed that the slower fluctuations were generated by rather general corrosion of the electrodes, whereas the shorter transients were very probably generated by the initiation of pits [44]. Similar electrode array methods have also been used to understand the thickness and compactness of concrete on rebar corrosion and found that the thicker and denser the protective layer of the concrete, the higher the natural corrosion potential and the higher the pH value [45]. Another study was to use the electrode array for studying the interaction between macrocell and microcell in the early corrosion process of reinforcing steel in cement mortar [46]. The corrosion potential and galvanic current of macrocell corrosion of the reinforcing steel in cement mortar were imaged by the electrode array technique during the corrosion initiation and propagation. It was found that the corrosion macrocell current is closely related with the difference of corrosion potentials between the anodic and cathodic areas. The interaction of corrosion macrocell and corrosion microcell of steel in concrete was directly sensed by the array electrode [46]. New insights into the corrosion mechanisms of steel in concrete were obtained from the characterization of steel corrosion in concrete at the macro- and micro-level using the electrode array in mortar exposed to periodic wetting and drying, in conjunction with steel corrosion damage assessment by micro X-ray computer tomography (CT) and scanning electron microscopy (SEM) [47]. Uneven distributions of galvanic current and open-circuit potential (OCP) on the WBE embedded mortar sample immersed in NaCl solution was observed, and oxygen depletion in mortar was found to facilitate the negative shift of the OCP and the smoothing of the current and potential distributions [48].

Electrode arrays with different geometries and dimensions were also designed for simulating and studying steel rebar corrosion in different reinforced concrete structures exposed to a wide variety of environmental conditions that may cause various damages to the reinforced concretes. A concrete embedding rebar electrodes array was designed and made based on the WBE to simulate the submerged, tidal, and splash marine environments and to monitor the carbonation and chloride ion penetration of concrete, the pH and chloride ion profiles of the concretes [49]. The vertical nonuniform corrosion of steel rebar was found to be caused by the difference of oxygen concentration, moisture content, free chloride content, and pH value in the vertical direction. The macroscopic galvanic corrosion between the splash zone rebars and the tidal zone rebars was inhibited and the galvanic corrosion between the adjacent rebars was promoted by the high resistivity of the splash zone concrete [49]. The electrode array has also been used as a part of integrated sensor unit for assessing the corrosion status of reinforcing steel rebar in concrete, includes an Ag/AgCl probe, a metal oxide probe, a multi-electrode array sensor, and a four-pin (Wenner) array stainless steel electrode for chloride content, pH, microcell corrosion current, and localized concrete resistivity measurements, respectively [50]. The application of electrode arrays were further extended to evaluate corrosion inhibitors and anti-corrosion coatings for protecting steel in concrete. A WBE was used to assess the inhibition of reinforcing bar corrosion in concrete with the addition of different inhibitors such as $NaNO_2$, $NaNO_2$ + Na_2CO_3, and $NaNO_2$ + aniline [51]. By measuring the open-circuit potential of an iron wire array, the impact of different inhibitors on the corrosion of rebar in the concretes was studied after immersion in 3% NaCl solution. It was found that best protection was achieved by adding $NaNO_2$ and aniline as combined inhibitors [51]. Electrode arrays were further used to evaluate the migration and inhibitive effect on rebar corrosion of migrating corrosion inhibitors in mortar. Migrating corrosion inhibitors were found to readily diffused through the mortar and increased the resistance to general corrosion much more than localized corrosion. This increase was mainly attributed to the sealing effect on mortar pores (pore-blocking effect) rather than on chemical absorption [52]. Multielectrode array sensors fabricated from carbon steel, stainless steel, and aluminum were utilized to assess the effectiveness of two corrosion inhibitors and one coating under cyclic deicer exposure conditions [53]. Both the sensor corrosion rates and mass loss results indicated that the inhibitors had a much greater inhibiting effect in protecting the metallic substrates during the cyclic exposure to $MgCl_2$ than to NaCl containing corrosive solutions [53]. The effect of concrete coating on corrosion of reinforced bar in concrete was also assessed by the WBE method [54]. It was found that coating the surface of concrete could keep the open-circuit potential of steel rebar within the passivation range, ensuring the integrity of the passivation [54]. More cases of designing and using electrode array-based corrosion probes for probing localized electrode processes occurring on steel buried in highly inhomogeneous and resistive soils, concrete, multiphase fluid media are described in References [38, 55, 56].

As shown in Figure 3.2c, an electrode array was designed for measuring and understanding corrosion of steel rebar in a reinforced concrete sewer pipe, where the nonuniform distribution of chlorides, the penetration of hydrogen sulfide (H_2S), and sulfuric acid (H_2SO_4) would create a corrosive environment [57–59]. The formation of local environments is affected by the concrete quality such as water/cement ratio, cement content, impurities in the concrete ingredients, presence of surface cracks, and by the external environment such as moisture, oxygen, humidity, temperature, bacterial attack, stray currents [60]. The monitoring of localized corrosion in concrete requires this ability to continuously measure the spatial distribution of local environmental conditions and associated localized corrosion activities together with their seasonal changes. Figure 3.2c also shows the electrode array used in conjunction with electrochemical noise analysis [56, 61] to enhance the detection sensitivity. When the electrode array is used in conjunction with electrochemical noise analysis (ENA) [56, 61–65], the detection sensitivity would be enhanced allowing the detection of early signs of localized corrosion initiation. Figure 3.2c also illustrates a localized corrosion probing video system that is designed by integrating a 100-element electrode array with 100 zero resistance ammeters in order to provide real-time recording and display of corrosion-related current flows between wire terminals in a WBE, displaying current flows over the electrode array surface as a "video." The images of localized corrosion initiation activities can be monitored instantaneously by displaying high-spatial and time-resolution "current flow video" in real time. Combined with the ENA, the monitoring system would allow both high temporal and spatial resolution detection of dynamically changing localized corrosion processes and mechanisms. Another method of obtaining localized corrosion video of reasonable resolution is to take WBE maps quickly and continuously, and display maps as a video [66] (sample video clips can be found online in Supplemental Material of an open access article Reference [66]). This approach was applied to better understand the mechanisms of corrosion under disbonded coatings and stray current corrosion [67, 68].

As illustrated in Figure 3.2d, a WBE probe is used to enable the *in situ* monitoring and visualization of electrochemical processes occurring on buried steel surfaces under cathodic protection (CP) and stray currents (anodic transient) conditions in soil. This setup has been used for probing the effects of stray current on pipeline corrosion, as described in the following sections of this chapter. For all these cases described above, good correlations were found between the corrosion patterns obtained based on electrochemical calculation using the electrode array corrosion monitoring data and the actual patterns visibly present on the array surface [65–69].

3.2 Steel Corrosion Behavior in Soil Under Disrupted Cathodic Protection

A case of major and dynamic environmental change that could significantly affect steel corrosion in soil is the disruption or failure of impressed current cathodic protection (ICCP) system. For instance the ICCP system could be disrupted due to severe anode deterioration or vandalism, or due to CP supply and control failure. Such disruption would certainly a concern to industry, and therefore knowledge of the potential effect of such CP disruption on steel corrosion is of interest.

3.2.1 Localized Corrosion on Steel in Soils Under Varying CP and Moisture Levels

Steel structures buried underground are commonly protected by CP, which induces changes to chemical environment around the steel surface, for instance, leading to a high pH on coating defects. When CP is disrupted or affected by stray currents, the environment surrounding a buried pipeline would change and therefore pipeline corrosion behavior and mechanism would also change with time. It was reported in the literature that the pH distribution on steel buried in soil would be affected by CP and the associated environmental conditions such as soil water saturation, soil texture, as well as coating defect size [25, 26]. Soil environment and corrosion behavior are expected to change when CP is malfunction, disrupted or affected by stray currents [27, 28]. Such change needs to be understood because it affects the validity of the modeling and prediction of localized corrosion under such complex and dynamically changing corrosion environments. Unfortunately the actual corrosion behavior of the pipeline under CP induced environmental change is generally unknown due to the lack of sufficient data, which poses problems in corrosion prediction and control. Current understanding of soil corrosion on CP protected steel is generally from the testing and modeling of homogeneous steel surfaces exposed to stable soil environmental conditions. In the pipeline industry, the corrosion behavior of pipelines is generally revealed through time-based inspection such as digging and pigging. Corrosion rates calculated from such inspection data are generally averaged and accumulated corrosion losses over a long period of time. Such data would be insufficient for corrosion prediction and management if the environmental conditions are varying and corrosion mechanisms are changing.

Over the past decades, extensive laboratory tests have been carried out to obtain corrosion data through evaluating the effects of a range of influencing factors such as soil moisture contents [2, 6–9], aeration [10, 11], electrical resistivity [12, 13], pH [14], ionic species [15], and microbiological activity [16] on corrosion in soil. Unfortunately in each soil corrosion test only one factor is varied, while all other soil condition parameters are kept consistent over the whole test period. This approach would need many individual tests in order to assess the effect of each influencing factor and its combinations with other factors on soil corrosion. It is very time consuming and is almost impossible to assess changes in localized corrosion under the effects of dynamically changing soil condition parameters such as moisture levels, CP malfunction, and fluctuating stray currents, and their combinations. To quantify the effect of soil variables on localized soil corrosion, there is a need for technology that is able to quickly acquire data under changing environmental conditions. An electrode array-based probe has been employed as a tool for this purpose. In a typical experimental study [67], as illustrated in Figure 3.3, Wang et al. monitored the localized corrosion initiation behavior of steel in soil with different moisture contents under the scenario that CP was disrupted. The electrode array probe enabled the visualization of dynamic localized electrode processes due to active changes in the soil corrosion environment and CP condition. Figures 3.4–3.7 show corrosion maps that reveal the behaviors of corrosion occurring on electrode array surfaces buried in sand. The changes in

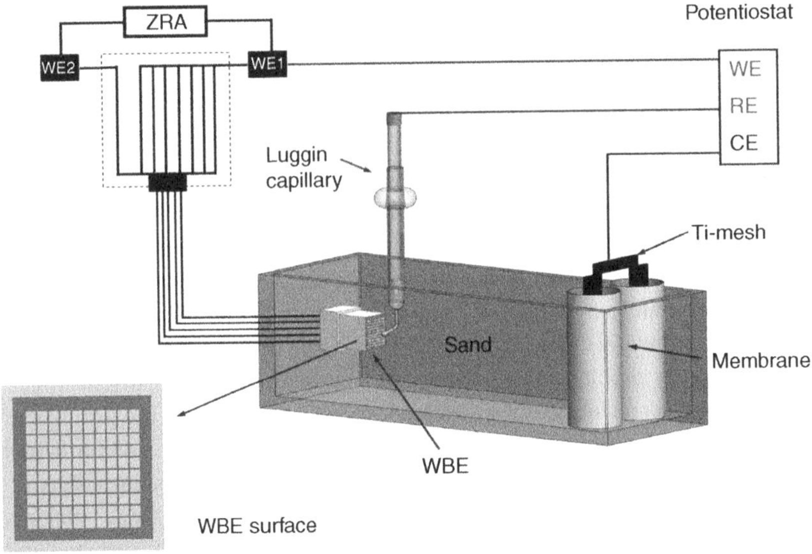

Figure 3.3 The schematic diagram of the test setup for probing localized corrosion initiation behavior of steel in soil with different moisture contents. *Source:* Wang et al. [67], Allen Press, CC BY.

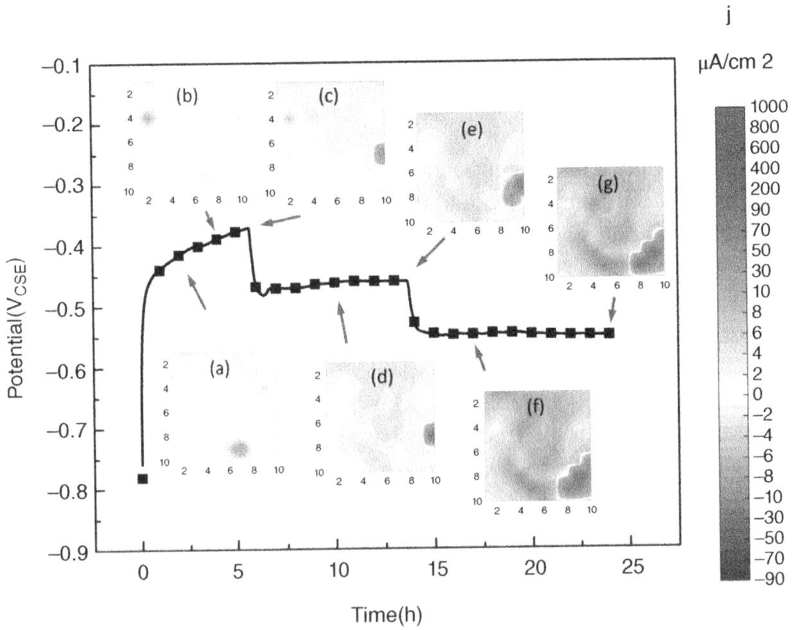

Figure 3.4 OCP recordings and the WBE current density distribution maps obtained from the WBE following 24 hours prior CP in the sand with 30% WHC. *Source:* Wang et al. [67], National Association of Corrosion Engineers.

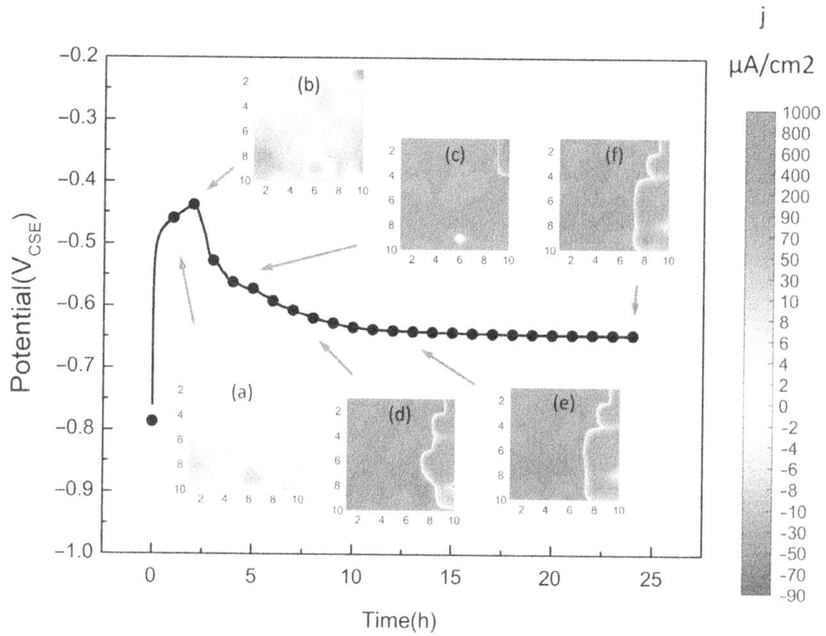

Figure 3.5 OCP recordings and the WBE current density distribution maps obtained from the WBE following 24 hours prior CP in the sand with 60% WHC. *Source:* Wang et al. [67]/ National Association of Corrosion Engineers.

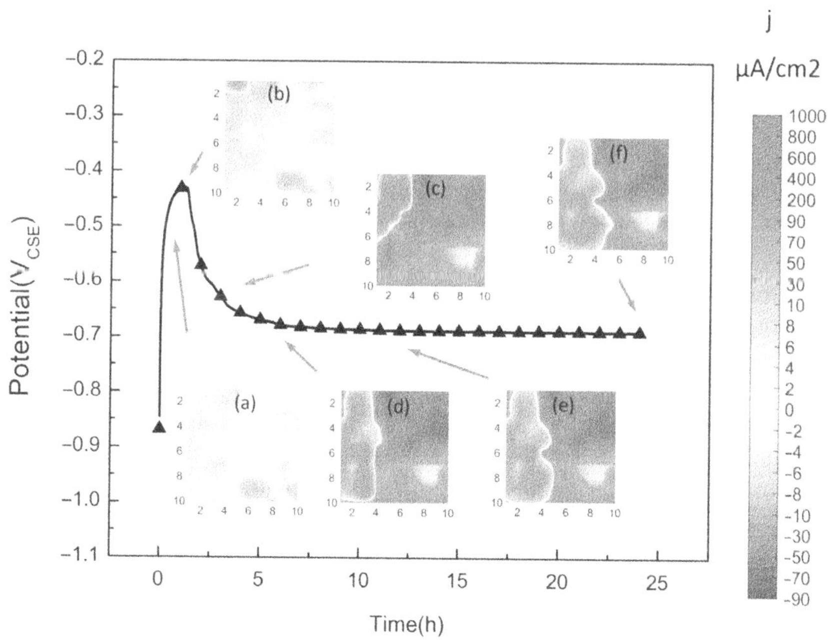

Figure 3.6 OCP recordings and the WBE current density distribution maps obtained from the WBE following 24 hours prior CP in the sand with 80% WHC. *Source:* Wang et al. [67]/ National Association of Corrosion Engineers.

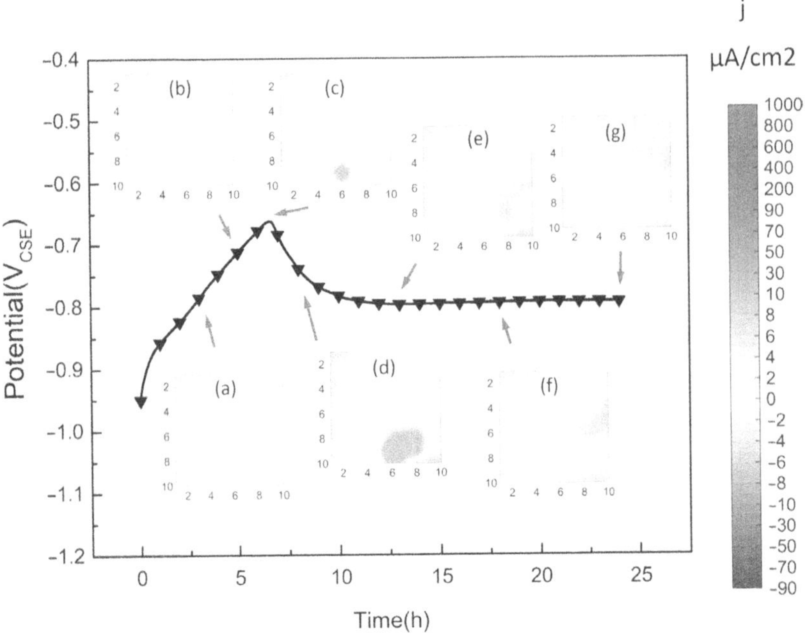

Figure 3.7 OCP recordings and the WBE current density distribution maps obtained from the WBE following 24 hours prior CP in the sand with 100% WHC. *Source:* Wang et al. [67]/ National Association of Corrosion Engineers.

open-circuit potential (OCP) and the current density distributions clearly show dynamic changes occurring on the WBE probe surface after CP was stopped, simulating the situation that CP was disrupted [21, 22, 28]. It is also clear that changes were different in soils with different moisture content (30, 60, 80, and 100%), suggesting different corrosion behavior in different soil conditions after CP disruption. Although only a few WBE current maps are shown in each of Figures 3.4–3.7, more WBE current maps have been taken and indeed corrosion can be visualized as a video if all these maps are shown in sequence (sample video clips showing corrosion video can be found online in Supplemental Material of an open-access article Reference [66]). This is another version of the WBE-video technique [66], in addition to the WBE-video device shown in Figure 3.2c that is based on a multichannel zero-resistance ammeter array, to provide real-time recording and display of corrosion-related current flows between wire terminals in a WBE.

A general trend observable in Figures 3.4–3.7 is that after CP disruption, the OCP changed rapidly. For instance, as shown in Figure 3.4, after CP was disrupted the OCP of the WBE gradually became nobler during the first five hours of exposure to the sand with 30% water holding capacity (WHC), which is consistent with the results presented in the literature [70, 71]. When the OCP gradually became nobler the anodic current densities (in gray color) registered on WBE probe surface (Figure 3.4a and b) decreased, reaching a small value of 1 μA/cm^2, suggesting some process occurred on the probe surface to reduce corrosion. The most likely cause is passivity that could occur on the WBE surface where the pH was higher than 10. The pH of the probe surface was higher than 10 due to prior CP. Subsequently OCP shifted rapidly to more negative value, as shown in Figure 3.4c. After the sharp decreasing of OCP at around 5.5 hours, localized corrosion accelerated immediately, as shown in Figures 3.4d and e. A large anodic current peak appeared at the edge of the WBE surface, indicating localized corrosion at the edge of the surface. The time required for the sharp decrease in OCP and increase in localized corrosion can be termed "critical duration" for corrosion initiation after CP disruption [21, 28]. This 'critical duration' is important to industry because it is the period that CP disruption is allowed without inducing corrosion. Corrosion would propagate after its initiation. As shown in Figure 3.4, with the extension of exposure of the WBE, the anode can be seen to propagate to neighboring locations over the WBE surface. This was accompanied by a second sharp decrease of OCP, until reaching the value of −550 mV$_{CSE}$. The maximum anodic current density over the WBE remained high (152 μA/cm^2) and the sum of anodic current increased from 8 to 9.7 μA. Subsequently, as shown in Figure 3.4f–g, the anodic areas propagated gradually to more neighboring locations over the WBE surface, with the increase of total anodic current to 23 μA. These data clearly show the dynamically changing nature of localized corrosion mechanisms, locations, and kinetics [67].

The 'critical duration' and corrosion behaviour were found to change with soil conditions especially its moisture contents. In a sand with higher moisture content of 60% of WHC, as shown in Figure 3.5, much larger anodic currents were recorded on the WBE surface and it required a shorter duration for OCP to decrease sharply. As shown in Figures 3.5a and b, a large anodic current peak appeared after the abrupt decreasing of OCP, indicating the initiation of localized corrosion at the edge of the WBE. With the extension of corrosion exposure, as shown in Figure 3.5c and d, corrosion anodes propagated quickly to neighboring electrodes over the WBE surface. The sum of anodic currents increased sharply to 100 μA and reached 135 μA, while the maximum anodic current density increased to 996 μA/cm^2 before stabilizing at approximately 310 μA/cm^2. These data clearly show not only the dynamically changing nature of localized corrosion mechanisms, locations, and kinetics but also rapidly propagating localized corrosion. When the moisture content of the sand further increased to 80% of WHC, as shown in Figure 3.6, an even shorter duration for localized corrosion initiation was observed although the general trend of the OCP and the WBE current density distribution patterns remained similar to those in Figures 3.4 and 3.5. When the moisture content of the sand further increased to saturated (100% WHC), as

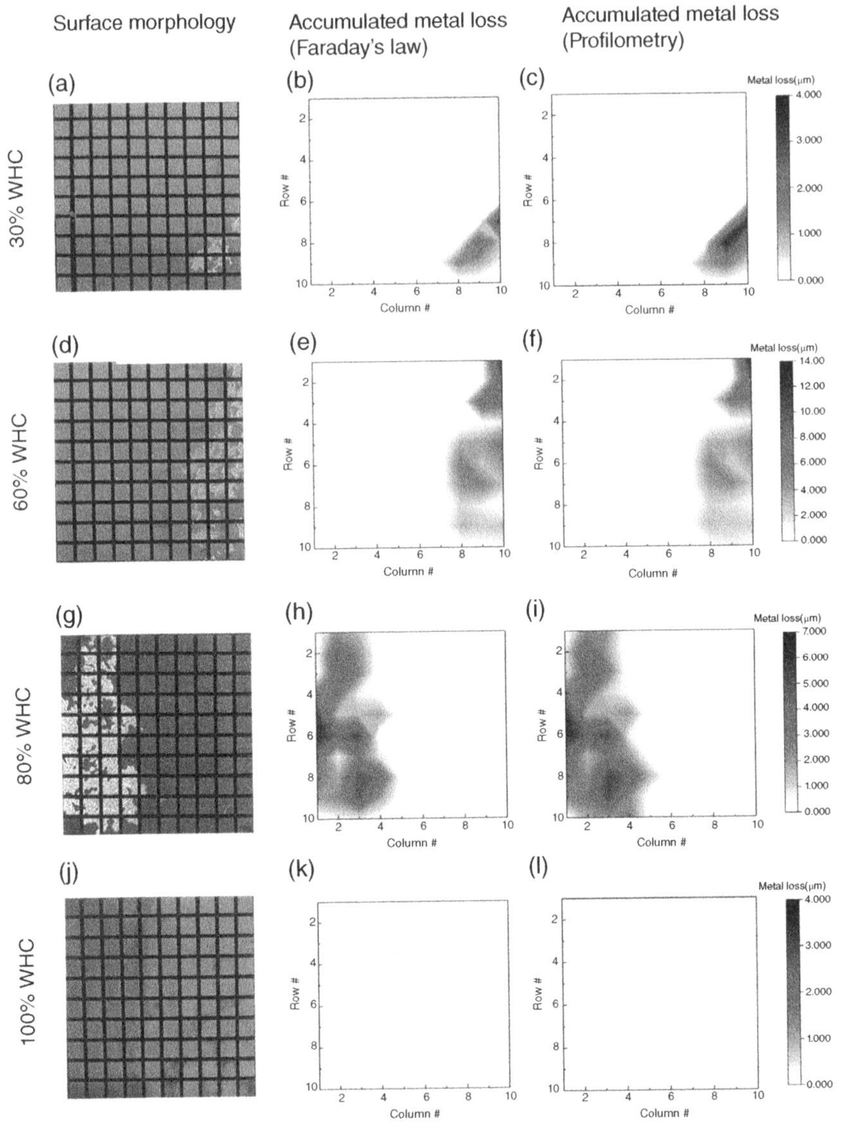

Figure 3.8 The surface morphologies of the WBE after rust removal with different moisture content soil: (a), (b) and (c) 30% of WHC; (d), (e) and (f) 60% of WHC; (g), (h) and (i) 80% of WHC; (j), (k) and (l) 100% of WHC. *Source:* Wang et al. [67], National Association of Corrosion Engineers.

shown in Figure 3.7, the potential values recorded was much more negative than those obtained in sand with lower moisture levels (see Figure 3.7a and b) although a similar trend of OCP was observable. In saturated sand cathodic reaction currents concentrated at the edges and corners of the WBE, which agrees with the simulation results presented in the literature [72]. These results suggest the significant effect of soil condition on corrosion in the case of CP disruption. It appears corrosion behaved differently under the effect of different combinations of environmental factors and CP conditions, suggesting difficulties in modelling and predicting corrosion based on only a few factors. For this reason, corrosion monitoring is needed although corrosion models are helpful corrosion management tools.

After corrosion experiments described above, the surface morphologies of the WBEs (after rust removal), as well as the electrochemically calculated and optically measured corrosion depth, are characterized. As shown in Figure 3.8, the highest corrosion metal loss occurred in soil with a moisture content of 60% of WHC although the total area of corrosion attack was the largest in soil with a moisture content of 80% of WHC. Localized corrosion damages and their distributions over the WBE surfaces determined based on Faraday's law and by surface profilometer, especially the extremely corroded sites, matched very well, indicating that the galvanic currents detected by the WBE current maps are mainly responsible for the localized corrosion damages. This confirms the accuracy of corrosion visualization shown in Figures 3.4–3.7. When the sand became saturated, although some oxide film was found on the WBE surface, no obvious corrosion was observed after the removal of corrosion products (Figure 3.8j) and the corrosion depth was too low to measure (see Figure 3.8k) [67].

In situ and site-specific corrosion data shown in Figures 3.4–3.7 provide lots of useful information about changing corrosion processes and mechanisms affected by soil moisture levels: (i) passivity of steel at an initial prior-CP generated high pH conditions; (ii) loss of passivity and localized corrosion initiation when CP was disrupted and after a 'critical duration'; (iii) localized corrosion growth and acceleration. From data shown in Figures 3.4–3.7, maximum anodic current density over a WBE surface can be determined that indicates the maximum corrosion dissolution rate, while the average anodic current density indicates the overall corrosion rate. It is clearly shown that the average cathodic current density on the WBE surface firstly increased significantly when the moisture content increased from 30 to 60%, while this value stabilized when the moisture content further increased to 80%. However, when the sand became saturated, rather small cathodic current was recorded on the WBE surface. The maximum and average anodic current densities shared similar trend, which reached the maximum value when the moisture content was 60% of WHC. Although these results are in agreement with the general corrosion behavior reported in the literature [3, 7], the localized corrosion behaviors shown in Figures 3.4–3.7 have enabled much more detailed understanding of localized corrosion on cathodically protected steel and corrosion after CP disruption. This knowledge would facilitate more accurate corrosion modeling and prediction of dynamically changing corrosion of steel pipeline in soil [67].

3.2.2 Localized Corrosion on Steel with Varying Coating Defect Sizes

Corrosion maps in Figures 3.4–3.7 only indicate localized corrosion behavior on simulated steel pipes with the same coating defect area of approximately 22.5 mm × 22.4 mm [67]. On practical pipelines, different sizes of coating damages or defects could exist on different areas of a pipeline, adding a further complication to the complexity of localized corrosion in soil [73]. It was reported in literature that the surface area of a metal surface can affect corrosion behavior such as pitting initiation [74, 75]. The initiation of localized corrosion on passive metals has often been considered to be a stochastic process that is attributed to the random distribution of defects or "active" spots in passive films [76], such as sulfide inclusions and grain boundaries [77, 78]. Since the total number of these "active" spots is proportional to the exposed metal surface area, naturally it is considerred that the chance for localized corrosion initiation on a larger metal surface would be higher than that on a smaller one [77, 78]. For instance, Aziz et al. [79] found that the pitting probability of an aluminum specimen in tap water increased with the increase of exposed specimen area (until it reached 60 cm^2). Li showed that the average pitting potential, which is related to the probability of pitting initiation, decreased significantly with increasing specimen size [78]. Similar behavior was found in concrete environment where a high pH usually keeps steel reinforcements passive. Angst et al. observed that the threshold chloride concentration decreases, which suggests higher localized corrosion susceptibility in concrete structures, with the increase in the exposed steel surface area [80]. All these suggest the strong influence of specimen size on experimental results. For this reason, Angst et al. suggested that results measured on certain sized samples can only be used to perform service life calculations and predictions of the same sized structures [80]. These results also suggest that the choice of electrode surface area would be an important factor affecting localized corrosion measurements and that results from laboratory corrosion testing using small-sized specimen could be misrepresentative when they are used to predict localized corrosion on larger structures such as underground pipelines. This is concern for the design of corrosion probes.

It should be noted that reports on the effect of electrode size on localized corrosion initiation were from experiments in relatively homogeneous and stable aqueous conditions [77–80]; little study was done in more complex environments such as underground pipelines that contain coating defects of various sizes due to various reasons such as manufacturing flaws, construction damages, and numerous forms of cracks and scratches. It has been recognized through experimental studies and mathematical modeling that the coating defect size affects the distributions of CP potential and currents [72, 81, 82] and the local chemical environment over a buried pipeline [25, 26]. However, there is a lack of understanding of the actual corrosion initiation behavior on differently sized coating defect areas on a buried pipeline buried in complex soil environments with coating defects of varied sizes and under dynamic environmental condition changes.

Wang et al. [73] carried out a study to examine the effect of electrode surface area on localized corrosion of X65 steel buried in soil under the scenario that CP was disrupted, mimicking a corrosion issue frequently observed on underground steel pipelines. A X65 pipeline steel multi-electrode array (the WBE probe) was used in a soil test cell similar to that shown in Figure 3.3. Current mapping using the probe was performed in conjunction with electrode potential monitoring to visualize dynamic corrosion initiation processes occurring on X65 steel coupon electrodes or WBEs of different surface area buried in unsaturated sandy soil under the scenario that CP was disrupted. After prior-CP was disrupted, as shown in Figure 3.9, OCP of steel electrode (or WBE) specimens of different sizes (5 mm × 5 mm, 10 mm × 10 mm, 20 mm × 20 mm, 30 mm × 30 mm) the OCP became nobler, and subsequently the OCP shifted rapidly to more negative values. These behaviours are rather similar to those shown in Figures 3.4-3.7 although the 'critical durations' are rather different. The initial positive movement of the OCP is most likely due to passivity of the WBE surface where the pH was higher than 10 due to prior CP, while the subsequent negative shift of the OCP should be due to surface chemistry changes leading to the local loss of passivity. This behavior is similar to that shown in Figures 3.4–3.7 and is in agreement with the results of Haleem et al. [71].

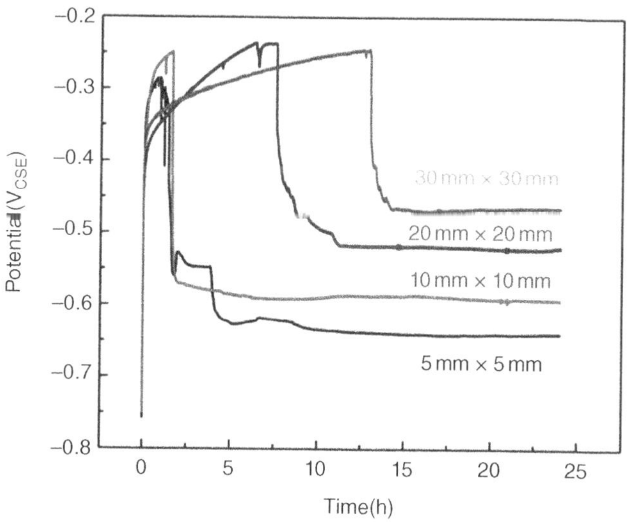

Figure 3.9 The variation of the open-circuit potential (OCP) of different size X65 steel specimen with time following 24 hours CP. *Source:* Wang et al. [73]/ With permission of Elsevier.

As shown in Figure 3.9, the duration at which the sharp decrease in OCP occurred was generally shorter with smaller sized electrodes. For instance, this sharp drop occurred on the 5 mm × 5 mm steel electrode approximately 1 hour after CP interruption, while it occurred on the 30 mm × 30 mm steel coupon after more than 13 hours. Obviously the 'critical duration' required for the OCP to drop sharply is directly related to the size of the X65 steel coupon [28, 68]. The sharp potential decreases shown in Figure 3.9 could be reasonably attributed to the initiation of localized corrosion, which can be directly verified by visualizing the dynamic electrode processes occurring on the electrode surface when its potential dropped sharply using the WBE. Figure 3.10 shows typical OCP recordings monitored from the WBE surface and current density distribution maps measured from the WBE probe at different points in time. From the anodic locations and values shown in the WBE current maps (inserted in Figure 3.10), it can be seen that initially when the OCP gradually became nobler, current densities on the WBE surface (Figure 3.10a–c) decreased until reaching a very small anodic current density value of 1 μA/cm^2, indicating corrosion was almost stopped after CP was disrupted. This behavior is similar to that shown in Figures 3.4–3.7 and that described by Huo et al. under anodic transits but before reaching the "critical duration" [21, 28, 68]. After the sharp decrease of the OCP, as shown in

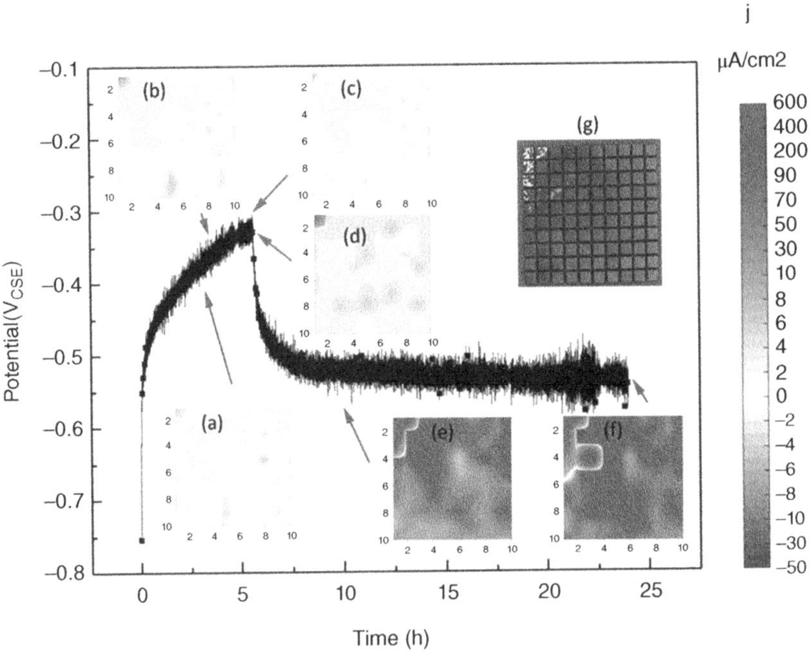

Figure 3.10 OCP recordings and current density distribution maps obtained with a WBE surface following 24 hours CP. *Source:* Wang et al. [73]/ With permission of Elsevier.

Figure 3.10d, a major anodic peak appeared and the anodic current (in gray color) increased significantly, indicating major localized corrosion occurred at the edge of the WBE surface. This clearly indicates that the sharp drop of the OCP was indeed due to the initiation of major corrosion anodes, leading to major increase in local anodic current value and rapid localized corrosion at the anodic location. This experiment confirms the validity of determining the "critical duration" by monitoring OCP. With the extension of exposure time, as shown in Figure 3.10e and f, the anodic areas propagated gradually to more neighboring electrodes over the WBE surface, suggesting the propagation of corrosion. It can also be seen in Figure 3.10d–f, these peak anodic areas were located at the edges of the WBE. Indeed, as shown in the inserted photo (Figure 3.10g), major corrosion occurred on the top left corner of the WBE surface. This is very similar to localized corrosion mechanisms that have been discussed previously in Reference [75].

These results shown in Figures 3.9 and 3.10 demonstrate that under the buried condition, the probability or susceptibility of localized corrosion initiation on a larger surface is lower than that on a smaller electrode. This behavior is opposite to that of conventional passive metals exposed to more homogeneous media reported in the References [77–80]. This suggests that for buried carbon steel and low-alloy steel, the controlling factor for corrosion initiation under CP interruption is not the randomness in local properties of the passive film, as in the case of conventional passive metals [77–80], and therefore the initiation of localized corrosion must be controlled by a different factor. To explain the behavior of buried X65 steel shown in Figures 3.9 and 3.10, pH monitoring was performed together with the OCP monitoring [73]. It was found that the steel surface became alkaline after the application of prior-CP and that the initial pH values on larger sized electrodes were higher than that on smaller size specimen. The pH decreased slower on larger sized electrode specimens, although the pH values decreased sharply after a period of gradual change in pH on all the specimen surfaces [73]. There appears a correlation between the critical duration for localized corrosion to occur and the duration for the pH to drop sharply, suggesting that the initiation of localized corrosion was controlled by the sharp decline of pH on the steel surface [73]. The importance of the steel surface pH on localized corrosion initiation is understandable given the significant effect of pH on the passivity of carbon steel and low-alloy steel in soil. These data in Figures 3.9 and 3.10 also clearly show the dynamically changing nature of localized corrosion mechanisms, locations, and kinetics. These results suggest again that accurate modelling and prediction of pipeline corrosion based on only a few factors is difficult, for this reason corrosion monitoring is needed if accurate understanding of corrosion behaviour is required for corrosion management.

The fact that X65 steel specimen size has a major effect on corrosion initiation is important because firstly it would affect the maintenance and CP requirement of underground carbon steel structures with varied coating defect sizes.

For instance, more CP current could be needed for pipeline with many small defects in order to increase its pH values and maintain its passivity when CP was interrupted or interfered. More importantly, the fact that electrode sizes affect corrosion initiation would significantly affect the design of corrosion testing coupons and probes for corrosion testing and monitoring. When a large steel coupon is employed in the laboratory to evaluate the "critical duration" for localized corrosion initiation in the case of CP system disruption, it would be misleading to use this "critical duration" as a guide for the whole buried structures that may have coating defects of various sizes. Therefore, when laboratory tests are designed, the size effect should be taken into consideration. In terms of corrosion monitoring, the choice of corrosion probe size also needs to be considered. If a corrosion probe is too large compared to practical coating defect areas on buried carbon steel structures, it would not be able to detect localized corrosion occurring on the smaller defects of the pipeline. In practice, since it is impossible to replicate all differently sized defects with the corrosion probe, corrosion probes should be designed to replicate typical or "worst-case scenario" cases that may occur on the steel structures in order to provide early warning of major corrosion issues. Overall, knowledge described above could be used to better devise CP and maintenance strategies for underground pipelines, as well as to improve the design of corrosion probes for pipeline corrosion monitoring and prediction.

3.3 Steel Corrosion Behavior in Soil Under Stray Currents

Corrosion data shown in Figures 3.4–3.7 as well as in Figures 3.9–3.10 clearly demonstrate the dynamically changing nature of localized corrosion mechanisms, locations, and kinetics when CP protected steels were affected by varying soil conditions, different coating defect sizes, and disrupted cathodic protection. In industry practical, such changes can occur even more dramatically when stray currents exist in the soil environment. It is well known that the existence of dynamic stray currents in the environment could nullify CP, leading to stray current corrosion of buried pipe. Due to the invisible and complex nature of buried pipelines, knowledge of stray current corrosion is generally from phenomena observed empirically on corroded pipelines through field inspection in the industry. There is a lack of reliable and reproducible experimental methodologies to systematically quantify the level and nature of damage as a result of various modes of dynamic stray current effects. This section describes new approaches to probing and quantifying various issues related to stray current corrosion using electrode array based corrosion probes.

3.3.1 Locating Stray Current Corrosion in Soil

A buried steel pipeline can be considered to be protected from soil corrosion by organic coatings and CP when a stable CP potential of −850 mV vs. $Cu/CuSO_4$ reference electrode (CSE) is applied on it. However, in practice, the validity of this standard CP potential has been under continuous debate [83–86] due primarily to the fact that the efficiency of CP can be significantly affected not only by the applied CP potential but also by several other major influencing factors, especially stray currents. It has been reported that anodic transients of varying durations (from a few seconds to a few minutes) from a tram line could push the actual potential of a pipeline to a level significantly more positive than the "safe" CP potential level, causing serious localized corrosion issues [68, 86]. When a buried pipeline is affected by stray currents, corrosion on the pipeline would not only change dynamically but also occur in some specific locations, leading to localised corrosion. As illustrated in Figure 3.11a [87], at the point where stray current enters a pipeline (the pickup point), the steel is catholically polarized and little corrosion would occur; while at the point where stray current exits the pipeline (the discharge point), the potential of the steel is shifted anodically and serious corrosion is expected. It is well known that usually coating defects on a coated pipeline form stray current pickup and discharge points when the pipeline is affected by stray currents. The location of the stray current pickup and discharge points could be affected by factors including the location and size of coating defects, as well as the soil resistivity distribution. For instance if only a small coating defect exists at the stray current discharge point, the corrosion attack could often be in a extremely localized form at the coating defect location [88]. These influencing factors clearly indicate the complexity of CP potential distribution over a stray current affected pipeline. In order to manage stray current corrosion, there is a need for monitoring not only the potential distribution, but also the location, the pattern and rates of stray current corrosion.

Unfortunately the determination of the location, the pattern and rates of stray current corrosion on a buried pipeline is not an easy task. Traditionally stray currents on pipeline are typically assessed by measuring the potential of a pipe by various data logging and survey methods [89], for instance the potential survey method is used for locating probable anodic locations on a stray current-affected pipe. Probable anodic sites, usually at stray current discharge point, are indicated at survey points where the most positive potential readings are detected. A major limitation of these potential measurement-based survey methods is that they are unable to determine the extent, pattern, and rates of pipeline corrosion because electrode potential is generally a thermodynamic parameter and is not an indicator of corrosion patterns nor kinetics [55]. This limitation could be overcome if a WBE-probe array system is used to directly locate, measure, and monitor pipeline

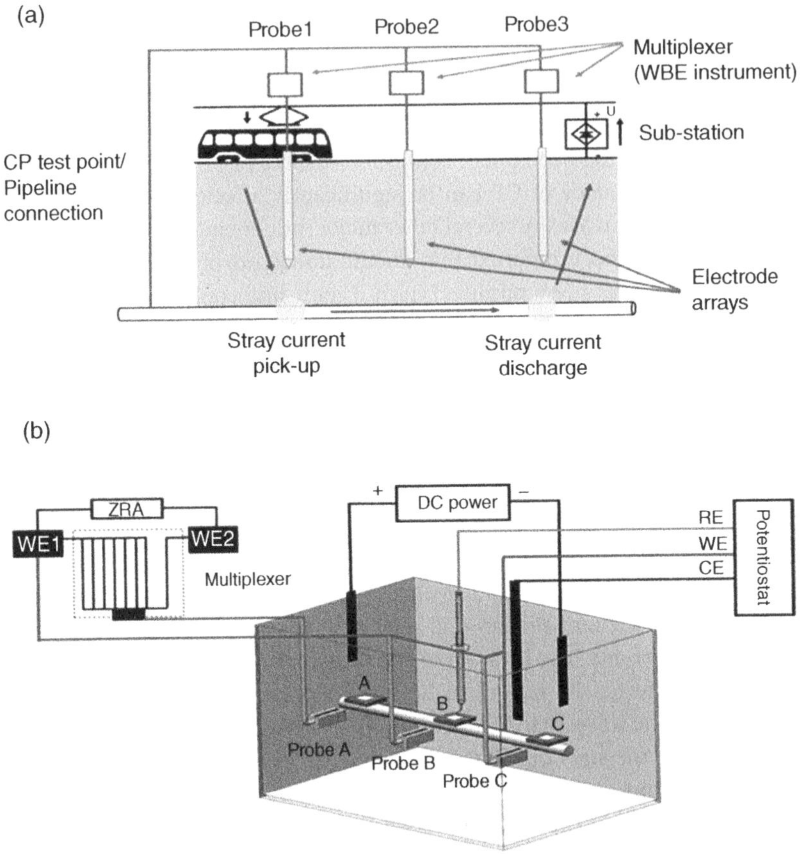

Figure 3.11 Schematic diagram of stray current and the test system (a), and schematic diagram of the experimental simulated pipe setup (b). *Source:* Wang et al. [87]/ National Association of Corrosion Engineers.

corrosion rates and patterns at survey points along a pipeline, as described by Wang et al. in reference [87]. As illustrated in Figure 3.11, the array system consists several WBE probes that are integrated by electrically coupling together to simulate a pipeline. This method was reported to be able to overcome major technical hurdle in identifying the stray current-affected pipeline location (e.g. the discharge point), as well as visualizing and monitoring stray current corrosion along a pipeline [87].

In a technical demonstration reported in Reference [87], this WBE-probe array system concept was demonstrated to survey corrosion due to DC stray currents. In order to mimic DC stray current effects on a pipeline section, as illustrated in Figure 3.11b, three steel coupons (A, B, and C) were mounted by epoxy resin and electrically connected together to simulate coating defects at different locations along the simulated pipe. Electrically "coupled" multi-WBE probes (only three are illustrated in Figure 3.11b) are placed next to the simulated coating defects on the pipe, and coupled together through a multiplexer. In this manner, this multi-WBE probe array system is integrated into the DC stray current-affected pipeline system, and each probe would behave differently depending upon the buried location of the probe [87]. Tests have been conducted to demonstrate the capability of the multi-WBE probe array system for (i) identifying the stray current affected pipeline location (e.g. the discharge point); (ii) measuring and monitoring the actual corrosion rates and patterns along a simulated pipeline that was affected by simulated DC stray currents; and (iii) assessing CP efficiency under the standard CP potential of $-850\,mV_{vs.CSE}$ [87]. Figures 3.12 and 3.13 show current distribution maps obtained from the WBE-probe array system buried in the sand box shown in Figure 3.11b under open-circuit potential (without CP protection) and under a CP potential of $-850\,mV_{CSE}$, respectively, with and without the effects of different levels of DC stray currents.

When no stray current was applied to the simulated pipe, as shown in Figure 3.12a–c, the current density distribution maps measured from the WBE-probes located at three different locations of the sand box display free corrosion characteristics: the distribution of anodes (gray color) and cathodes (dark gray color) over the probes' surfaces is random and the sum of all anodic and cathodic currents is approximately zero. This is in agreement with a buried pipeline without CP. When a small DC stray current of 1 mA was applied to the simulated pipe, as shown in Figure 3.12d–f, for the probe A, the number and magnitude of positive current peaks decreased, suggesting a decrease in anodic corrosion reaction at the stray current "pickup" point, while the number and magnitude of negative peaks increased, suggesting an increased cathodic polarization and cathodic reactions. When the applied stray current further increased to 2, 4 and 6 mA, as shown in Figure 3.12a, d, g, j and m, the positive currents at the "pickup"point gradually disappeared, suggesting the prevention of anodic corrosion reaction at this point due to stray current "induced" cathodic protection. The cathodic current at this point gradually increased and reached $-50\,\mu A/cm^2$. These behaviors are in good agreement with the behavior of stray current "pick-up" point of a pipeline reported in the literature. This suggests that the probe "A" successfully simulates and measures corrosion of stray current "pick-up" point of a pipeline. For the probe C, however, the number and magnitude of positive current peaks increased with increasing stray current values, as shown in Figure 3.12c, f, i, l, and o, suggesting an increase in anodic corrosion reaction at the stray current "discharge" point. These behaviors are in agreement with the behavior of stray current "discharging" point of a pipeline reported in the literature. This suggests that the probe C successfully simulated and measured corrosion of stray current "discharging" point of a pipeline. At the middle of the sand box, as shown in Figure 3.12b, e, h, k, and n, the applied stray current had little effect on corrosion. This is expected since

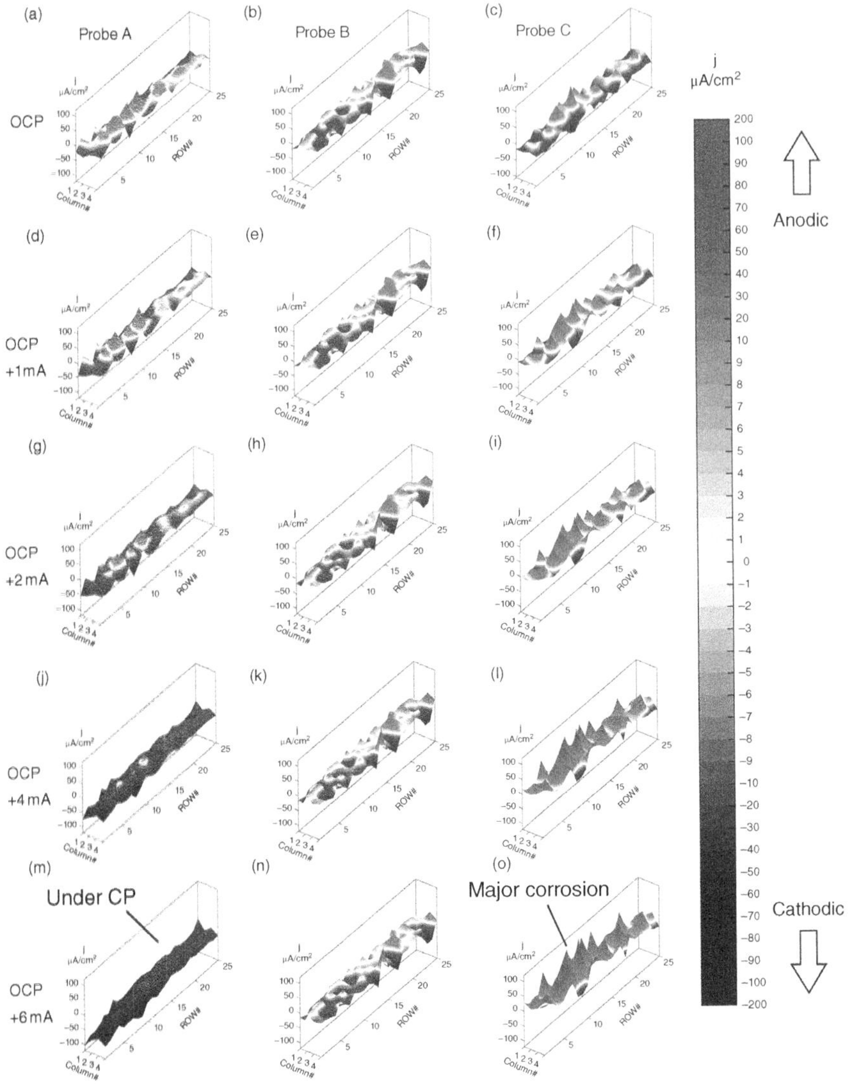

Figure 3.12 The current distribution maps obtained when probe buried in different locations and different stray current applied at OCP. *Source:* Wang et al. [87]/ National Association of Corrosion Engineers.

probe B is probably at the stray current "neutral" location [87]. Please note that more probes can be added to the simulated pipe to increase the detection resolution of stray currents along a pipeline.

When the standard CP potential of $-850\,\mathrm{mV_{vs.CSE}}$ was applied to the simulated pipe for preventing stray current corrosion, as shown in Figure 3.13, significant

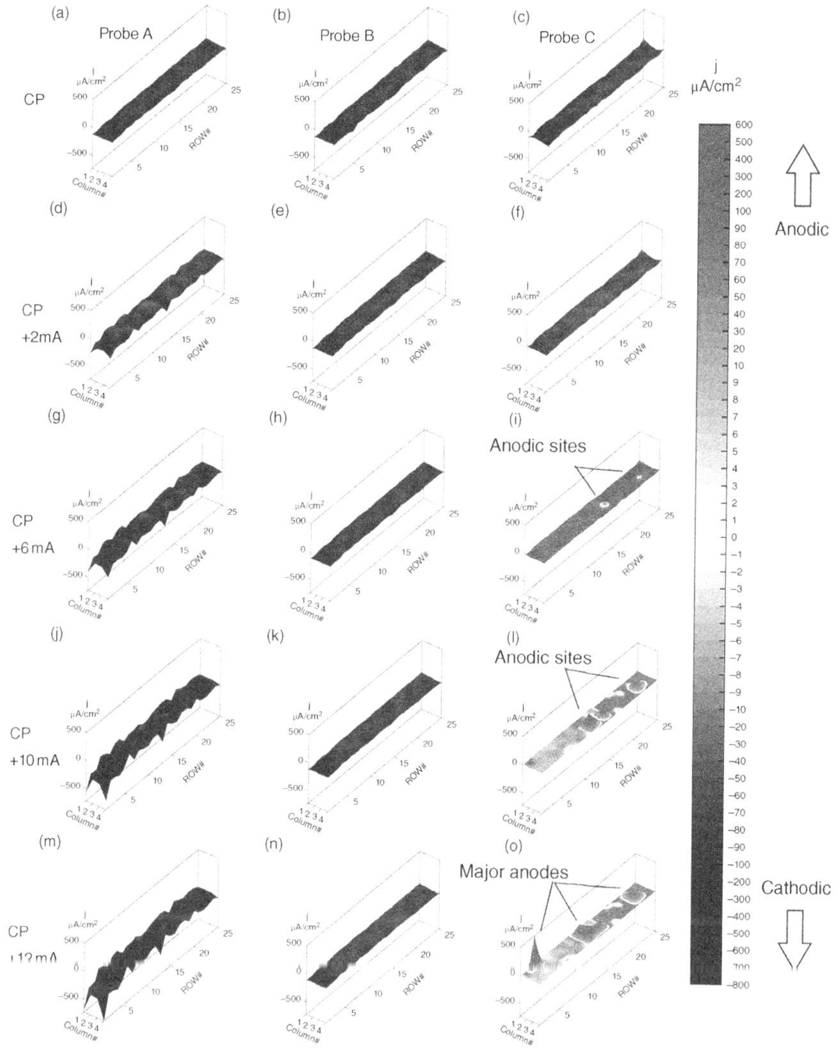

Figure 3.13 The current distribution maps obtained when probes are buried in different locations of a sand box at the CP potential of $-850\,mV_{CSE}$ with different stray currents applied. *Source:* Wang et al. [87]/ National Association of Corrosion Engineers.

changes to current density distribution maps can be observed. When no stray current was applied, as shown in Figure 3.13a–c, all steel electrodes were protected by CP as indicated by the overall distribution of cathodes (dark gray color) and negative CP currents flowing over all probes' surfaces. However when DC stray currents were applied, as is shown in Figure 3.13i, some anodic corrosion sites initiated on the current distribution map obtained from the probe C. With the further increase of stray current value to 10 and 12 mA, as shown in Figure 3.13l and o, major increases in anodic currents and sites occurred. The maximum anodic current density increased to approximately 500 uA/cm^2. Critically it should be noted that corrosion was highly localized with anodic current peaks increased when the applied DC stray current increased from 6 to 12 mA. Similar to the maps shown in Figure 3.12, the probe B located at the middle position of the sand box was almost not affected by the stray current since it is most likely at the stray current "neutral" location, as is presented in Figure 3.13b, e, h, i, and n. These results indicate that the standard CP potential of −850 mV$_{vs.CSE}$ was insufficient for preventing corrosion at stray current "discharging" point when the applied stray current value was over 6 mA [87]. These results suggest that the standard CP potential would be insufficient to protect pipelines under significant stray current influence.

All these behaviors shown in Figures 3.12 and 3.13 are in good agreement with the behavior of stray current on steel corrosion reported in the literature [90–92] and are also in good agreement of the corrosion observable from photos taken from coupons after corrosion experiment was completed. As shown in optical

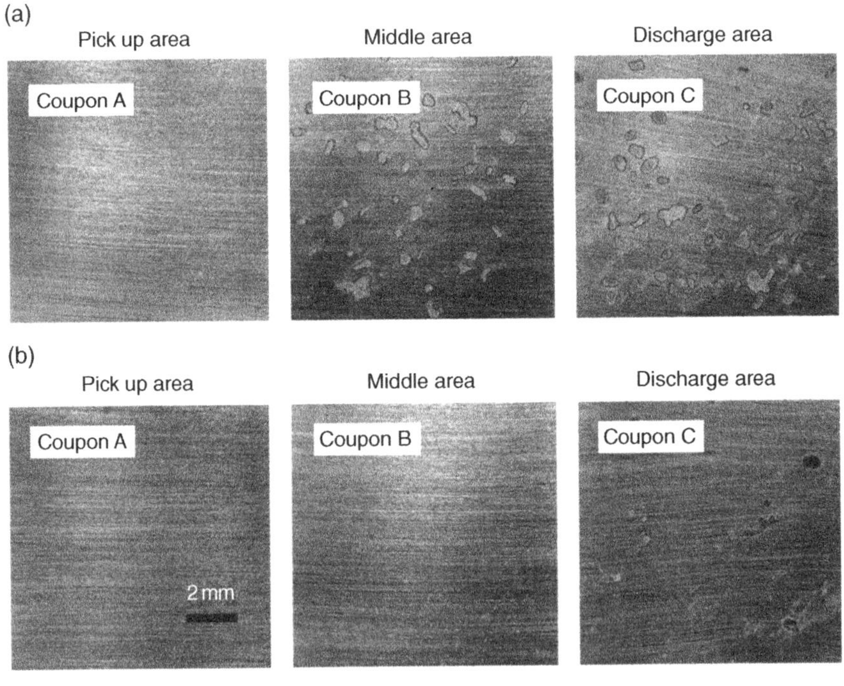

Figure 3.14 (a) Photos of coupons located at different locations after one hour exposure to 6 mA DC stray current at OCP; (b) photos of coupons located at different locations after one hour exposure to 12 mA DC stray current under CP potential of −850 mV$_{CSE}$. *Source:* Wang et al. [87], Allen Press.

microscope images in Figure 3.14a, little corrosion occurred on the "coupon A" located at the stray current "pick-up" point, minor corrosion occurred on the "coupon B," while significant corrosion occurred on the "coupon C" located at the stray current "discharging" point. In the CP protected condition, as shown in Figure 3.14b, little corrosion occurred on the coupon as shown on both the "coupon A" and "coupon B" photos, while significant corrosion occurred on the "coupon C" that was located at the stray current "discharging" point. These results clearly illustrate the failure of preventing steel from stray current corrosion using a CP system that supplied a standard protection potential. Again these results demonstrate the capability of the WBE-probe array in locating the stray current "discharging" point where serious corrosion occurs and visualizing and quantifying the effects of stray currents on corrosion at different locations of a simulated buried pipeline [87].

Based on results described above, two applications of this WBE-probe array system have been be proposed [87]. Firstly, this array can become a "permanent monitoring tool" for monitoring pipeline corrosion by installing probes permanently at selected and regularly spaced locations over a buried pipeline. Such array could provide early warning of the pipeline system because these installed probes are able to detect not only the existence of stray current corrosion but also corrosion rates and their distribution over a pipeline surface. Different corrosion probes could be placed at different locations of a pipeline to detect different stray current effects. For instance, a corrosion probe should be installed at the stray current "discharging" point where the maximum corrosion rate could occur, while a coating disbondment probe such as the type described by Mahdavi et al. [93] could be installed at the stray current "pick-up" point to detect accelerated coating disbondment due to a stray current caused very negative potential. In practical industry applications, if multiple WBE probes were used in separate locations of a stray current affected pipeline, these probes could be connected and coupled through pipe rather than directly connected to the same point through an external wire, which could simplify the probe application and allow the probe array to function the same as those shown in Figure 3.11. These probes can detect stray current affected areas, such as "pick-up" point for "permanent monitoring tool" installation. Another application of this array is that it could be used as a "corrosion survey tool" to identify the possible discharge and pickup points over a stray current affected pipeline. When the electrode array is used as survey tool, probes can be designed as "portable probe sticks" that can be inserted into small holes drilled on the ground near a stray current affected pipeline, while the other end of the probes is connected to the pipeline. If multiple of such coupled arrays are used in separate locations of the stray current affected pipeline, these probes can detect stray current effects and the "pick-up" point where "permanent monitoring tool" can be installed afterwards [87].

3.3.2 Dynamic Stray Current Corrosion Under Anodic Potential Transients

Although corrosion maps shown in Figures 3.12 and 3.13 clearly exhibit the effects of DC stray currents on buried steel corrosion over different locations, the DC stray current applied in the experiment was at constant values and therefore corrosion at each location was under relatively steady conditions. In engineering practice, however, significant and dynamic pipeline potential transients could occur at a location of a pipeline due to various forms of external electrical interference usually referred to as stray currents [83–86], therefore corrosion at the location would vary with the dynamic potential changes. A common form of external electrical interference is anodic transient that causes the steel pipeline potential to shift rapidly toward the anodic direction. Anodic transients are believed to significantly reduce CP efficiency [86]; however, the exact effect of either the magnitude or duration of these transients on the likelihood of steel corrosion was unknown, remaining a uncertainty of industrial concern for a long time. Over the past several decades many attempts have been made to develop "rules" for quantifying the effect of anodic transients on the corrosion of cathodically protected buried steel pipelines. Unfortunately these "rules" remain empirical because they are based only on field inspection data that are generally accumulated results after long periods of pipeline corrosion under the effects of stray currents of unknown characteristics [83, 86, 94]. For instance Peabody [94] described a "rule" that an anodic transient of one hour duration would lead to a 27% increase in buried pipeline corrosion risk while an anodic transient of one day's duration would increase the "chance" of corrosion by 50%. Although this "rule" clearly points out the importance of the duration of anodic transients, it does not give any clear justification or fundamental understanding for the effects of these durations. A better understanding of the influence of anodic transients on corrosion of CP protected structures would be of significance for pipelines and other buried infrastructure.

The ability to directly probe corrosion currents on cathodically protected metal structures under the effects of different stray currents would allow better understanding of stray current influence on corrosion. A major difficulty is the lack of reliable and reproducible experimental methodologies that are able to systematically categories and quantify the level and nature of corrosion damage as a result of various modes of CP excursions. Currently potential recording is the most commonly used method for inspecting stray current activities in the pipeline industry; however potential recording does not provide sufficient information about corrosion rates and patterns. Weight-loss coupons have been used to determine corrosion rates of steel buried in soil; however, weight-loss coupons are unable to provide in situ corrosion rate data required for quantifying the effects of relatively short-duration CP potential excursions. In order to address these issues, as shown in Figure 3.2d, a soil corrosion cell with electrode array WBE probes was designed and used by Huo et al. to simulate dynamic and complex localized corrosion in soil under the effect of stray current induced potential transients [21]. This new method has shown necessary temporal and spatial resolutions required for the visualization of rapidly changing electrode processes and the effective quantification of the effects of dynamically changing potential excursions on buried steel corrosion. This capability has been demonstrated by mapping vigorously changing localized corrosion processes occurring on buried steel surfaces under the effect of anodic transients and has been used to achieve better understanding of stray current corrosion phenomena [21, 28, 68]. Details of measurement procedures can be found in Reference [21].

Figure 3.15 shows typical current density recordings and current maps obtained using the WBE probe under the effects of anodic step changes of different amplitudes (+500, +800, and +1000 mV) [21]. The current densities were recorded from the electrode array under the prior CP potential and three different anodic step change levels, while the currents shown in the maps are measured from each of the 100 individual electrodes in the WBE probe. All experiments were started by applying a CP potential of $-950\,mV_{vs.CSE}$ on the WBE for a 24-hour period in order to establish a stable prior CP condition on buried WBE surfaces, simulating a CP protected pipeline. Under prior CP, as shown in the relevant WBE map in Figure 4.15, only cathodic currents (in negative values) were detected over the WBE surface, indicating that the probe was under CP and no corrosion would be expected. Immediately after the anodic transient signals were superimposed on the CP potential, the steel potential was shifted from $-950\,mV_{vs.CSE}$ to $+50\,mV_{vs.CSE}$, −150 mV vs. CSE and $-450\,mV_{vs.CSE}$ upon application of either 1000 mV, 800 mV, or 500 mV, respectively (see Figure 3.2d). As shown in Figure 3.15, an initial anodic current peak accompanying this shift in potential can be observed in all cases. This is believed to be due to an anodic current spike charging the double-layer capacitance over the electrode surface in addition to metal dissolution current occurring over the anodically polarized WBE surface at these anodic transient potentials [21]. As shown in map (a) of Figure 3.15, an area with high local anodic current is observable on the background of a smaller uniform anodic current, suggesting that local anodic corrosion dissolution was occurring (0.1 mm/y) on the WBE surface immediately after the anodic step change was applied. After approximately five minutes, as shown in map (b), the anodic currents dropped dramatically and the corrosion rate dropped from 0.1 mm/y to approximately 0.01 mm/y. This low corrosion rate was then maintained, as shown in map (c), with no clear anodic site observable in the WBE map recorded up to nine minutes. If we consider that, under an initial CP-induced high pH environment (local pH was measured using a pH indicator paper to be approximately pH 12), the steel surface should be under passive conditions [95], a short anodic transient is not sufficient to disrupt this passivity. This result suggests that under the sandy soil box cell and CP conditions, the application of an anodic transient of +1000 mV for less than

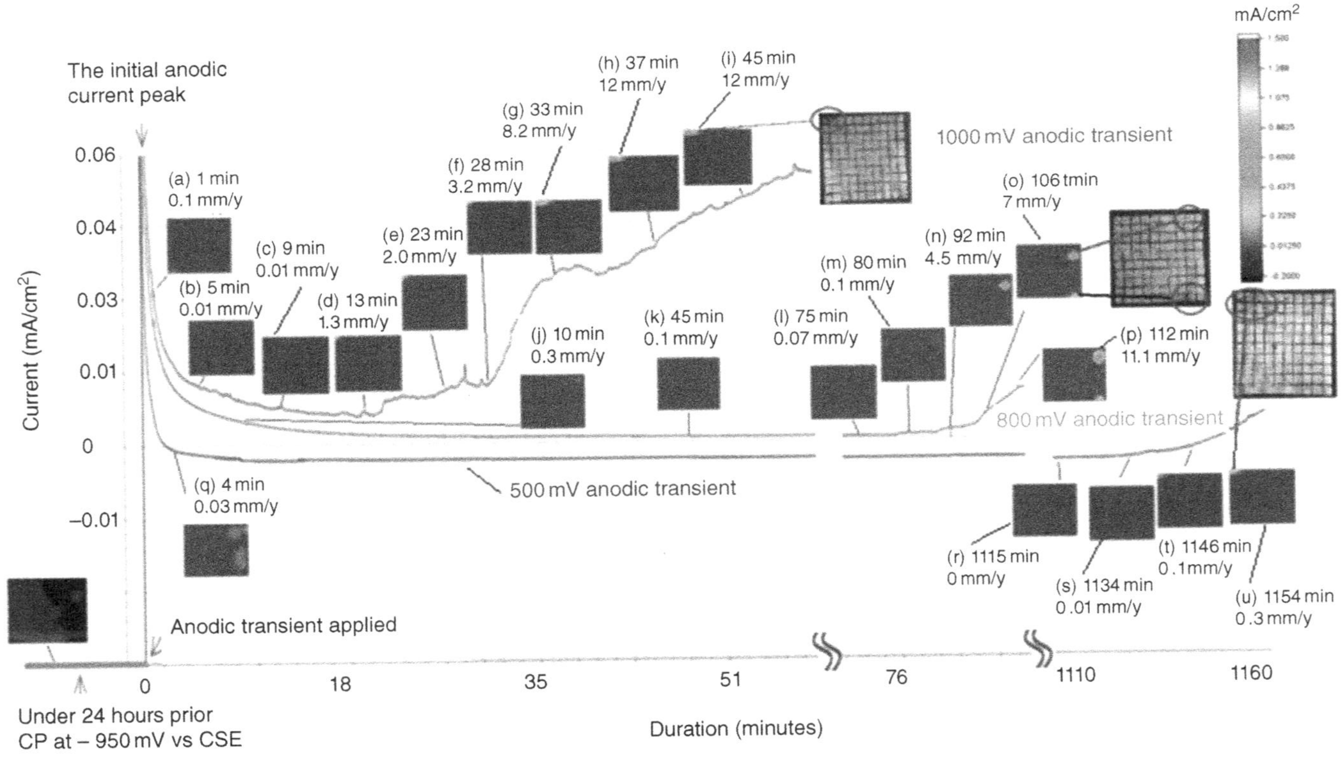

Figure 3.15 Current recordings and WBE maps obtained using a WBE probe under three different CP and anodic transient conditions. *Source:* Huo et al. [21], ELSEVIER.

nine minutes would not result in significant steel corrosion. Map (d) in Figure 3.15 shows that passivity is disrupted, at least locally, after 15 minutes with a major anode indicating a significantly high local corrosion rate of 1.3 mm/y appearing on the WBE surface, despite the fact that the overall current recording shown in Figure 3.15 is still low (approximately 0.021 mA/cm^2) and the overall corrosion rate is small (approximately 0.07 mm/y) at this time. Thus, the WBE probe indicates that localized anodic sites initiate upon extending the anodic transient duration, leading to a large increase in localized corrosion clearly visible in the WBE maps. This result also suggests that the visualization of the localized corrosion using the WBE maps is necessary for detecting the critical anodic transient durations. These results suggest that, under these conditions, there exists a critical anodic transient duration of approximately nine minutes for the buried steel surface under a 1000 mV anodic step change (note: the pipeline industry considers the acceptable corrosion rate is less than 0.1 mm/y). This was the first time that a critical anodic transient duration has been observed and it may have a significant impact on standards used in industry in the future [21].

Under changing environment surrounding the steel and the amplitude of the transient, the critical anodic transient duration was found to vary, as might be expected given the changes in passivity at the steel interface. As shown on map (m) in Figure 3.15, the critical anodic transient duration increased to approximately 80 minutes when a lesser anodic excursion amplitude of 800 mV was applied to the WBE surface buried in soil. As shown on map (t) in Figure 3.15, the critical anodic transient duration increases to approximately 1146 minutes when the anodic excursion amplitude was decreased to 500 mV under similar conditions. On the other hand, when the steel was immersed in an aqueous NaCl solution of the same resistivity and under the same CP and anodic transient conditions (i.e. an initial CP potential of $-950\,mV_{vs.CSE}$ for a 24 hours period and then application of a 1000 mV anodic transient), the critical anodic transient duration was found to be zero, i.e. passivity is never maintained under such aqueous conditions. This result suggests significant differences in the effect of anodic transients in aqueous systems, indicating the need for a different approach for protecting freshwater, marine and underground infrastructure assets under the threat of stray currents. The propagation of corrosion under further extended anodic transient is also clearly visualized in Figure 3.15. It can be seen in maps (e)–(i) that the anodic current increased rapidly after approximately 28 minutes and the localized corrosion rate reached and maintained a very high level, leading to significant localized corrosion on the WBE surface (the photo of the excavated WBE surface is also shown in Figure 3.15). Similar behaviors are observable on maps measured from the WBE probes exposed to different levels of anodic transients (e.g. maps (i)–(p)). These results suggest a major breakdown of passivity under the continued effect of the anodic transient due to changes in steel surface pH [95]. For instance significant anodic transients could lead to potentials where the water oxidization reaction readily occurs, generating acid on the electrode surface. This would immediately depassivate the steel surface. Furthermore, with the oxygen reduction reaction not as prominent at the more anodic potentials, OH^- ions are not being generated and those initially present from the CP condition may diffuse away from the steel surface, leading to a lower pH and a more corrosive steel surface condition. A smaller anodic transient amplitude and a lower diffusion soil medium would decrease the rate at which these pH changes take place, leading to a longer critical anodic transient duration. This is consistent with our experimental observations that the critical anodic transient duration changed with the anodic transient amplitudes and the corrosion media [21].

This critical anodic transient duration was also found to be influenced by the anodic transient amplitude, with or without pre-applied CP, test media (soil or aqueous media) and their conductivity, the cathodic cycle during the anodic transient and the amplitude of cathodic transient [21, 28, 68]. As shown in Figure 3.16, an increase of the anodic excursion amplitude from +1000 mV (potential at +50 $mV_{vs.CSE}$) to +1950 mV (potential at +1 $V_{vs.CSE}$) would change the current vs. time behavior significantly. This is clear by comparing the anodic current traces in Figures 3.15 and 3.16. In Figure 3.16, anodic current remains at a rather high level (approximately 1 mA in Figure 3.16 or 285 mA/m^2 in Figure 3.17), although it dropped from a very high initial anodic current peak value. Such a large anodic current suggests a corrosion rate of more than 3 mm/y. As a consequence, the critical anodic transient duration in this condition was zero. The anodic transient duration was also found to be influenced by the testing cell environment. Under much harsher conditions in an aqueous cell, zero critical anodic transient duration could be identified when an anodic excursion of +1000 mV (to +50 $mV_{vs.CSE}$) (see Figure 3.17) was applied, with the anodic current remaining at a very high level) [28].

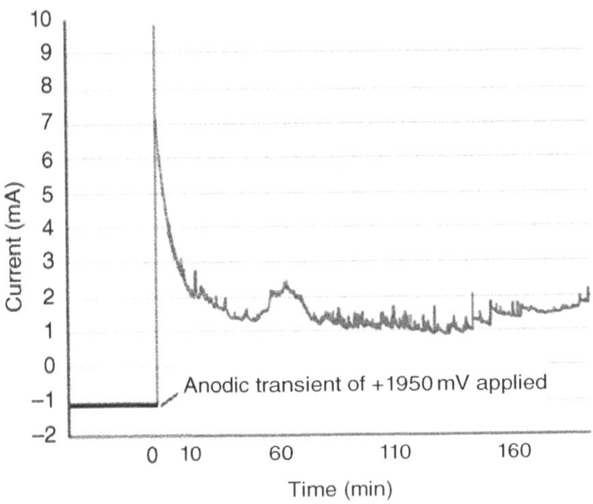

Figure 3.16 The monitored CP (Black line) and anodic transient (gray line) current from experiment of constant +1950 mV anodic transient signal (to +1 $V_{vs.CSE}$) applied on the buried steel coupon after 24 hours CP −950 $mV_{vs.CSE}$ in sandy soil. *Source:* Huo and Tan [28]/ With permission of Taylor & Francis.

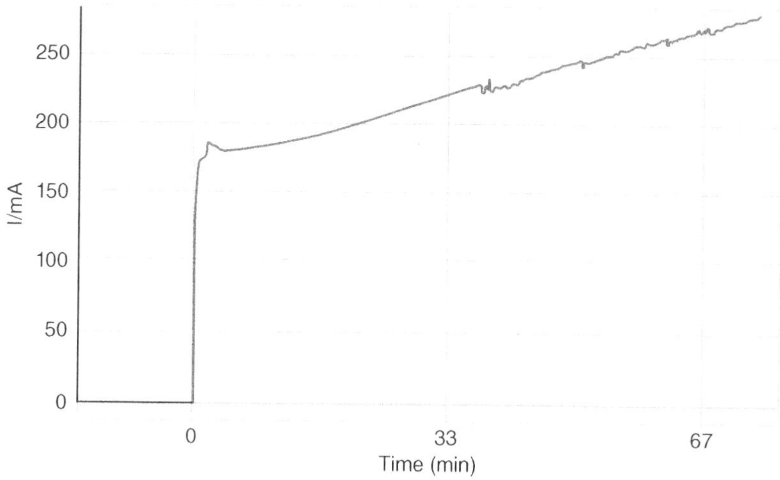

Figure 3.17 The monitored CP (Black line) and anodic transient (Gray line) current of constant +1000 mV (+50 $mV_{vs.CSE}$) anodic transient signal applied on the steel coupon after 24 hours CP −950 $mV_{vs.CSE}$ (pH at 9.96) in aqueous cell. *Source:* Huo and Tan [28]/ With permission of Taylor & Francis.

The phenomena associated with critical anodic transient duration can be explained by the passivity of steel surface under CP-induced high pH condition, and the loss of passivity or the breakdown of passivity under the effects of anodic transient-induced pH decrease and surface chemistry changes [28, 68]. The Pourbaix diagram shown in Figure 3.18 can be used to explain the corrosion behaviors shown in the WBE maps in Figures 3.15–3.17. Figure 3.18 includes the effect of chloride ions on creating a pitting zone and demonstrates that the decreasing pH leads to localized corrosion initiation on the steel surface. This figure shows that passivity is disrupted, at least locally, beyond this time frame with a major anode indicating a significantly high local corrosion rate of 1.3 mm/y (see Figure 3.18, G), appearing on the WBE probe surface, despite the fact that the overall monitored current recording (see Figure 3.15) at 15 minutes is still low and the overall corrosion rate is small (approximately 0.07 mm/y) under the applied anodic transients. As shown in Figure 3.18, under the effects of anodic transients, the pH of the steel surface would decrease from the initial CP-induced high pH (up to 14) due to natural diffusion of OH^- from the steel surface and also the generation of H^+ at the steel surface under the anodic transient-induced anodic reaction. Decrease in pH was commonly observed when an anodic transient was applied on steel electrodes buried in sand. This pH drop also occurs on the steel surface exposed in aqueous cells where the diffusion rates are much higher than that in the sandy soil environment, thus allowing the OH^- ions to diffuse away from the steel surface. With the pH decreasing, the steel would eventually lose passivity [28, 68].

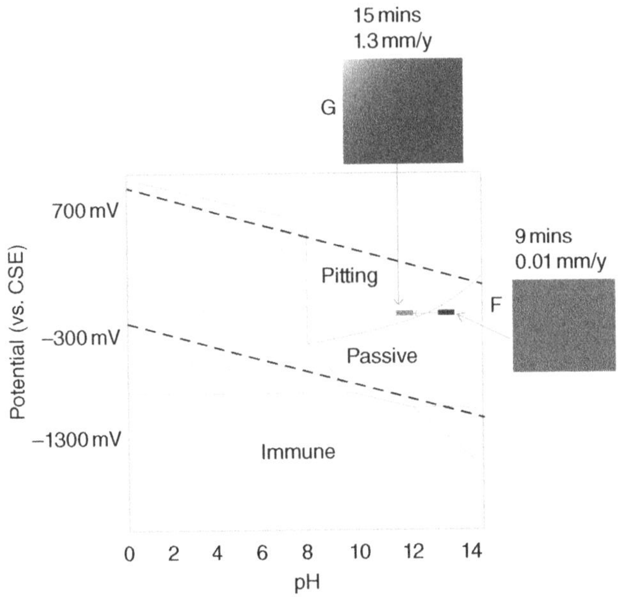

Figure 3.18 Pourbaix diagram in the presence of chloride ions to explain the passive layer disruption due to the pH changing (Adapted from Reference [95] for [Cl^-] = 0.1 M.): F. current map after nine minutes anodic transient applied; G. current map after 15 minutes' anodic transient applied. *Source:* [28, 68]/ With permission of Taylor & Francis.

In the buried steel–CP–anodic transient system studied in this work [28, 68], the initiation of localized corrosion is actually a deterministic event because steel shows passive behavior only when the environment is alkaline (e.g. higher than pH 10). The initiation of localized corrosion is profoundly dependent on the pH of the local environment and therefore it is a deterministic event that is predictable based on environmental pH values. This work suggests that passivity can be achieved on cathodically protected buried steel if the CP-generated local pH at the steel surface is high (usually above pH 10), and passivity can be maintained if such a high pH can be maintained. In the case of anodic transients that push the potential into the passive region and beyond, the environmental pH would reduce with time due to natural diffusion of OH− away from the steel surface and also the generation of H+ at the steel surface under the anodic transient-induced anodic reaction. Breakdown of such passivity is a time-dependent process and occurs due to the decrease of pH and the presence of chloride and other ions. This suggests that not all the anodic transients cause buried pipeline corrosion initiation. This is the first time that a critical anodic transient duration has been clearly determined for steel based on the initiation of localized corrosion. This work contributed to the understanding towards a unified consensus on how "big" CP excursion (either in magnitude, frequency, or duration) needs to be in order to cause major corrosion problems, and it may have a significant impact on standards used in industry in the future [21, 28, 68].

3.3.3 Stray Current Corrosion Under the Effects of Cyclic Potential Transients

A series of corrosion maps in Figures 3.15 clearly show that single-step anodic transients of different amplitudes (+500, +800, and +1000 mV) can significantly affect the corrosion behavior of buried steels, causing dynamic localized corrosion processes to initiate and propagate [21]. A limitation of the critical anodic transient duration and amplitude values reported previously [21, 28] is that they are obtained from electrodes under a single anodic transient with very long anodic transient durations (e.g. 24 hours) or with a large amplitudes (e.g. 1950 mV) [28]. It should be noted that although long duration and large amplitude anodic transients were recorded in industry with large amplitudes of several voltages and with long durations of almost 30 minutes, in most cases practical potential excursions are much more complex. In typical traction affected stray current corrosion systems, potential excursion can consist of both anodic and cathodic transients of 10- and 30-seconds in duration and vary in frequency from 2 to 3 minutes [83–86]. Corrosion behavior of buried steel under such dynamically and cyclically changing external electrical interference is extremely difficult to quantify in engineering practice because of the rapidly changing nature of the corrosion process.

The simulation of cyclic potential excursions is difficult because of the numerous forms of electrical interference signals that are likely to occur in industry. The soil corrosion cell, as shown in Figure 3.2d, was also used again by Huo et al. [68] to simulate such complex corrosion conditions under the effect of periodic and cyclic stray currents such as in wave forms (c) and (d) in Figure 3.2d because this method has shown necessary temporal and spatial resolutions required for the visualization of dynamically changing electrode processes and the effective quantification of the effects of dynamically changing potential excursions on buried steel corrosion [21, 28, 68]. Square and sine waves were selected to simulate potential excursion signals because the low-frequency sine wave could be used to simulate slow moving underline potential shifts while the square wave can simulate "spike" types of potential transients. When an anodic transient was applied to a buried steel electrode, a common behavior reported previously [21, 29] is that there is a period of time for which anodic current becomes very small, suggesting passivation, before the anodic current eventually increases rapidly. This period has been defined as the "critical anodic transient duration" because it indicates the maximum allowable anodic transient duration before major corrosion occurs. Analysis of the critical anodic transient duration shows that it is due to the incubation period required for pH drop to allow the breakdown of passivity and the initiation of localized corrosion under the effect of anodic transients [21]. Significant corrosion would occur if the anodic transient extends beyond the critical anodic transient duration with localized corrosion rates reaching as high as 12 mm/y [21, 29]. Although long-term anodic transients such as those reported previously [21, 29] do exist on some pipeline conditions, in most industrial cases, however the stray currents are relatively short-term "cyclic" potential fluctuations.

In the case of cyclic transients, the situation is more complex due to the extremely dynamic nature of the combination of anodic/cathodic transients of varying durations and amplitudes. Figure 3.19 shows typical corrosion behavior after (a, b) two typical cyclic forms of transients in both anodic and cathodic directions were applied, and (c, d) only anodic transients were applied [68]. As shown in Figure 3.19, significantly more corrosion occurred when a cyclic potential excursion, in both anodic and cathodic directions, was applied (±250 and ±500 mV potential excursion from the baseline CP of $-850\,mV_{VS.CSE}$) compared with only anodic transients (+250 mV and +500 mV from the same baseline CP of $-850\,mV_{VS.CSE}$). Another observation is that corrosion patterns in Figures 3.19a and b were spread over a large area, while the corrosion patterns in Figure 3.19c and d were small pits. This indicates that the cathodic transients actually affect the nature of steel corrosion, not only on its rates, but also on its pattern. This could be due to the fact that very negative potential cycles could lead to damage of a passive film on the steel surface, as discussed in sections below. These results suggest that

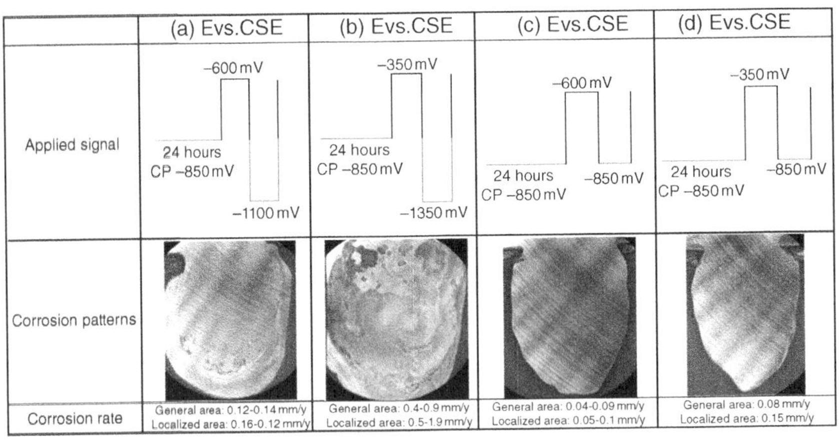

Figure 3.19 Coupon electrodes corroded under the cyclic potential fluctuation (30 seconds anodic and cathodic cycles) imposed after 24 hours CP at $-850\,mV_{VS.CSE}$ in sandy soil, and photos of the corroded surface after 15 days (24 hours CP + 14 days transients). *Source:* Huo and Tan [68]/ With permission of Taylor & Francis.

short duration (30 seconds) potential excursions only in the anodic direction (e.g. Figure 3.19c and d) would not cause major corrosion compared to a similar anodic shift, which follows a deep negative excursion (e.g. Figure 3.19a and b). Similar behavior was found in experiments with single-direction and double-direction sine waves as potential excursion signals. In the case of single-direction sine wave test (+500 mV amplitude on CP $-850\,mV_{vs.CSE}$), little corrosion occurred with the highest localized corrosion rate <0.1 mm/y. In the double-direction sine wave test (±500 mV amplitude on CP $-850\,mV_{vs.CSE}$), major localized corrosion occurred with the highest localized corrosion rate of 1.34 mm/y [68].

Figure 3.20 shows results from experiments with the WBE exposed to the same condition as the coupon electrode shown in the photo of Figure 3.19b where serious corrosion is visible. As shown in Figure 3.20, the selected form of "cyclic" potential fluctuations was rather dynamic with 30 seconds anodic excursion at $-350\,mV_{vs.CSE}$ and then 30 seconds at $-1350\,mV_{vs.CSE}$. The WBE current maps in Figure 3.20 show that a low corrosion rate of around 0.01 mm/y was recorded after the first two minutes of experiments. This could suggest that the buried steel WBE surface was passive after 24 hours prior CP. However, after a few cycles of the potential fluctuation, at 10 minutes, the WBE current map shows that the WBE surface became slightly active at the anodic cycle and the corrosion rate increased to 0.06 mm/y. The whole surface became more active after the signal had been applied for 18 hours, and the whole surface corrosion rate reached 0.5 mm/y. This corrosion rate kept increasing. After 24 hours, the WBE map shows that under the anodic cycle, the whole surface corrosion rate reached 0.6 mm/y. The WBE surface photo after test is also shown in Figure 3.20. The whole surface corrosion rate (measured by surface profilometry) is 0.58 mm/y and the highest localized corrosion rate is 1.3 mm/y. These are similar to those shown in Figure 3.19b. These results show that corrosion can initiate on the WBE surface that was under ±500 mV anodic and cathodic cyclic transients fluctuating around a baseline CP level of $-850\,mV_{vs.CSE}$. The corrosion pattern recorded by the WBE maps shows that the ±500 mV anodic and cathodic transients fluctuating around a CP value of $-850\,mV_{vs.CSE}$ lead to a general corrosion pattern, which is similar to the corrosion pattern shown in Figure 3.19b. However under the constant anodic transient, as shown in Figure 3.19d, the corrosion initiation is mainly from a localized area. This is in agreement with the behavior reported previously in References [21, 29].

These results clearly suggest that cathodic transients has quite dramatic effect on corrosion of buried steel under cyclic potential transients [29, 68]. This has implications for predicting the likelihood of corrosion under the influence of stray currents and therefore more detailed studies were carried out by Huo et al. [68]. Figure 3.21 shows results from experiments with the WBE exposed to a different form of "cyclic" potential fluctuations [68]. In this case, the applied potential

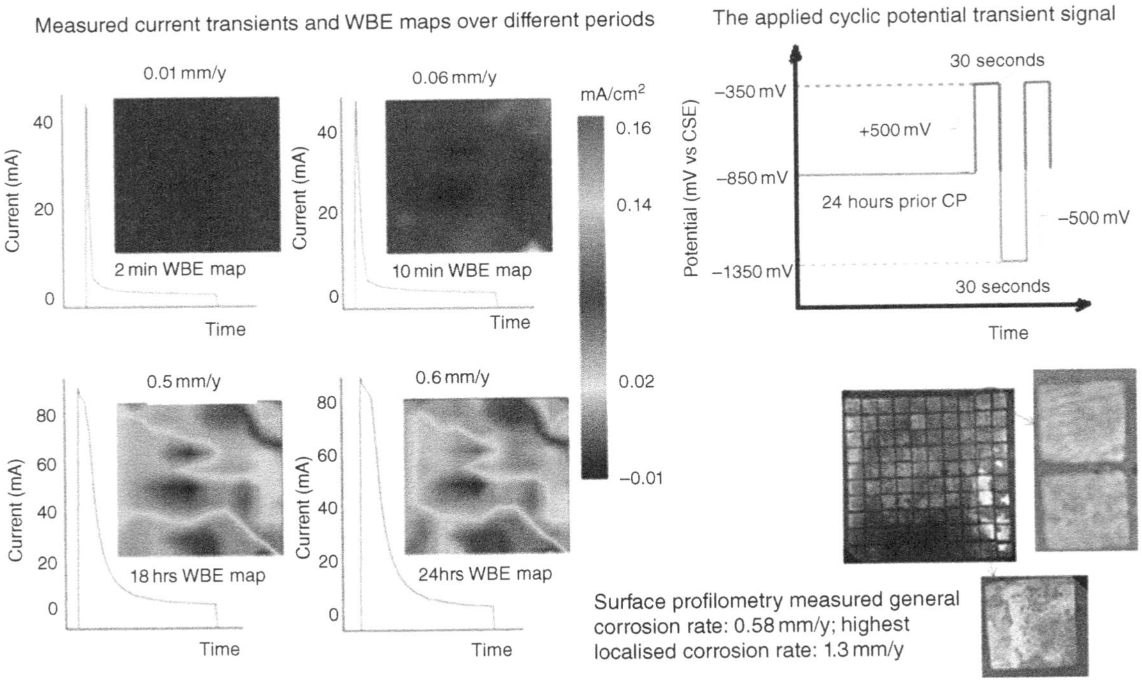

Figure 3.20 Current recording and WBE current maps measured under the cyclic potential fluctuation (30 seconds anodic and cathodic cycles) imposed after 24 hours CP at −850 $mV_{vs.CSE}$ in sandy soil, and photos of the corroded WBE surface after one day transient test. *Source:* Huo and Tan [68]/ With permission of Taylor & Francis.

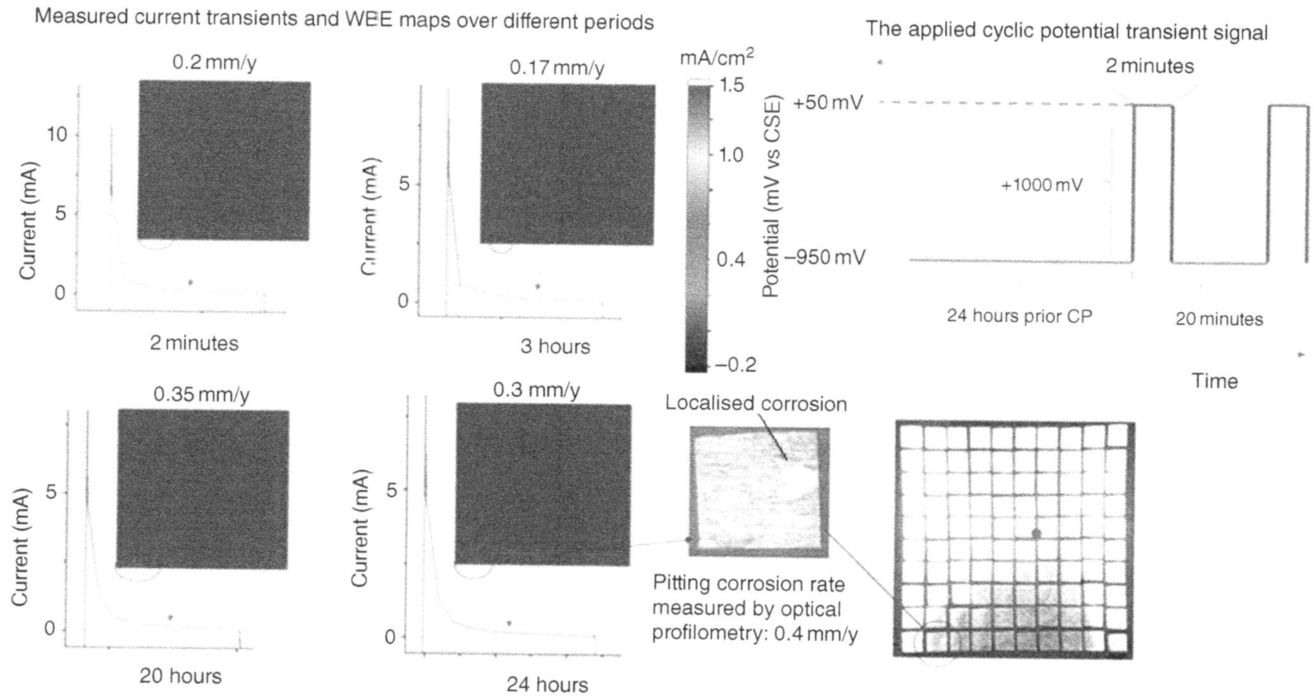

Figure 3.21 Current recording and WBE current maps measured from a WBE under the cyclic potential fluctuation imposed after 24 hours CP at $-950\,mV_{vs.CSE}$ in sandy soi., and photos of the corroded WBE surface after 24 hours exposure. *Source:* Huo and Tan [68]/ With permission of Taylor & Francis.

changed dynamically with two minutes of anodic excursion of +1000 mV from CP −950 mV$_{VS.CSE}$ (+50 mV$_{VS.CSE}$) and then 18 minutes at CP −950 mV$_{VS.CSE}$ applied on the WBE sensor that was pre-conditioned by 24 hours CP −950 mV$_{VS.CSE}$. As shown in Figure 3.21, when the cyclic anodic potential excursion starts to apply on the WBE sensor, the initial anodic cycle currents were small, following the same trend and behavior as the case when the constant anodic transients were applied in experiment reported previously in Reference [29]. The corrosion rate during the first two minutes was around 0.2 mm/y. After two minutes of an anodic cycle, the cyclic signal shifted back to baseline CP potential level of −950 mV$_{VS.CSE}$ for 18 minutes. WBE maps recorded after three hours of potential cycling, the monitored localized corrosion rate is 0.17 mm/y, as shown in Figure 3.21 (at the left bottom area of the WBE map). This relatively low corrosion rate increased to 0.3 mm/y, at a localized area, after one day of this cyclic anodic transient testing. After two days, the WBE sensor was removed from the soil, cleaned by Clark acid solution and the surface was imaged with an optical microscope. The images confirmed that slight localized corrosion occurred at the corresponding area where the WBE maps showed an anode. By measuring the corrosion rate through optical profilometry, as shown in Figure 3.21, this localized area corrosion rate is around 0.4 mm/y. This corrosion rate is much lower than that when constant anodic transient was applied to the WBE, as reported previously in Reference [29]. This could be due to the fact that the cathodic cycle would generate OH^- for maintaining the high pH environment on the steel surface. In other words, the cathodic cycle at CP −950 mV$_{VS.CSE}$ regenerated OH^- ions and thus reduced the buried steel corrosion rate during the anodic cycle at +50 mV$_{VS.CSE}$.

This hypothesis has been tested by extending the duration of the anodic transient and reducing the duration of the cathodic transient [68]. As shown in Figure 3.22, once the anodic cycle duration was extended to 10 minutes, the passivation behavior during the anodic cycle was more obvious during the first few cycles. The localized corrosion rate at the beginning was low, 0.07 mm/y, which was in line with the constant anodic transient at +50 mV$_{VS.CSE}$ test results reported previously in Reference [29]. Even after 2 hours, this low corrosion rate was still kept at 0.09 mm/y as shown in the WBE map. However, two stable anodic areas with high corrosion rates, 5.1 and 16 mm/y, appeared on the WBE maps after 23 hours. The appearance of high anodic current sites suggest that the steel passive layer broke down during the cyclic anodic transient at +50 mV$_{VS.CSE}$. This could be due to the steel surface pH decrease due to the shorter duration of the cathodic cycle (five minutes) at −950 mV$_{VS.CSE}$ than anodic cycle (10 minutes) at +50 mV$_{VS.CSE}$ cycle duration. Therefore, the cathodic reaction could not generate enough OH^- to neutralize the H^+ generated from the anodic reaction during the anodic cycle and compensate the OH^- diffused away from the steel surface. The surface

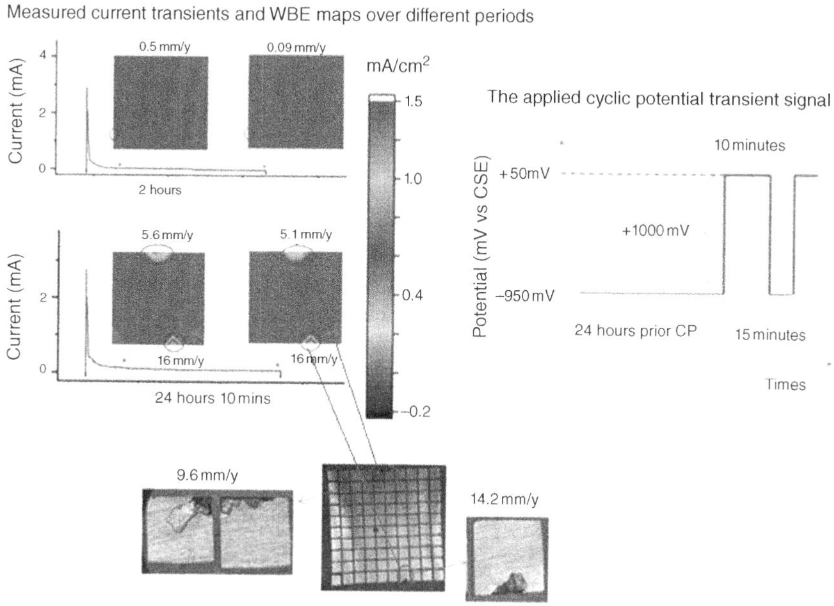

Figure 3.22 Current recording and WBE current maps measured from a WBE under the cyclic potential fluctuation imposed after 24 hours CP at −950 $mV_{vs.CSE}$ in sandy soil, and photos of the corroded WBE surface after 24 hours exposure. *Source:* Huo and Tan [68]/ With permission of Taylor & Francis.

photo of the WBE is also shown in Figure 3.22 after being removed from sandy soil. The obvious corrosion areas on the WBE are in agreement areas recorded high anodic currents. By optical profilometry measurement, the highest localized corrosion rate at the top area is 9.6 and 14.2 mm/y at the bottom one. The pH at the interface of steel/soil was 5, as measured by pH indicator paper [68].

Under cyclic ±500 mV anodic and cathodic transients fluctuating around a baseline CP potential of −850 $mV_{vs.CSE}$, a general corrosion pattern is observed, which could be due to the large negative transient amplitude cyclically removing the whole surface passive layer which formed during the 24 hours CP and probably also at every anodic cycle. At the beginning of the anodic cycle, because of the high pH environment induced by the 24 hours prior CP, the steel surface becomes passive. As shown in the Pourbaix diagram of iron in Figure 3.23, when the anodic transient begins on the cathodically protected steel surface, the steel surface state reaches the passive region (shown as the orange bar in Figure 3.23). At this stage, the WBE current map recorded at two minutes showed the whole surface had a very low corrosion rate, around 0.01 mm/y, suggesting that the passive layer on the steel surface was protective. When the signal shifted to a more negative potential, i.e. $-1350\,mV_{vs.CSE}$, the whole surface state entered the immune region (as shown by the dark gray bar in Figure 3.23). During this very negative potential, the WBE current maps recorded the average whole surface current density as $-1.6\,mA/cm^2$. In the immune region, the passive layer cannot exist at such a negative potential and begins to be reduced by the high cathodic current density [96]. After a few such cathodic cycles, the passive layer becomes weaker, as shown in the 10 minutes WBE current map, suggesting the passive layer on the whole WBE surface was gradually consumed by the large cathodic transient current. The passive layer formation rate during the anodic cycle cannot compensate for the passive layer reduction rate during the cathodic cycle. After 18–24 hours, most of the WBE surface becomes active during the anodic cycle, suggesting that the WBE surface cannot maintain a passive layer and rapid corrosion occurred [68].

These results appear in agreement with a theory presented by Markus [96] that was used to explain the a.c. corrosion processes under cyclic anodic transient and cyclic high cathodic current density transient, over $-0.1\,mA/cm^2$. Markus suggested that under very negative potential, the passive layer formed during the anodic cycle would become a rust layer which can no longer protect the substrate metal. A new passive layer would then be formed under the rust layer at the next anodic cycle. This suggests that under cyclic anodic transients and a very negative potential transient, corrosion is not directly caused by the anodic transient itself but results from the unstable passive film formed during the anodic cycle and removed by the large negative transient cycle.

Overall, an electrochemically integrated electrode array has been employed to probe dynamic localized corrosion on CP-protected buried steel under the

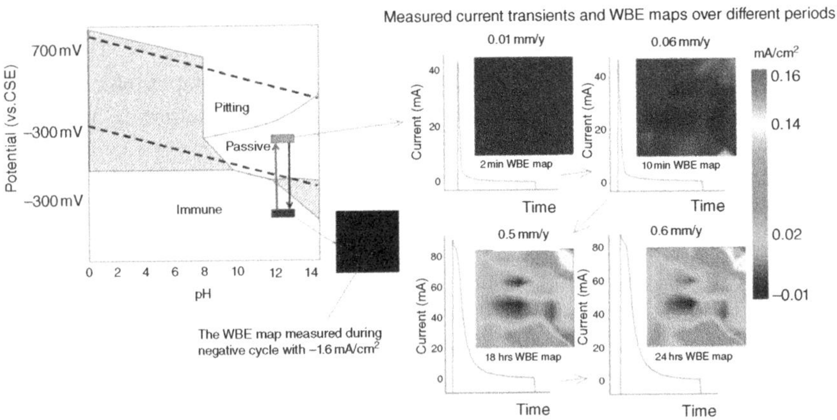

Figure 3.23 Pourbaix diagram of iron describes the corrosion behavior under ±500 mV anodic and cathodic transients fluctuate surround CP $-850\,mV_{vs.CSE}$ test. *Source:* Huo and Tan [68]/ With permission of Taylor & Francis.

influence of cyclic stray currents, with particular focus on the effect of cathodic transients, by visualizing the effects of variously designed anodic excursions on the passivity, corrosion initiation, and propagation of buried steel. Cathodic transients are found to affect corrosion in different ways, depending upon the amplitude and duration of the cathodic transients applied. If the cathodic transient leads to a very negative potential, e.g. $-1350\,mV_{vs.CSE}$, significantly rapid corrosion would occur due to the breakdown of the passive film. If the cathodic transient leads to a potential of around CP $-950\,mV_{vs.CSE}$, less corrosion would occur probably due to the fact that passive condition was re-established at the cathodic cycle [68]. These results are in general agreement with longer-term coupon tests and surface profilometry measurements of corrosion damage.

References

1 The Straits Times news, Why collapse happened? *The Straits Times* (26 October 2002).

2 Jiang, J., Wang, J., Wang, W., and Zhang, W. (2009). Modeling influence of gas/liquid/solid three-phase boundary zone on cathodic process of soil corrosion. *Electrochimica Acta* 54: 3623–3629.

3 Cole, I.S. and Marney, D. (2012). The science of pipe corrosion: a review of the literature on the corrosion of ferrous metals in soils. *Corrosion Science* 56: 5–16.

4 Tan, M.Y.J. (2020). A critical overview of monitoring infrastructural health using corrosion probes. *Corrosion Engineering, Science and Technology* 55: 103–117. https://doi.org/10.1080/1478422X.2019.1695390.

5 Tan, M.Y.J., Varela, F., Huo, Y. et al. (2017) Acquisition, visualisation and interpretation of pipeline corrosion monitoring data. *EUROCORR 2017 – The Annual Congress of the European Federation of Corrosion, 20th International Corrosion Congress and Process Safety Congress 2017*, Czech Association of Corrosion Engineers. Prague, Czech Republic (3–7 September 2017).

6 Murray, J. and Moran, P. (1989). Influence of moisture on corrosion of pipeline steel in soils using in situ impedance spectroscopy. *Corrosion* 45: 34–43.

7 Gupta, S. and Gupta, B. (1979). The critical soil moisture content in the underground corrosion of mild steel. *Corrosion Science* 19: 171–178.

8 Fu, J., Pei, F., Zhu, Z. et al. (2013). Influence of moisture on corrosion behaviour of steel ground rods in mildly desertified soil. *Anti-Corrosion Methods and Materials* 60: 148–152.

9 Hendi, R., Saifi, H., Belmokre, K. et al. (2018). Effect of black clay soil moisture on the electrochemical behavior of API X70 pipeline steel. *Materials Research Express* 5: 036523.

10 Levlin, E. (1996). Aeration cell corrosion of carbon steel in soil: in situ monitoring cell current and potential. *Corrosion Science* 38: 2083–2090.

11 Azoor, R., Deo, R., Birbilis, N., and Kodikara, J. (2018). A coupled electro-chemical-soil model to evaluate the influence of soil aeration on underground metal pipe corrosion. *Corrosion* 74: 1177–1191.

12 Schwerdtfeger, W. (1965). Soil resistivity as related to underground corrosion and cathodic protection. *Journal of Research of the National Bureau of Standards- Co Engineering and Instrumentation* 69C: 20–21.

13 Arzola, S., Palomar-Pardave, M., and Genesca, J. (2003). Effect of resistivity on the corrosion mechanism of mild steel in sodium sulfate solutions. *Journal of Applied Electrochemistry* 33: 1233–1237.

14 Song, G.-L., Cao, C.-N., and Lin, H.-C. (1995). A new technique for correlation of underground corrosion. *Corrosion* 51: 491–497.

15 Yan, M., Sun, C., Xu, J. et al. (2014). Role of Fe oxides in corrosion of pipeline steel in a red clay soil. *Corrosion Science* 80: 309–317.

16 Mehanna, M., Basséguy, R., Délia, M.-L., and Bergel, A. (2009). Effect of Geobacter sulfurreducens on the microbial corrosion of mild steel, ferritic and austenitic stainless steels. *Corrosion Science* 51: 2596–2604.

17 Romanoff, M. (1957). *Underground Corrosion*, National Bureau of Standards Circular 579, vol. 25. Washington: US Government Printing Office.

18 Fitzgerald, J. (1993). Evaluating soil corrosivity – then and now. *Materials Performance United States* 32.

19 Scully, J. and Bundy, K. (1985). Electrochemical methods for measurement of steel pipe corrosion rates in soil. *Materials Performance* 24: 18–25.

20 Nie, X., Li, X., Du, C., and Cheng, Y. (2009). Temperature dependence of the electrochemical corrosion characteristics of carbon steel in a salty soil. *Journal of Applied Electrochemistry* 39: 277–282.

21 Huo, Y., Tan, M.Y., and Forsyth, M. (2016). Visualising dynamic passivation and localised corrosion processes occurring on buried steel surfaces under the effect of anodic transients. *Electrochemistry Communications* 66: 21–24.

22 Varela, F., Tan, M.Y.J., and Forsyth, M. (2015). Understanding the effectiveness of cathodic protection under disbonded coatings. *Electrochimica Acta* 186: 377–390.

23 Dang, D.N., Lanarde, L., Jeannin, M. et al. (2015). Influence of soil moisture on the residual corrosion rates of buried carbon steel structures under cathodic protection. *Electrochimica Acta* 176: 1410–1419.

24 Akkouche, R., Rémazeilles, C., Jeannin, M. et al. (2016). Influence of soil moisture on the corrosion processes of carbon steel in artificial soil: active area and differential aeration cells. *Electrochimica Acta* 213: 698–708.

25 Kodým, R., Šnita, D., Fíla, V. et al. (2017). Investigation of processes occurring at cathodically protected underground installations: mathematical modeling of reaction transport processes in soil. *Corrosion Science* 120: 28–41.

26 Kodým, R., Šnita, D., Fíla, V. et al. (2017). Investigation of processes occurring at cathodically protected underground installations: experimental study of pH alteration and mathematical modeling of oxygen transport in soil. *Corrosion Science* 120: 14–27.

27 Hoseinieh, S.M., Homborg, A.M., Shahrabi, T. et al. (2016). A novel approach for the evaluation of under deposit corrosion in marine environments using combined analysis by electrochemical impedance spectroscopy and electrochemical noise. *Electrochimica Acta* 217: 226–241.

28 Huo, Y. and Tan, M.Y. (2017). Measuring and understanding the critical duration and amplitude of anodic transients. *Corrosion Engineering, Science and Technology* 1–8.

29 Sun, M., Staszewski, W.J., and Swamy, R.N. (2010), Article ID 724962). Review article: smart sensing technologies for structural health monitoring of civil engineering structures. *Advances in Civil Engineering* 2010: 13.

30 Chang, F.K. (1997) Structural health monitoring: a summary report on the first international workshop on structural health monitoring. *Proceedings of the 2rd International Workshop on Structural Health Monitoring* (ed. F.K. Chang), pp. 3–11. Technomic Publishing Company, Lancaster, UK (September 1997).

31 Zeng, X., Bao, X., Chhoa, C.Y. et al. (2002). Strain measurement of a concrete beam.... *Applied Optics* 41: 5105–5114.

32 Chen, G., Xu, B., McDaniel, R.D. et al. (2005) Distributed strain measurement of a large-scale reinforced concrete beam-column assembly under cyclic loading. *Smart Structures and Materials: Sensors and Smart Structures Technologies for Civil, Mechanical, and Aerospace Systems, Proceedings of SPIE*, vol. 5765, article 55, pp. 516–527, San Diego, CA, USA (7–10 March 2005).

33 Fuhr, P.L. and Huston, D.R. (1998). Corrosion detection in reinforced concrete roadways and bridges via embedded fibre optic sensors. *Smart Materials and Structures* 7: 217–228.

34 Ayres, J.W., Lalande, F., Chaudhry, Z., and Rogers, C.A. (1998). Qualitative impedance-based health monitoring of civil infrastructures. *Smart Materials and Structures* 7: 599–605.

35 Ay, A.M., Wang, Y., Khoo, S.Y., and Li, A.J. (2013) Vibration based damage identification of a scale-model steel frame structure subjected to bolt connection failures. *The 6th International Conference on Structural Health Monitoring of Intelligent Infrastructure*, Hong Kong, China (9–11 December 2013).

36 Tiedemann, W., Newman, J., and Bennion, D.N. (1973). Errors in measurements of electrode kinetics caused by nonuniform ohmic-potential drop to a disk electrode. *Journal of the Electrochemical Society* 120: 256–258.

37 Prentice, G.A. and Tobias, C.W. (1982). A survey of numerical methods and solutions for current distribution problems. *Journal of the Electrochemical Society* 129: 72–78.

38 Tan, Y. (2013). *Heterogeneous Electrode Processes and Localized Corrosion*, 1–246. Hoboken, NJ: Wiley.

39 Glass, R.S., Clarke, J., Willis, L., and Ciarlo, D.R. (1995). Method for monitoring environment and corrosion. US Patent 5,437,773.

40 Raupach, M. and Schießl, P. (2001). Macrocell sensor systems for monitoring of the corrosion risk of the reinforcement in concrete structures. *NDT and E International* 34: 435–442.

41 Aung, N.N. and Tan, Y.J. (2004). A new method of studying buried steel corrosion and its inhibition using the wire beam electrode. *Corrosion Science* 46: 3057–3067.

42 Tan, Y.J. (2011). Experimental methods designed for measuring corrosion in highly resistive and inhomogeneous media. *Corrosion Science* 53: 1145.

43 Aung, N.N., Wai, W.K., and Tan, M.Y. (2006). A novel electrochemical method for monitoring corrosion under insulation. *Anti-Corrosion Methods and Materials* 53: 175–179.

44 Legat, A. (2007). Monitoring of steel corrosion in concrete by electrode arrays and electrical resistance probes. *Electrochimica Acta* 52: 7590–7598.

45 Hu, J., Li, G., Deng, K., and Liu, Y. (2005). Study on non-uniform corrosion of rebar embedded in concrete by wire – beam electrode. *Journal of the Chinese Society of Corrosion and Protection* 25 (2): 88–92.

46 Li, L.Q., Dong, S.G., Wang, W. et al. (2010). Study on interaction between macrocell and microcell in the early corrosion process of reinforcing steel in concrete. *Science China Technological Sciences* 53 (5): 1285–1289.

47 Česen, A., Kosec, T., and Legat, A. (2013). Characterization of steel corrosion in mortar by various electrochemical and physical techniques. *Corrosion Science* 75: 47–57.

48 Shi, W., Dong, Z.H., Kong, D.J., and Guo, X.P. (2013). Application of wire beam electrode technique to investigate initiation and propagation of rebar corrosion. *Cement and Concrete Research* 48: 25–33.

49 Hu, J., Deng, P., Li, X. et al. (2018). The vertical Non-uniform corrosion of reinforced concrete exposed to the marine environments. *Construction and Building Materials* 183: 180–188.

50 Yu, H. and Caseres, L. (2012). An embedded multi-parameter corrosion sensor for reinforced concrete structures. *Materials and Corrosion* 63 (11): 1011–1016.

51 Liubin, S., Daowu, Y., Sanjun, P. et al. (2011). Electrochemical study of inhibitors to improve the anti-corrosion performance of reinforced bar in the concrete. *Anti-Corrosion Methods and Materials* 58 (1): 22–25.

52 Shi, W., Wang, T.-Z., Dong, Z.-H., and Guo, X.-P. (2017). Application of wire beam electrode technique to investigate the migrating behavior of corrosion inhibitors in mortar. *Construction and Building Materials* 34: 167–175.

53 Wu, J., Yu, H., and Shi, X. (2016). Effectiveness of products in managing metallic corrosion induced by cyclic deicer exposure: laboratory study using multielectrode array sensors, electrochemical impedance, and laser profilometer. *Journal of Materials in Civil Engineering* 28 (5): 04015186.

54 Yang, D., Song, L., Wang, C., and Xiao, Z. (2009). Electrochemistry study on epoxy resin to anti-corrosion performance of reinforced bar in the concrete. *TMS Annual Meeting* 3: 735–740.

55 Tan, Y.J. (2009). Sensing electrode inhomogeneity and electrochemical heterogeneity using an electrochemically integrated multielectrode array. *Journal of the Electrochemical Society* 156: C195.

56 Tan, Y.J. (2009). Sensing localised corrosion by means of electrochemical noise detection and analysis. *Sensors and Actuators B: Chemical* 139: 688–698.

57 John McCarter, W. and Vennesland, Ø. (2004). Review: sensor systems for use in reinforced concrete structures. *Construction and Building Materials* 18: 351–358.

58 Melchers, R.E. and Wells, P.A. (2010) Modelling the long term corrosion of reinforced concrete sewers. *Proceedings of the 7th Leading-Edge Conference on Water and Wastewater Treatment Technologies*, Phoenix, AZ, USA (2–4 June 2010).

59 Wells, P.A., Melchers, R.E., and Bond, P. (2009). Factors involved in the long term corrosion of concrete sewers. *Australasian Corrosion Association Proceedings of Corrosion and Prevention 2009*, Coffs Harbour, Australia (15–19 November 2009).

60 Ahmad, S. (2003). Reinforcement corrosion in concrete structures, its monitoring and service life prediction – a review. *Cement and Concrete Composites* 25: 459–471.

61 Cottis, R.A. (2001). The interpretation of electrochemical noise data. *Corrosion (NACE)* 57: 265.

62 Hladky, K. and Dawson, J.L. (1981). The measurement of localised corrosion using electrochemical noise. *Corrosion Science* 21: 317.

63 Mansfeld, F. and Xiao, H. (1994). Evaluation of coating degradation with electrochemical impedance spectroscopy and electrochemical noise analysis. *Journal of the Electrochemical Society* 141: 1403.

64 Bertocci, U., Gabrielli, C., Huet, F. et al. (1997). Noise resistance applied to corrosion measurements. *Journal of the Electrochemical Society* 144: 37.

65 Liu, T., Tan, Y.J., BZM, L. et al. (2006). Novel corrosion experiments using the WBE: (I)-(IV). *Corrosion Science* 48: 23–78.

66 Varela, F., Tan, M.Y.J., and Forsyth, M. (2015). An electrochemical method for measuring localized corrosion under cathodic protection. *ECS Electrochemistry Letters* 4 (1): C1–C4. Video in Supplemental Material: https://iopscience.iop.org/article/10.1149/2.0091501eel.

67 Wang, K., Varela, F.B., and Tan, M.Y. (2019). Visualizing dynamic and localized corrosion processes on cathodically protected steel exposed to soil with different moisture contents. *Corrosion* 75 (4): 398–407.

68 Huo, Y. and Tan, M.Y. (2018). Localised corrosion of cathodically protected pipeline steel under the effects of cyclic potential transients. *Corrosion Engineering, Science and Technology* 53: 348–354.

69 Varela, F., Tan, M.Y.J., and Forsyth, M. (2015). Electrochemical method for studying localized corrosion beneath disbonded coatings under cathodic protection. *Journal of the Electrochemical Society* 162: C515–C527.
70 Abd El Haleem, S.M., Abd El Aal, E.E., Abd El Wanees, S., and Diab, A. (2010). Environmental factors affecting the corrosion behaviour of reinforcing steel: I. The early stage of passive film formation in $Ca(OH)_2$ solutions. *Corrosion Science* 52 (12): 3875–3882.
71 Abd El Haleem, S.M., Abd El Wanees, S., Abd El Aal, E.E., and Diab, A. (2010). Environmental factors affecting the corrosion behavior of reinforcing steel II. Role of some anions in the initiation and inhibition of pitting corrosion of steel in $Ca(OH)_2$ solutions. *Corrosion Science* 52: 292–302.
72 Orazem, M.E., Esteban, J.M., Kennelley, K.J., and Degerstedt, R.M. (1997). Mathematical models for cathodic protection of an underground pipeline with coating holidays: part 1 – theoretical development. *Corrosion* 53 (4): 264–272.
73 Wang, K., Varela, F.B., and Tan, M.Y. (2019). The effect of electrode surface area on corrosion initiation monitoring of X65 steel in soil. *Corrosion Science* 152: 218–225.
74 Frankel, G. (1998). Pitting corrosion of metals a review of the critical factors. *Journal of the Electrochemical Society* 145: 2186–2198.
75 Tan, Y. (2011). Understanding the effects of electrode inhomogeneity and electrochemical heterogeneity on pitting corrosion initiation on bare electrode surfaces. *Corrosion Science* 53: 1845–1864.
76 Frankel, G.S. and Sridhar, N. (2008). Understanding localized corrosion. *Materials Today* 11: 38–44.
77 Angst, U., Rønnquist, A., Elsener, B. et al. (2011). Probabilistic considerations on the effect of specimen size on the critical chloride content in reinforced concrete. *Corrosion Science* 53: 177–187.
78 Li, L. and Sagues, A. (2004). Chloride corrosion threshold of reinforcing steel in alkaline solutions – effect of specimen size. *Corrosion* 60: 195–202.
79 Aziz, P. and Godard, H.P. (1955). Influence of specimen area on the pitting probability of aluminum. *Journal of the Electrochemical Society* 102: 577–579.
80 Angst, U.M. and Elsener, B. (2017). The size effect in corrosion greatly influences the predicted life span of concrete infrastructures. *Science Advances* 3: e1700751.
81 Orazem, M., Esteban, J., Kennelley, K., and Degerstedt, R. (1997). Mathematical models for cathodic protection of an underground pipeline with coating holidays: part 2 – case studies of parallel anode cathodic protection systems. *Corrosion* 53: 427–436.
82 Orazem, M., Kennelley, K., and Bone, L. (1993). Current and potential distribution on a coated pipeline with holidays part II – comparison of the effects of discrete and distributed holidays. *Corrosion* 49: 211–219.
83 Buchler, M. (2015). A new perspective on cathodic protection criteria promotes discussion. *Materials Performance* 54: 44–48.

84 Gummow, R.A. (2010). Examining the controversy surrounding the −850 mV CP criteria. *Pipeline and Gas Journal* 237: 85–87.

85 Angst, U., Büchler, M., Martin, B. et al. (2016). Cathodic protection of soil buried steel pipelines – a critical discussion of protection criteria and threshold values. *Materials and Corrosion* 67: 1135–1142.

86 Martin, B. (2006). A history of stray current corrosion. *Corrosion & Materials* 31: 12–14.

87 Wang, K., Varela, F.B., and Tan, M.Y.J. (2018). An array of multielectrodes for locating, visualizing, and quantifying stray current corrosion. *Corrosion* 74 (10): 1093–1101.

88 Bertolini, L., Carsana, M., and Pedeferri, P. (2007). Corrosion behaviour of steel in concrete in the presence of stray current. *Corrosion Science* 49: 1056–1068.

89 Varela, F., Tan, M.Y.J., and Forsyth, M. (2015). An overview of major methods for inspecting and monitoring external corrosion of on-shore transportation pipelines. *Corrosion Engineering, Science and Technology* 50: 226–235.

90 Zakowski, K. and Darowicki, K. (2000). Methods of evaluation of the corrosion hazard caused by stray currents to metal structures containing aggressive media. *Polish Journal of Environmental Studies* 9: 237–242.

91 Ormellese, M., Lazzari, L., Goidanich, S., and Sesia, V. (2008) CP criteria assessment in the presence of AC interference. *Proceedings of the Corrosion 08*, NACE International, New Orleans, Louisiana, USA (16–20 March 2008).

92 Gummow, R.A., Wakelin, R.G., and Segall, S.M. (1998) AC corrosion – a new challenge to pipeline integrity. *Proceedings of the Corrosion 98*, NACE International, San Diego, USA (22–27 March 1998).

93 Mahdavi, F., Tan, M.Y.J., and Forsyth, M. (2016). Communication – an approach to measuring local electrochemical impedance for monitoring cathodic disbondment of coatings. *Journal of the Electrochemical Society* 163: C228–C231.

94 Peabody, A.W. (1979). Corrosion aspects of arctic pipelines. *Materials Performance* 18: 27–31.

95 Pourbaix, M. (1972). Significance of protection potential in pitting, intergranular corrosion and stress-corrosion cracking. *Journal of the Less-Common Metals* 28: 51–65.

96 Büchler, M., Voute, C.H., and Joos, D. (2012). Effect of variation of cathodic protection level over time on the a.c. corrosion process, Ceocor International Congress and Exhibition, Lucerne, Switzerland.

4

Localized Corrosion Influenced by Changing Mechanisms

In practical engineering structures, the mechanisms of corrosion and materials degradation can change with environmental variations and the extension of exposure, significantly altering the nature, rates, and patterns of corrosion. The severity of corrosion and the efficiency of corrosion control are often determined by corrosion mechanisms that can change due to time-based variations in environmental parameters; therefore, it is important to identify these mechanism changes in order to better understand and prevent corrosion and materials degradation. An example is buried steel pipeline corrosion and materials degradation that can start from coating cracking and damage, soil corrosion at bare metal areas, coating disbondment, and corrosion under disbonded coatings (CUDC). Each of these mechanisms could occur one after another and become the dominant and "worst-case scenario" mechanism of corrosion and material degradation over different periods of exposure. Such mechanism changes can also occur due to many other factors, such as temperature changes, local pH changes, aeration, flow velocity changes, corrosion product layer formation, the presence of wear or abrasion, and the existence of crevices and local stress. An example is the formation of protective iron carbonate scales inside oil and gas pipelines at a temperature above 70 °C that can significantly affect the rates and mechanism of carbon dioxide corrosion of steel.

The fact that corrosion mechanism change with time suggests that corrosion data from short exposure testing could falsely represent corrosion mechanisms and kinetics over longer periods, posing major challenges on the prediction and control of corrosion and materials degradation. It is evident that most existing corrosion predictive models and control methods are designed based on a single corrosion mechanism and are unable to tackle localized corrosion affected by multiple and varying mechanisms. It is essential to detect such mechanism changes if accurate and appropriate corrosion prediction, modeling, and control

Localized Corrosion in Complex Environments, First Edition. Mike Yongjun Tan.

is required. Electrochemical monitoring is probably the only practical approach to achieving this because it has the temporal resolution required for detecting such dynamic mechanism changes. This chapter presents electrochemical methods and tools including the WBE-video technique and integrated three-dimensional (3D) corrosion probes for *in situ* monitoring of corrosion and materials degradation affected by changing mechanisms. Coating damage, disbondment and subsequent corrosion of underground steel pipelines are discussed as typical examples. It is shown that corrosion mechanism changes are usually due to accumulated changes in local chemical conditions, in particular the local pH, oxygen contents, and humidity. The design of corrosion probes should focus on the simulation of such local chemistry changes in order to effectively mimic and monitor the dominating mechanisms over every period of corrosion and materials degradation processes.

4.1 Localized Corrosion and Materials Degradation with Varying Mechanisms

4.1.1 Issues with Changing Corrosion Mechanisms on Buried Pipelines

Buried metallic structures such as underground steel pipelines are usually protected against corrosion by organic coatings in conjunction with cathodic protection (CP). Organic coatings protect metallic substrate primarily by acting as a barrier for corrosive species such as water, ions, and oxygen. Unfortunately, coatings usually contain various forms of defects such as pores and could crack or degrade under mechanical impacts and environmental effects, resulting in the formation of pathways for corrosive species to reach metal substrate. Figure 4.1 shows examples of mechanically induced pipeline coating damages following hydrostatic test and horizontal directional drilling (HDD) installation of pipeline. In order to protect exposed metallic areas under damaged coatings, CP is used by applying appropriate cathodic polarizing potential. The potential of $-850\,mV_{CSE}$ is the commonly applied "safe" CP potential in industry criterion [1, 2] for protecting steel pipelines exposed to normal underground conditions. However this "safe" CP level may not be sufficient in some sections of the pipelines if they are influenced by various complex factors such as stray currents, cathodic shielding, and microorganisms. Under the influence of these factors, multiple corrosion mechanisms including CDUC and stray current corrosion could occur [3, 4], as illustrated in Figure 4.2.

Coatings on buried pipelines could degrade in various ways and mechanisms such as water sorption and ions transportation, blistering, anodic undermining, and cathodic disbondment. Coating blistering is a common form of localized

Figure 4.1 Sample field joint coating damage following hydrostatic test and coating damages observed on pipelines installed by horizontal directional drilling (HDD).

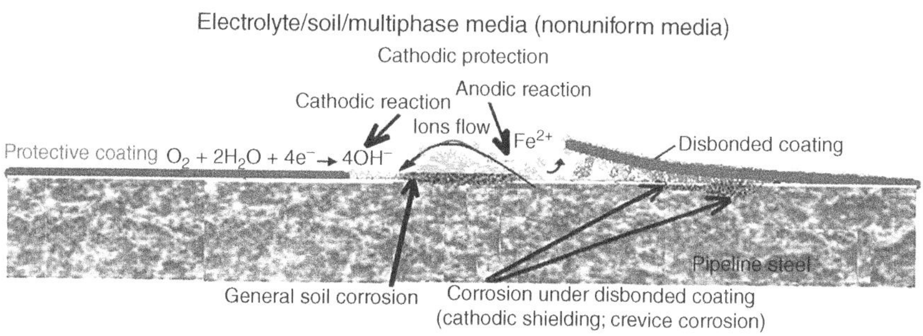

Figure 4.2 Illustrating multiple mechanisms that could occur over a steel surface with damaged and disbonded coating.

adhesion loss, which happens through various mechanisms such as coating's swelling and volume expansion due to water absorption, gas inclusion, electroosmotic, and osmotic influences [5, 6]. Cathodic disbondment is a more common form of coating disbondment that occurs frequently on coated metal structures protected by cathodic potential and is considered to be a major mechanism of coating failure occurring on steel structures such as buried steel pipelines that are protected by coating and CP. Cathodic disbondment takes place at the coating and metal interface, causing the loss of adhesion between a coating and its metal substrate due to local chemical and physical changes [5–10]. At the beginning of a pipeline's exposure to soil, coating defect areas on steel would experience slow general soil corrosion under CP-induced passive conditions, in similar ways as various forms of soil corrosion described in Chapter 3. Cathodic disbondment usually occurs at locations where CP overprotection exists such as at the stray current "pickup" points where the potential could be significantly more negative than other areas of the pipeline (see Figures 3.12 and 3.13). The products of CP-induced oxygen reduction or hydrogen evolution reactions such as hydroxyl ions, hydrogen gas, and short-lived superoxide intermediates could form at coating's defect sites, causing cathodic disbondment of coatings either through dissolving metal oxide layer, saponification of polymer coating, or displacing the coating at metal/coating interface [5–10]. Cathodic disbondment creates "crevices," as illustrated in Figure 4.2, over the surfaces of "overprotected" pipelines.

Coating disbondment is often followed by localized corrosion initiation and propagation, as illustrated in Figure 4.2, within the crevice between the metal surface and the disbonded coating layer. The crevice provides easy lateral paths for corrosive species to reach the metal surface while simultaneously shielding cathodic protection currents from reaching the metal substrate. This causes corrosion mechanism changing from the initial general soil corrosion to much more complex localized corrosion usually referred to as corrosion under disbonded coatings (CUDC). CUDC is the worst-case scenario form of electrochemical corrosion that occurs frequently on the external surface of buried steel structures such as underground pipelines, causing many types of corrosion damage such as crevice corrosion, pitting, microbiologically induced corrosion, and even stress corrosion cracking. Figure 4.3 shows some photos of pipeline samples with coating disbondment and corrosion under disbonded coatings. Some more cases of corrosion under disbonded coatings on pipelines are shown in Figure 1.18.

Currently, cathodic disbondment and the subsequent CUDC remain as major failure modes affecting the integrity of civil and industrial infrastructures such as buried and submerged pipelines [3]. It was commented by industry that currently some 90% corrosion occurred on buried gas pipelines was due to corrosion under disbonded coatings or heat shrink sleeve (behaving like shielding coating) [3]. For this reason, aged pipelines with disbonded shielding coatings are identified as

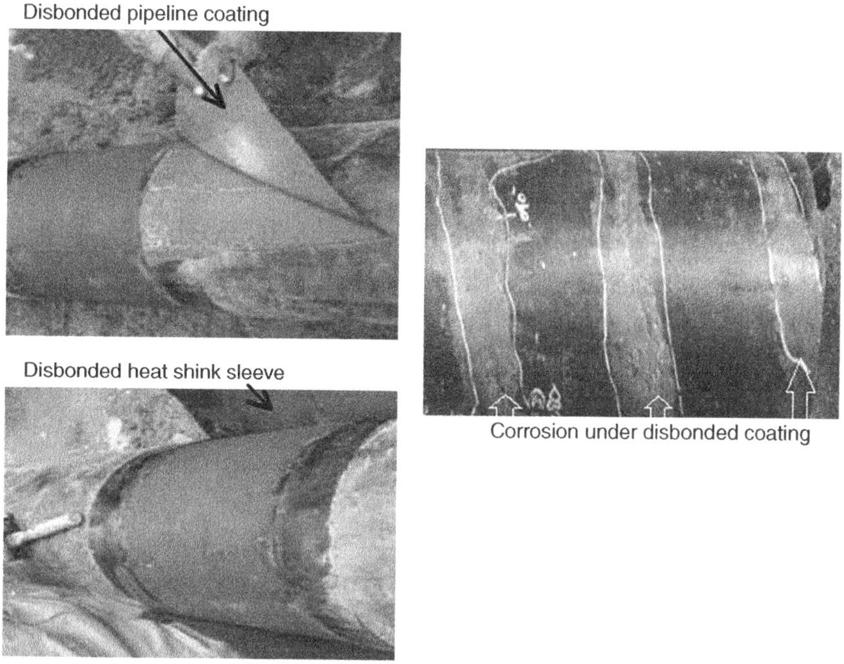

Figure 4.3 Cases of coating disbondment and corrosion under disbonded coatings on pipelines.

prime "high-risk" pipelines because the industry is concerned that this type of corrosion cannot be mitigated by conventional CP or be detected through normal inspection and survey means due to current shielding, leading to continuous localized corrosion occurring on metal surface unnoticed [3–7]. Cathodic shielding under disbonded coatings on underground and submerged engineering structures meets all conditions leading to complexity of localized corrosion: Invisibility; dynamically changing mechanisms with exposure time and environmental conditions; corrosion initiation and propagation over multi-time and length scales.

4.1.2 Assessing Coating Disbondment and Corrosion Under Disbonded Coatings

A prerequisite to effective prediction and control of coating disbondment and associated corrosion is techniques capable of detecting and monitoring the occurrence of coating disbondment and CUDC. Practical methods used in industry for assessing and managing coating disbondment and CUDC include (i) the excavation and visual inspection of coated metal structures such as buried pipelines; (ii) pre-assessment of the cathodic disbondment resistance of coatings in simulated exposure environment; (iii) *in situ* monitoring of the cathodic disbondment of coatings as well as CUDC. Traditional excavation and visual inspection of coated metal structures is useful, but a slow and expensive way of detecting and evaluating cathodic disbondment of coatings and CUDC. A more effective approach to providing data useful for minimizing cathodic disbondment of coatings and CUDC is to perform laboratory tests that pre-assess coating's resistance against cathodic disbondment before being applied on metallic structures. Over the past decades major innovative efforts have been made to overcome technical obstacles in assessing the cathodic disbondment of coatings. The advantages and limitations of various techniques have been critically reviewed and discussed in the literature such as references [4, 11]. Currently the most widely used laboratory testing method for pre-assessing coating's ability to resist cathodic disbondment is standard cathodic disbondment test (CD test) [12, 13] that is based on ex-situ observation of test samples subjected to the accelerated CD test conditions after a specific time period. These tests require a coated sample with an artificial defect exposed to a simulated CP condition. After a specified period of exposure time, the test sample is removed from the test cell for visual assessment of the radial disbondment of the coating film by striping the coating using a knife. The CD test is an accelerated test in nature, operating at a controlled current or potential significantly higher than those experienced in normal industry practice, therefore the test results may not be representative of the long-term coating performance under normal CP operating conditions [14]. The test results may not correlate well with field observed behavior [15]. Despite of these shortcomings, CD tests are still widely used for pre-assessing the cathodic disbondment resistance of coatings, and for investigating the effects of some environmental parameters on coating disbondment rates such as coating thickness [14, 16, 17], cathodic potential level [14, 17–20], solution type and concentration [14, 17, 18, 20], testing temperature [14, 17, 19–21], substrate topography [21], and exposure time [20, 21].

Pre-assessment of coating can assist the appropriate selection of coatings; however it does not ensure the avoidance of coating disbondment because the practical environment can be very different form the accelerated testing condition. In industry practice, monitoring is desirable for the timely detection, prediction, and repair of coating disbondment and CUDC. Techniques that could provide *in situ* monitoring of cathodic disbondment of coatings and CUDC is therefore needed. One such technique is electrochemical impedance spectroscope (EIS) that could reveal organic coating degradation through *in situ* measurement of interfacial properties such as the resistance and capacitance of coated electrodes [22–25]. EIS has also been evaluated for studying the cathodic disbondment of defective coatings after the coated electrode was exposed to solution under CP for various periods of time [26, 27]. Unfortunately no correlation between the derived EIS parameters and the coating disbonded area was found [26–30] because global EIS measurements data are dominated by the small impedance of the defective coating area that "masked" the larger impedance at disbonded area. A method to address the issue in global EIS is the localized EIS (LEIS) technique that measures the a.c. current density in solution very near the polarized working electrode to detect the local impedance over different areas of the electrode [31–33]. The LEIS scanning can detect changes at a coating defect under CP potentials [34]; however, there are significant obstacles for the techniques being practically used to monitor in-situ disbondment of thick pipeline coatings (e.g. 500–1000 μm thick) under cathodic protection conditions because a thick coating would "shield" the current from reaching the disbonded area. Therefore, LEIS could provide little information from higher impedance regions deep in the coating disbondment areas.

Scanning Kelvin probe (SKP) is another scanning technique that could be used in the laboratory to detect electrochemical changes at coating/metal interface due to cathodic disbondment of coatings [35]. It measures the differences of Volta-potential between a vibrating probe and the coated sample without contacting it and indicates the corrosion potential and its changes underneath the intact polymer layers [36]. The SKP has been used extensively to study coatings delamination at OCP [36–38]; however, little information is available on the suitability of the SKP technique for probing coating disbondment under cathodic protection condition [39, 40]. Non-electrochemical techniques such as scanning acoustic microscopy (SAM) have also been employed for imaging the commencement and propagation of cathodic disbondment of coatings [41–44]. The SAM uses ultrasound signals to visualize coatings degradation, delamination, and cathodic disbondment [42–44], unfortunately this technique was reported to be unsuccessful in the detection of coating disbondment because of the limited vertical resolution of the microscope [45–47] and limitations in measuring thick polymer films [48].

A probe designed based on an electrochemically integrated multi-electrode array, often referred to as the wire beam electrode (WBE), is an unconventional approach in localized electrochemical measurement for probing cathodic disbondment. Le Thu et al. [49] took the advantage of the WBE technique in monitoring electrochemical parameters from localized areas under a disbonded coating in order to enable the observation of the coating disbondment. Wang et al. also used the WBE to measure and map the current distributions over carbon steel under an alkyd primer exposed to 3.5% sodium chloride solution [50]. Zhang et al. further used the combination of the WBE and EIS method to evaluate metal corrosion under defective coatings [51]. These reports suggest that the WBE approach possesses the potential for *in situ* measurement and monitoring of cathodic disbondment of coatings [49–52]. However there are two issues with the WBE probes used in these reported work: Firstly the gaps between electrodes/wires in WBEs appear to be too big. These large gaps could prevent the movement of disbondment front from one electrode/wire to another on the WBE surface. It is well known that the WBE surface needs to simulate the chemistry and its distribution over a large electrode surface. This is especially important since cathodic disbondment is believed to be surface chemistry driven. Secondly, the coating appears to be very thin and therefore lots of corrosion occurred in experiments reported work. The current or impedance measured from such complex surface might not be from "underneath" coating film, but under the corrosion products. A more recent application of the WBE method by Mahdavi et al. has suggested the potential to address major difficulties in the *in situ* measurement and monitoring of cathodic disbondment of coatings [53].

On the other hand, CUDC has also been extensively studied as the worst-case scenario form of electrochemical corrosion that occurs frequently on the external surface of buried steel structures such as underground pipelines, causing many types of corrosion damage such as pitting [54], microbiologically induced corrosion (MIC) [55, 56], and stress corrosion cracking (SCC) [57, 58]. Unfortunately buried steel pipelines that are usually protected by organic coatings and cathodic protection (CP) [59, 60] can suffer coating disbondment, leading to the formation of crevices of various geometries between the coating and the pipeline steel surface [61–63]. CP current would be difficult to penetrate into the crevice under the disbonded coating (usually referred to as cathodic shielding) [61, 64] and thus create a serious localized corrosion environment [65]. CUDC has been reported to be critically affected by the disbondment gap size and the environment. For instance, it was reported in the industry that if a disbonded coating was tightly bound to a steel pipe, CUDC would not normally be an issue, while corrosion issues were observed on disbonded coatings loosely bound on pipes [66]. Over the past decades, many attempts have been made to understand and quantify CUDC in aqueous solutions under the influence of disbondment crevice geometry by measuring the potential gradient [54, 56, 64, 67–75], pH distribution [54, 64, 67, 68, 70–75], current profile [64, 69, 74, 75], and oxygen concentration [54, 64, 67, 68, 72] within the crevices of disbonded coatings. However the in-situ measurement and monitoring of CUDC has been a difficult in industry because of the lack of suitable CUDC probes. The key to designing corrosion probes for probing CUDC is that corrosion probes need to simulate localized corrosion mechanisms and their changes with time, as well as to monitor CUDC and its changes with localized corrosion mechanisms, with special attention on the interactions between different mechanisms. A major effort of *in situ* monitoring of corrosion under disbonded coating by an electrochemically integrated multi-electrode array was made by Varela et al. [76, 77] who firstly applied the probe to map the spatial and temporal current density distributions along the coating disbondment crevice. More details of these technical advancements are described in the following sections.

4.2 Probing Localized Coating Degradation and Disbondment

Since coating disbondment is a critical step leading to CUDC, ideally coating disbondment could be prevented or significantly reduced in order to protect a buried pipeline from CUDC. This could be achieved by monitoring the occurrence of coating degradation and disbondment, and by adjusting CP potential and environmental conditions in order to control coating disbondment. Monitoring coating disbondment is also useful for investigating factors affecting coating damages and for predicting CUDC, unfortunately measuring and probing coating degradation and disbondment is a technological difficulty.

4.2.1 Probes for Monitoring Cathodic Disbondment

In order to overcome technological limitations in monitoring the cathodic disbondment of coatings, a method has been developed by Mahdavi et al. for *in situ* probing of coating disbondment based on an electrochemically integrated multi-electrode array probe and local electrochemical impedance spectroscopy (EIS) measurements [53, 63]. Figure 4.4 shows a multi-electrode array probe employed

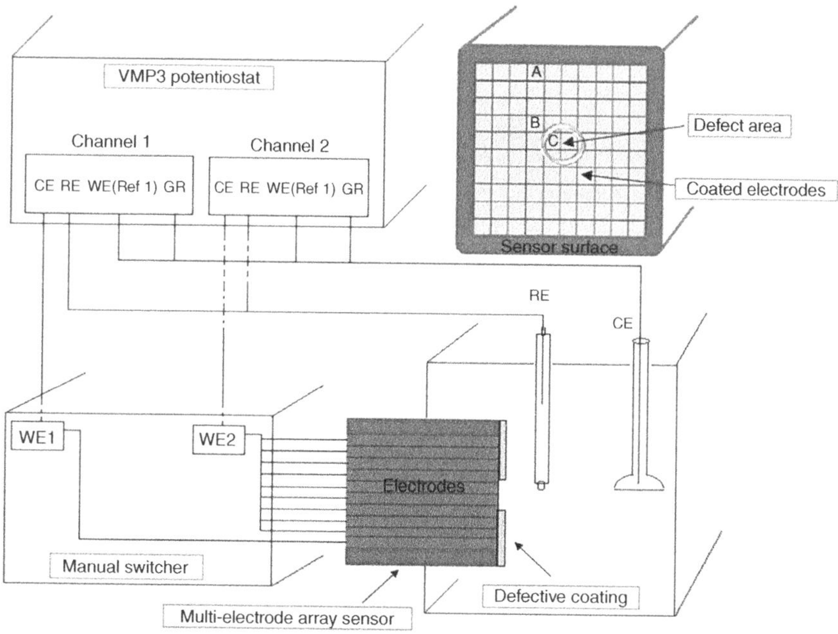

Figure 4.4 Schematic diagram of the experimental setup for local EIS mapping under CP. *Source:* Mahdavi et al. [53]/ IOP Publishing / CC BY.

to perform local EIS measurements over a surface covered with a defective coating. In a typical experiment [53], transparent polyester coating was applied on the surface of an electrode array consisting of 100 closely packed but isolated square-shaped electrodes of carbon steel (2.44 mm × 2.44 mm) embedded in epoxy resin. The gaps between neighboring electrodes were kept very small (0.10 ± 0.05 mm). The coating was dried under room temperature for five days to form a film with dry film thickness of 1000 ± 20 μm. An artificial defect of 5 mm diameter was introduced at the center of the coated probe surface to simulate a coating defect or damage. Figure 4.5 shows typical local EIS data measured on selected areas of the electrode array surface after 624 hours exposure to solution under a CP potential of $-1.40\,V_{Ag/AgCl}$. It can be seen that different electrode areas exhibit very different EIS characteristics. Most of the electrodes located far away from the defect area (typically location "A") show a capacitive behavior with a high magnitude of impedance ($>5 \times 10^7$ ohm), which indicates that coating at the location was intact

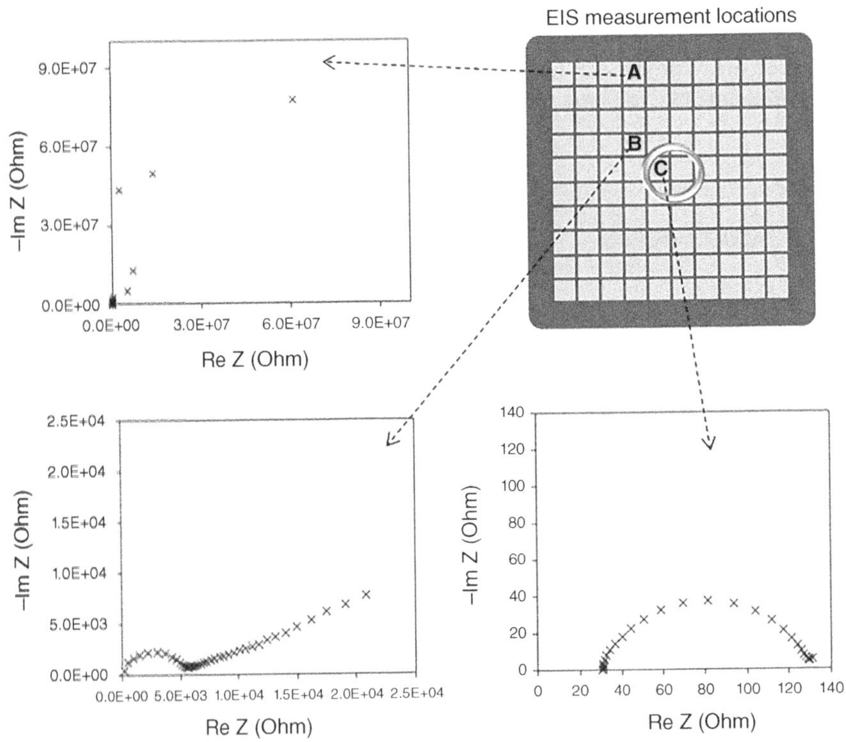

Figure 4.5 The Nyqusit plots recorded over locations "A", "B", and "C" over multi-electrode array sensor surface after 624 hours exposure to testing solution under a CP potential of $-1.40\,V_{Ag/AgCl}$. *Source:* Adapted from Mahdavi et al. [53].

and disbondment did not occur at these electrodes even after 624 hours exposure. Electrodes located near the defect area including the electrode at location "B" are characterized with two semi-circles in Nyquist plots. The semi-circle that appears in the high frequency is believed to be related to coating properties and the second one is a depressed semi-circle at lower frequency related to electrochemical reactions at metal surface under the disbonded coating at electrode. The impedance of these electrodes is 3–4 orders of magnitude lower than that of the electrodes at location "A," although it is still much larger than that at location "C" where an artificial coating defect existed. The electrodes at the defect area exhibit one semi-circle relating to electrochemical reactions at metal surface [53]. Figure 4.6 shows maps of local impedance amplitude (at |Z| 300 mHz) measured after different periods of exposure. It is clearly shown that the impedance amplitude values over the electrodes surrounding the defect area decreased continuously over a 624 hours exposure period, and that the low impedance area expanded with the extension of the exposure time. The decrease in the impedance should be indicating the permeation of the testing solution through the coating/metal interface under the disbonded coating film because the impedance of the coating itself did not change much. In this way, coating disbondment could be effectively monitored in the laboratory and similar approach could also be taken in field conditions [53]. It should be noted that in this application the gaps between neighbouring electrodes in the electrode array probe should be kept very small in order to simulate the local surface chemistry changes over the probe surface, as well as the electrochemical interactions among electrodes in the array, otherwise the accuracy of simulating a real cathodic disbondment process on a pipeline would be compromised.

4.2.2 Examining Factors Affecting the Rate of Coating Cathodic Disbondment

Using the experimental setup shown in Figure 4.4, the effects of various factors affecting coating disbondment could be examined. An example is probing the effect of CP potential on cathodic disbondment by monitoring the coating disbondment processes under different CP potentials [63]. The rates of cathodic disbondment was found to be highly dependent upon the applied CP potential. As shown in Figure 4.7a–f, little or only slight coating disbondment occurred at open-circuit potential (OCP) and at low CP potentials of $-760\,mV_{Ag/AgCl}$ and $-950\,mV_{Ag/AgCl}$. When the CP potential reached more negative levels, such as $-1200\,mV_{Ag/AgCl}$, and $-1400\,mV_{Ag/AgCl}$, as shown in Figure 4.7g–l, the rates of disbondment increased significantly. These results show that current mapping over the WBE is able to probe the disbondment processes under different CP testing conditions. Figures 4.8a-l show the electrochemical impedance maps collected

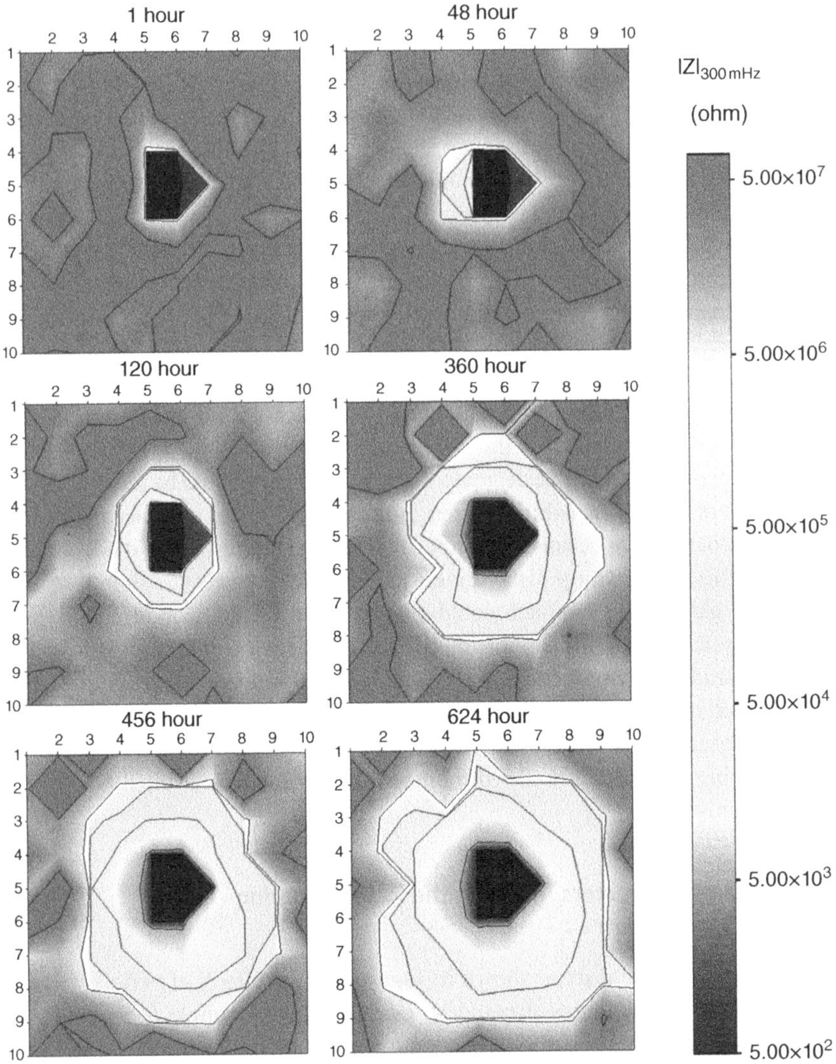

Figure 4.6 Maps of amplitude of impedance (|Z| 300 mHz) over a coated multi-electrode array sensor with an artificial defect after different periods of exposure time to testing solution under CP potential of $-1.40V_{Ag/AgCl}$. *Source:* Mahdavi et al. [53]/ IOP Publishing / CC BY.

over the multi-electrode array surfaces exposed to different potentials: OCP (a, b, c), −950 mVAg/AgCl (d, e, f), −1200 mVAg/AgCl (g, h, i) and −1400 mVAg/AgCl (j, k, l). The impedance maps for all samples after the first hour of exposure are characterized with an area at the center with low impedance values ($<1\times10^7$ ohm) and surrounding areas with much larger impedance values. However, as shown in Figure 4.8h and k, the electrode arrays exposed at potentials

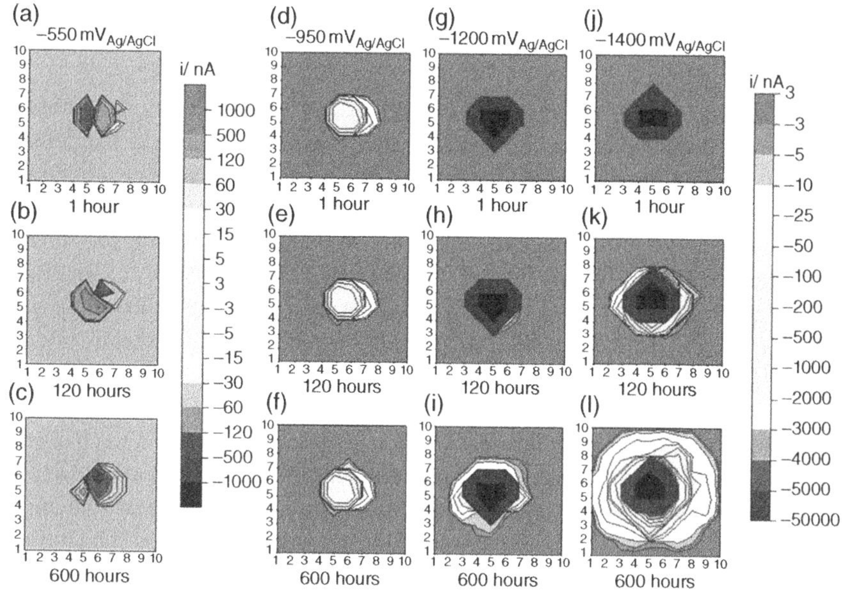

Figure 4.7 Direct current (i) maps obtained on coated multi-electrode array samples under different potential levels measured after 1, 120, and 600 hours exposure under OCP (a, b, c), −950 mVAg/AgCl (d, e, f), −1200 mVAg/AgCl (g, h, i) and −1400 mVAg/AgCl (j, k, l). *Source:* Mahdavi et al. [63]/ With permission of Elsevier.

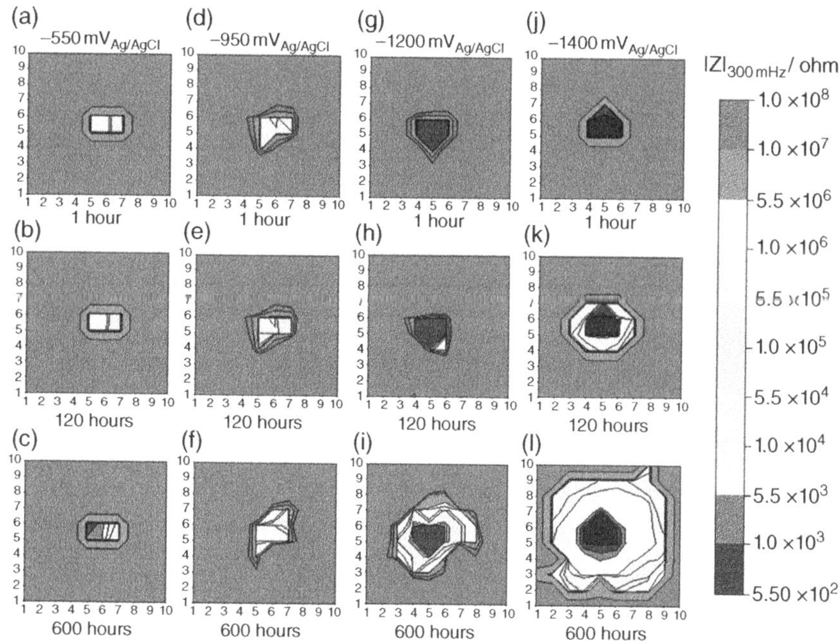

Figure 4.8 Amplitude of impedance (|Z| 300 mHz) maps obtained on coated multi-electrode array samples under different potential levels measured after 1, 120, and 600 hours exposure to OCP (a, b, c), −950 $mV_{Ag/AgCl}$ (d, e, f), −1200 $mV_{Ag/AgCl}$ (g, h, i) and −1400 $mV_{Ag/AgCl}$ (j, k, l). *Source:* Mahdavi et al. [63]/ With permission of Elsevier.

in the hydrogen evolution region (−1200 $mV_{Ag/AgCl}$ and 1400 $mV_{Ag/AgCl}$) show areas with low impedance values (<1×10^7 ohm) expanding around the defect after 120 hours cathodic disbondment testing. The impedance maps obtained after 600 hours (Figure 4.8c) show that coating disbondment did not occur at OCP, as the EIS plots still are characterized with intact coating behavior. A very small area of disbondment is observable on the map measured at −950 $mV_{Ag/AgCl}$ after 600 hours. The electrode arrays tested at −1200 $mV_{Ag/AgCl}$ and −1400 $mV_{Ag/AgCl}$ (Figure 4.8i and l) show the growth in disbonded area around the defect by extension of exposure time to 600 hours. Compared with the current maps shown in Figure 4.7, data from EIS measurements contain more informative on coating and its disbondment in localized area [63]. These results clearly show the changed behaviour and rates of coating disbondment under different CP potential levels.

Figure 4.9 shows the disbonded distance from the defect area, measured after 600 hours, against the six different levels of applied CP potential. It is shown that the relationship between the applied potential and disbonded distance is nonlinear. For instance, no disbondment was observed on electrode array samples exposed under CP potential of −760 $mV_{Ag/AgCl}$ and significant cathodic disbondment occurred at potentials at or more negative than −1100 $mV_{Ag/AgCl}$. The fact that hydrogen evolution would start at a potential level close to −1100 $mV_{Ag/AgCl}$, while accelerated disbondment also occurs at potentials more negative than −1100 $mV_{Ag/AgCl}$, suggests a correlation between hydrogen evolution and coating disbondment. This CP potential of approximately −1100 $mV_{Ag/AgCl}$ appears to be a "critical potential" that could be associated with the 'pushing' or 'lifting' pressure from hydrogen generated from the hydrogen evolution reaction, i.e. $2H_2O + 2e^- \rightarrow H_2 + 2OH^-$. This potential is also in the similar range as the most negative potential level allowed in industry standards such as the Australian standard [78] for preventing overprotection of pipelines that specifies that CP potential should

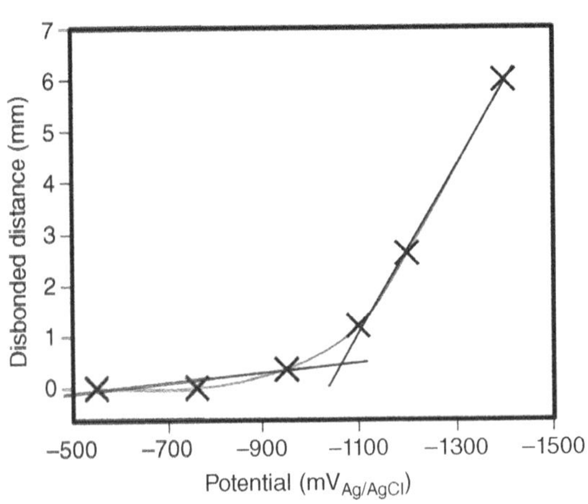

Figure 4.9 Disbonded distance against the potential level measured after 600 hours of exposure. Note the potential level at the intercept of secant lines located between −950 and −1100 $mV_{Ag/AgCl}$ can be considered as the critical potential for accelerating the cathodic disbondment. *Source:* Mahdavi et al. [63]/ With permission of Elsevier.

not be more negative than $-1.2\,V_{SCE}$ [78], which is based on industry experience and observation. These results provide evidences supporting the suggestion that hydrogen evolution caused accelerated cathodic disbondment. Previously it has been suggested that accelerated cathodic disbondment is due to the mechanical lifting of the coating by the build-up of hydrogen gas at the interface of the coating and the steel substrate [79–82]. In order to further verify the effect of hydrogen evolution on cathodic disbondment at elevated CP potentials, tests have been carried out under different pH and oxygen conditions. As shown in Figure 4.10, for the sample tested under a CP potential of $-1400\,mV_{Ag/AgCl}$, cathodic

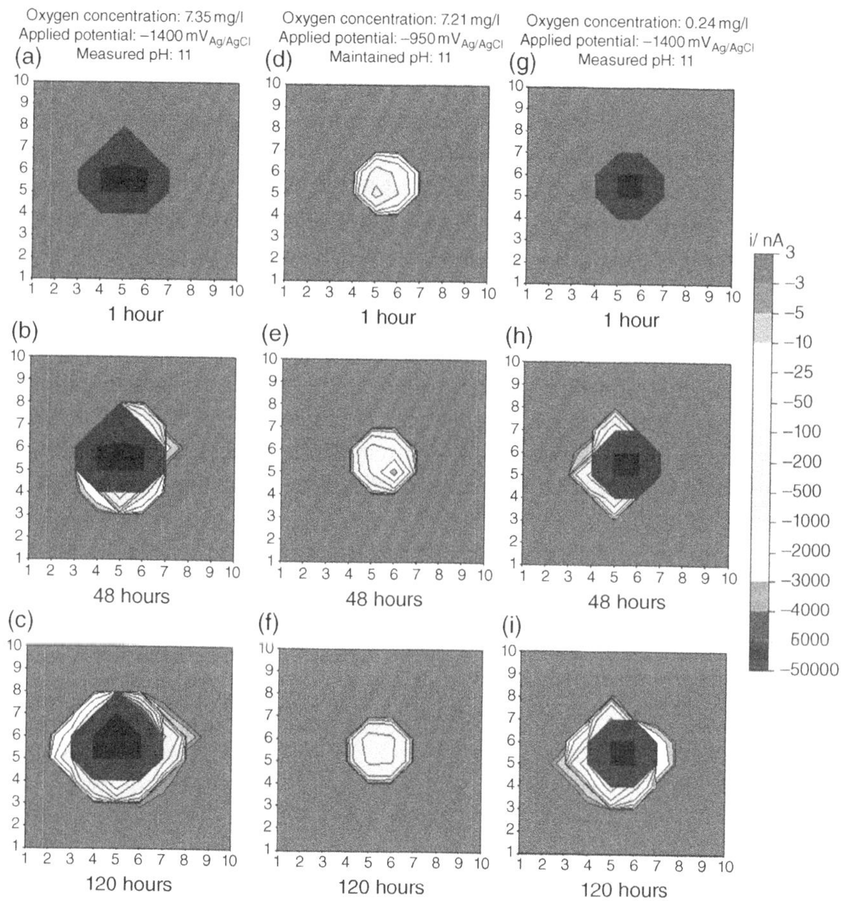

Figure 4.10 Direct current (i) maps obtained on coated array electrode samples during 120 hours exposure to: (a–c) $-1400\,mV_{Ag/AgCl}$ in 3 wt.% NaCl solution open to air; (d–f) $-950\,mV_{Ag/AgCl}$ in 3 wt.% NaCl solution and maintained pH of 11 open to air; (g–i): $-1400\,mV_{Ag/AgCl}$ in deaerated 3 wt.% NaCl solution. *Source:* Mahdavi et al. [63]/ With permission of Elsevier.

disbondment commences after 48 hours of exposure and increases further as the exposure duration is extended to 120 hours (see Figure 4.10a–c), while no disbondment occurred on the sample tested under $-950\,mV_{Ag/AgCl}$ (see Figure 4.10d–f). The pH of solutions in both test cells were high (measured as pH 11) and the oxygen levels in both cells were similar. The only difference is that no hydrogen gas was produced in the cell under $-950\,mV_{Ag/AgCl}$ CP, while the evolution of a large amount of hydrogen bubbles was observed at the coating defect site in the cell under $-1400\,mV_{Ag/AgCl}$ CP. Similar findings have also been reported by Betz et al [79]. These results might be also an evidence on the role of hydrogen evolution as a critical driving force of cathodic disbondment. The results of the experiment on the sample exposed to deaerated solution at $-1400\,mV_{Ag/AgCl}$ (Figure 4.10g–i) indicate that, at this level of applied potential, the hydrogen evolution occurs not only at the coating defect area but also under the disbonded coating, providing additional evidence indicating hydrogen evolution as a critical driving force of cathodic disbondment [63].

It was also found that once the cathodic disbondment initiated under very negative CP potentials, the disbonded distance would spread at an almost constant rate. These results suggest that a practical method of controlling coating disbondment is to ensure that the applied CP potential is below the "critical potential" [63]. The effect of the duration of environmental exposure on cathodic disbondment is shown in Figure 4.11 that presents changes in coating disbonded distance with exposure time measured on coated multi-electrode array probes after different periods of exposure to test solution under different CP potentials. It is evident that cathodic disbondment occurred after an induction period of 240, 120, 120, and, less than 48 hours to probes exposed to test cells that were under CP potential of $-950\,mV_{Ag/AgCl}$, $-1100\,mV_{Ag/AgCl}$, $-1200\,mV_{Ag/AgCl}$, and $-1400\,mV_{Ag/AgCl}$, respectively. The significantly shorter induction period under more negative CP

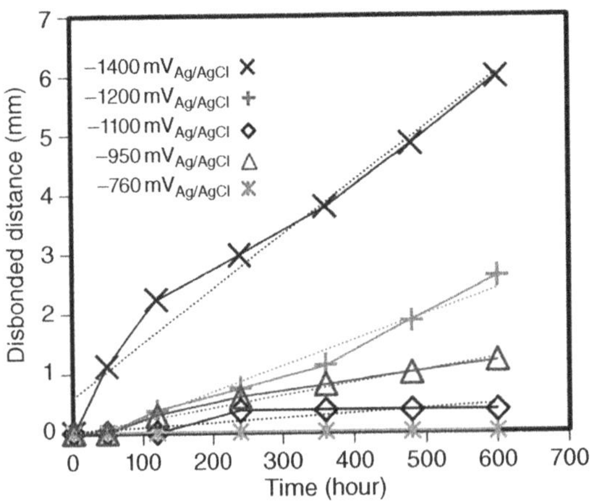

Figure 4.11 Disbondment behavior of coated multi-electrode samples with extension of time of exposure to solution under different levels of CP potentials. *Source:* Mahdavi et al. [63]/ With permission of Elsevier.

potentials (e.g. $-1200\,mV_{Ag/AgCl}$, and $-1400\,mV_{Ag/AgCl}$) could be due to the significant role of build-up hydrogen in accelerating the coating disbondment. After the induction period, as shown in Figure 4.11, at very negative CP potentials the coating's disbondment progressed continuously at a constant rate throughout the exposure period. This linear relationship between the disbonded distance and exposure time is in good agreement with the findings reported in the literature [14, 17, 18, 20]. These results indicate that in this highly conductive testing media (3 wt.% NaCl), the elevated CP potential applied at the defect area could penetrate and reach the metal surface under the disbonded coating and continuously provide sufficient cathodic reaction products to maintain a constant rate of cathodic disbondment during 600 hours test period. Overall these sample experiments described above clearly show cathodic disbondment of a thick coating containing a defect under the effect of CP potential [63]. It should be noted that the coating disbondment behaviour could be different in highly resistive soil conditions.

4.3 Probing Localized Corrosion Under Disbonded Coatings

4.3.1 Integrated 3D Probes for Probing Corrosion Under Disbonded Coatings

The occurrence of cathodic disbondment of coatings leads to the formation of crevices between the metal surface and the disbonded coating layer. Corrosive electrolyte from the surrounding environment can access these crevices, resulting in an enclosed corrosive environment within the crevice and corrosion under disbonded coatings (CUDC). This local crevice environment is particularly problematic for pipelines buried in soil because the resistive soil could prevent CP current from reaching deep into the crevice, causing a CP problem often referred to as cathodic shielding [61, 64]. Industrial experience has shown that various localized corrosion issues and mechanisms such as crevice and pitting corrosion [54], microbiologically induced corrosion (MIC) [55, 56], and stress corrosion cracking (SCC) [57, 58] can occur under coating disbondment, creating a "worst-case" pipeline corrosion scenario affected by environment and mechanism changes. Early detection of such corrosion due to cathodic shielding of CP within the crevice under coating disbondment is critical to ensure appropriate protection of pipelines with disbonded coatings [76, 77]. Unfortunately, as explained in Chapter 2, inspection and timely detection of CUDC is a difficult task. Conventional potential survey methods, currently used by the pipeline industry, are not able to provide information about the CP effectiveness and corrosion under a disbonded coating. For instance pipeline survey techniques such as direct current voltage gradient (DCVG) (see Figure 2.2e) [83, 84] that is an effective tool to locate coating defects cannot detect disbonded coatings and CUDC because of the shielding of currents by the coating film [85]. While the use of in-line metal loss inspection pigging tools can detect pipeline corrosion including corrosion losses under disbonded coatings, they are sensitive only when considerable metal-losses occurred and therefore may not provide corrosion initiation information. On the other hand, they are applied only once every few years due to high costs and interruptions to operation and therefore they may not provide timely corrosion management and CP control information.

Using sensors and probes to simulate, detect and measure early stages of corrosion, or corrosion susceptibility under disbonded coatings could provide a valuable means of pro-actively managing pipeline corrosion. However, several limitations need to be overcome before a sensor or probe (e.g. a metal coupon, an electrical resistance probe or an electrochemical probe) could be practically applied to monitor CP performance and corrosion rates under a disbonded coatings in buried structures [85, 86]. Since the mechanisms of soil corrosion, coating disbondment, and CUDC are different and each mechanism is dominating or the "worst-case scenario" mechanism over different period of time, a probe needs to effectively simulate and *in situ* monitor each of these mechanisms. A approach is to integrate differently designed probes including a probe for simulating and monitoring bare steel corrosion (such as those shown in Figures 2.4–2.7), a probe for simulating and monitoring coating disbondment (such as the electrode array shown in Figure 4.4) and a probe for simulating and monitoring CUDC. Another approach to simulating and probing CUDC in three-dimensional crevices is the design of 3D electrode array probes and the use of WBE-video technique that can visualize the dynamic CUDC processes with rates and patterns changing with the time.

Over the past decade, Varela et al. [76, 77] firstly applied an electrochemically integrated multi-electrode array to map the spatial and temporal current density distributions along the crevice in dilute sodium sulfate solutions and in soil [87–89]. It is pointed out that the key to designing corrosion probes for probing corrosion under disbonded coatings is to simulate and monitor localized corrosion mechanisms and their changes with special attention on the interactions between different mechanisms [86]. An electrochemical probe denoted differential aeration sensor (DAS) was designed to measure corrosion rates and their distribution on metals subjected to CP, as illustrated in Figure 4.12, based on a high spatial resolution electrochemically integrated electrode array under disbonded coatings. As shown in Figure 4.12, square- and rectangle-shaped corrosion probes were designed using pipeline steel wires that are closely packed into a 10-by-10 array (Figure 4.12a) and 4-by-25 array (Figure 4.12b) and mounted on epoxy resin avoiding electrical contact between them. To simulate the disbonded coating, a plastic film (e.g. polymethyl methacrylate, PMMA) or epoxy coating cover with an

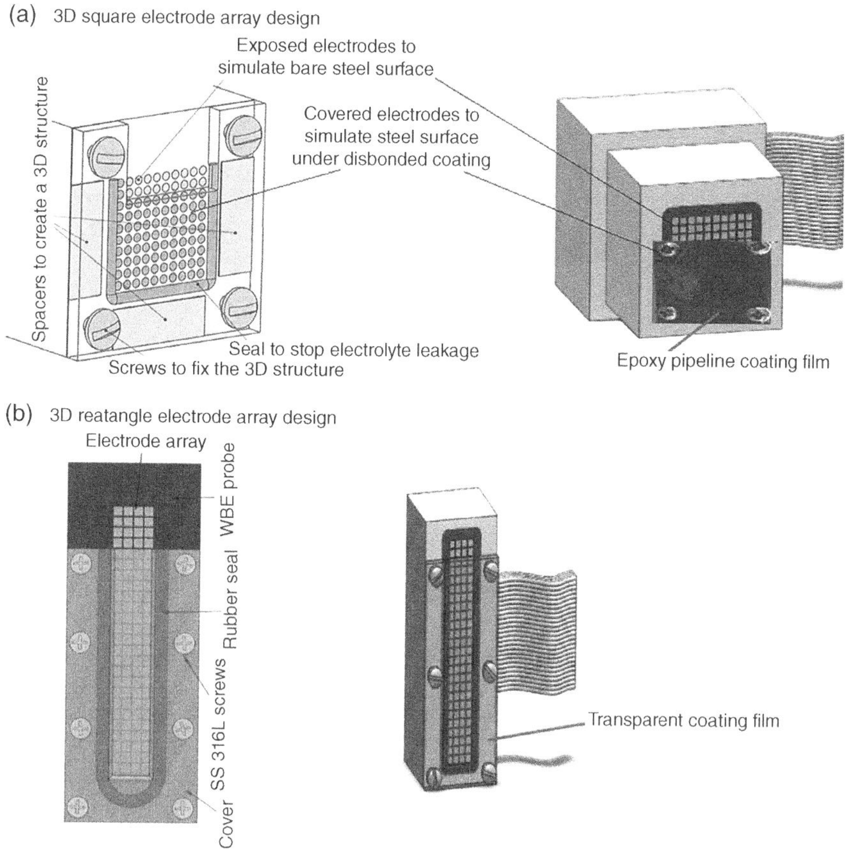

Figure 4.12 Schematic diagram of the differential aeration corrosion probe with a 3D designed surface simulating corrosion under disbonded coating environment: (a) a square-shaped 3D electrode array probe [76, 77], and (b) rectangular-shaped 3D electrode array probes. *Source:* Wang et al. [89], ELSEVIER.

incorporated rubber seal was used. The crevice gap between the cover and the array's surface was fixed at 0.25 mm but can be fixed at other sizes by introducing spacers at the areas indicated in Figure 4.12. In this specific probe design, two rows of electrodes were directly exposed to the bulk solution outside of the crevice to simulate a coating holiday, and the eight remaining rows were under the cover to simulate a coating disbondment. Figure 4.13 illustrates the basic geometry of the electrochemical cell and the electrical connections used to measure current density maps under CP. A Luggin capillary was used to minimize IR drops. More details on experimental details can be found in references [11, 77, 89]. Current densities across the probe surface can be mapped regularly or continuously at

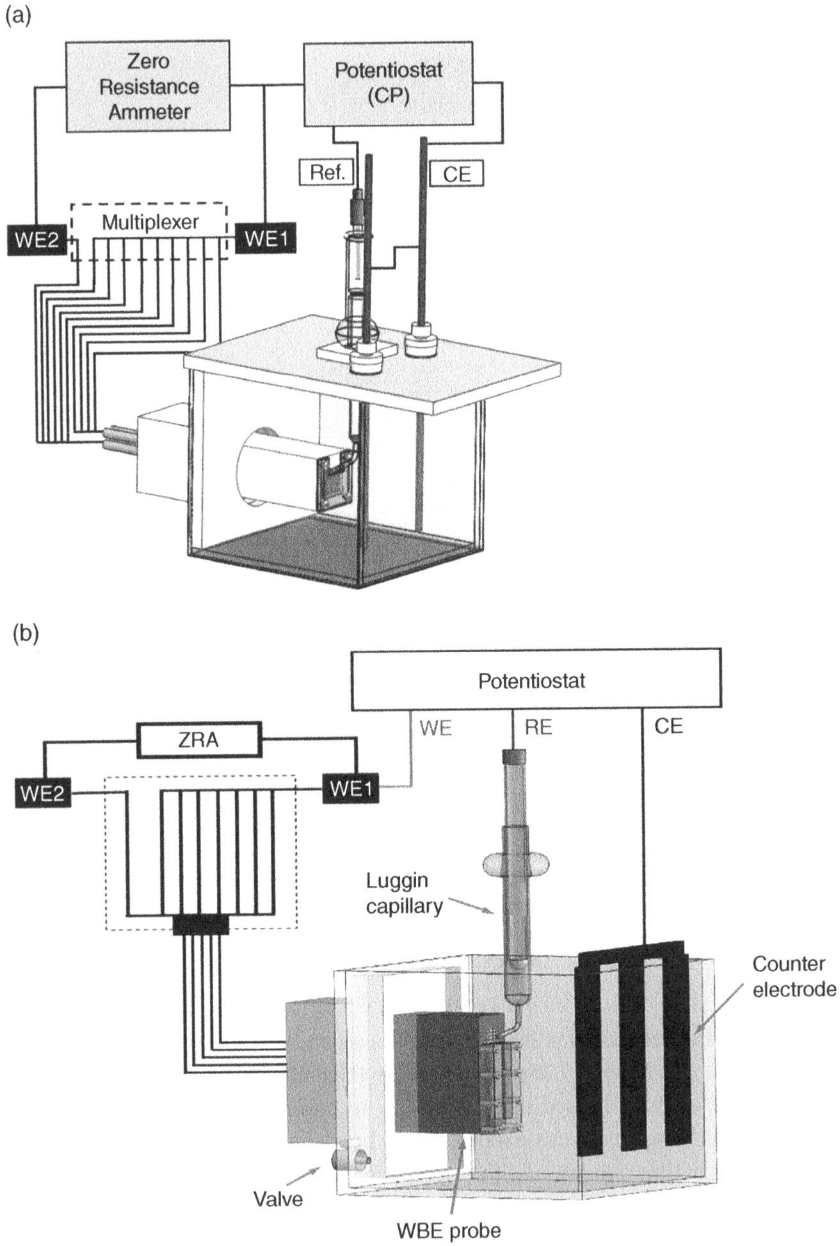

Figure 4.13 Schematic diagram of the experimental arrangement for the determination of current density maps under CP using (a) a square-shaped 3D electrode array probe [76, 77], and (b) a rectangular-shaped 3D electrode array probe. *Source:* Wang et al. [89]/ With permission of Elsevier.

open-circuit potential (OCP) or at CP potential. Raw data obtained from the measuring instrument can be used to generate current density surface maps. A series of these maps can be used to generate high-resolution videos if measurements are repeated quickly [77]. These data can be presented as the WBE-video. An sample video is shown in Supplemental Material: https://iopscience.iop.org/article/10.1149/2.0091501eel in reference [77].

Figures 4.14 and 4.15 show sample current density maps measured from a square-shaped DAS array (see Figure 4.12a) at OCP and at a CP potential of $-760\,mV_{Ag/AgCl}$, respectively (videos showing dynamic behavior in greater detail are presented as a supplementary material of reference [77]). At OCP, as shown in Figures 4.14, corrosion shows a typical differential aeration controlled free corrosion pattern. Initially cathodic sites were mainly located outside the crevice where high concentration of oxygen existed, while within the crevice large anodic currents were found especially at the third and fourth rows of the electrode array that were close to the cathodes. With the extension of the test, the anodic region that was initially located in the third and fourth electrode rows moved toward the mouth of the crevice and began to cover part of the directly exposed sensor area. The pH measurements of the test solution within the crevice at the end of the test recorded a value of approximately 7, i.e. no significant changes from the initial value. When the DAS array was under cathodic protection (CP) by cathodically polarizing it to $-760\,mV_{Ag/AgCl}$, current density maps (see Figure 4.15) shows very different features from that shown in Figure 4.14. Large cathodic currents (CP and cathodic reaction currents) were registered outside the crevice and along the third row of electrodes that were close to the crevice month and had good accessibility to oxygen. Deeper into the crevice, relatively low magnitude anodic current densities were measured, indicating slow corrosion inside the crevice area. These anodic current densities reached a maximum of about $5\,\mu A/cm^2$ and gradually decreased to extremely low values. The pH inside the crevice increased from 6 to 11 that may have passivated the steel and caused the decrease of anodic current densities. Differences in Figures 4.14 and 4.15 indicate corrosion mechanism changes when CP is applied on a steel pipeline with disbonded coating. Results shown in Figure 4.15 indicate effective CP in a conductive media, however it should be noted that this CP effectiveness is expected to change with the conductivity of the media, the crevice gap size and other factors affecting CP current penetration into the crevice areas.

The current density distribution maps obtained from the DAS under both OCP (Figure 4.14) and CP (Figure 4.15) conditions are in general agreement with previous research findings reported in the literature [62, 64, 69, 90, 91]. For instance, in the OCP case, the current distribution profile along the crevice measured at the beginning of the test agrees with Song's model, which predicts the formation of cathodes outside and anodes inside the crevice area [91]. However, Song's model

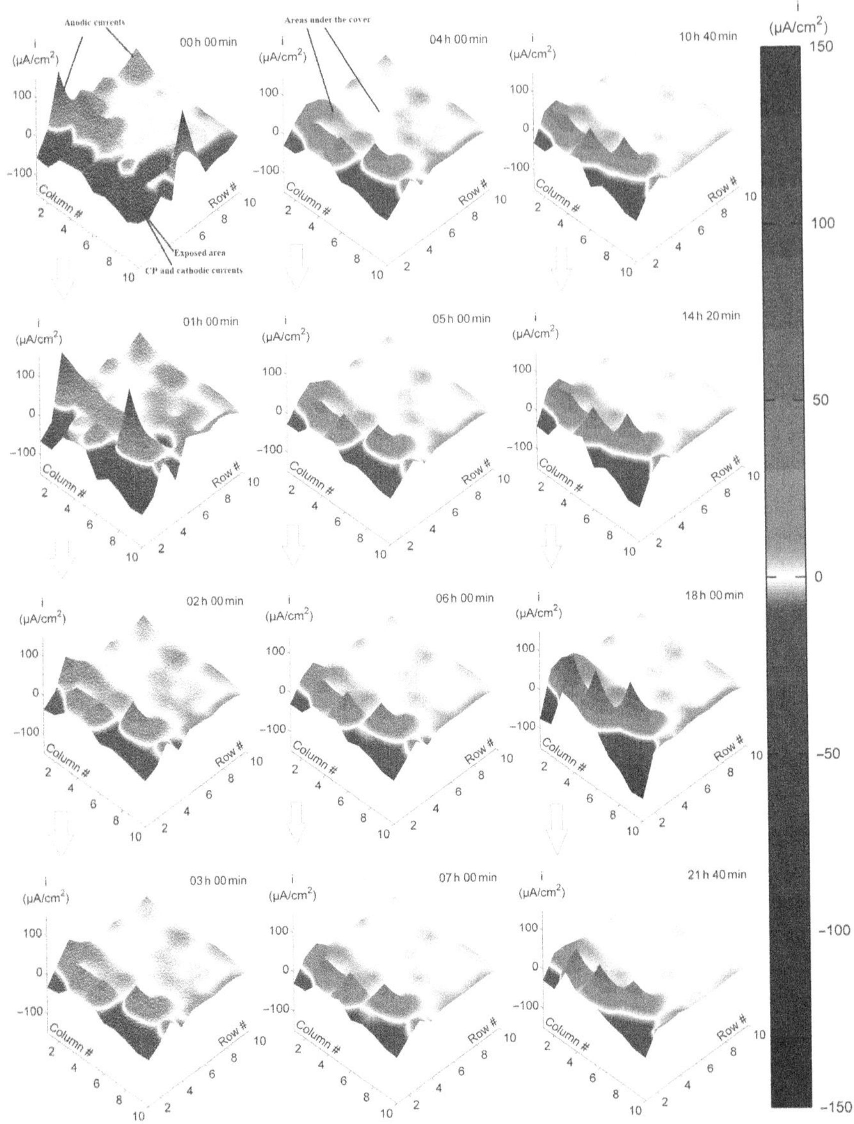

Figure 4.14 Current density maps at various immersion times in 0.1 M NaCl solutions without CP. *Source:* Varela et al. [76]/ IOP Publishing / CC BY.

is unable to predict the further development of corrosion. As shown in Figure 4.14 (01 hour 00 minute), anodes started to grow outside the crevice area. This is understandable since carbon steel would not develop a passive film at neutral pH, and consequently the anode could move outside the crevice in order to minimize

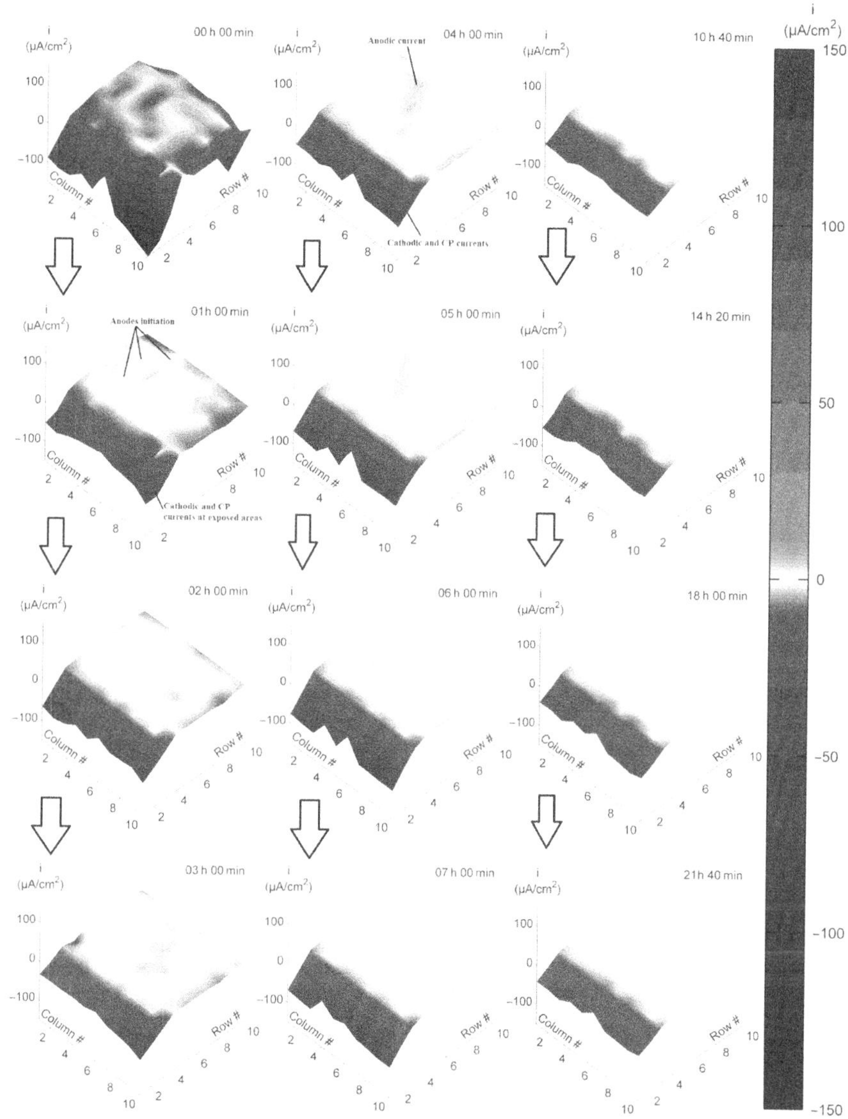

Figure 4.15 Current density maps at various immersion times in 0.1 M NaCl solutions with CP. CP potential: −760 $mV_{Ag/AgCl}$. *Source:* Adapted from Varela et al. [77].

the solution resistance between it and the main cathode. In the case when a CP potential was applied, the observed current density profiles were also, in general, similar to those previously reported in the literature [62, 64, 69, 90, 91]. The anodic current densities registered in Figures 4.14 and 4.15 were used to calculate the accumulated metal thickness losses at each electrode over the whole corrosion period using the Faraday's Law [76, 77]. Figures 4.16 and 4.17 show a comparison between the electrode array surfaces at the end of the tests, and the calculated metal losses at the OCP and CP, respectively. The pattern of calculated thickness losses correlates well with the observed corrosion damage, with thickness losses

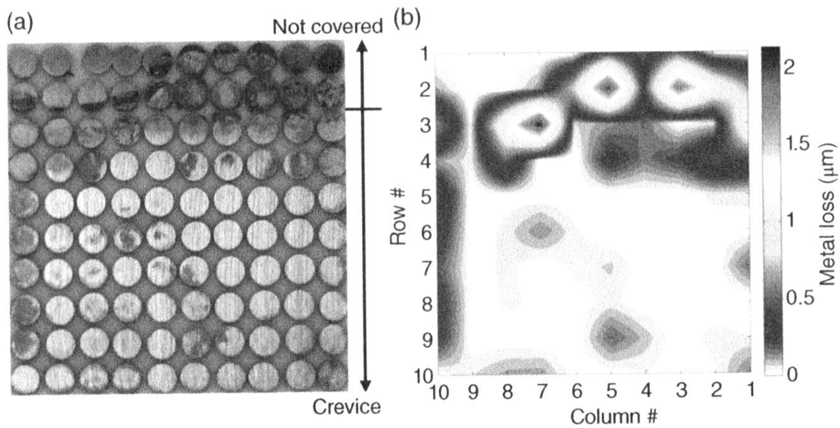

Figure 4.16 Comparison of the sensor's array surface after 23 hours immersion in 0.1 M NaCl solutions without CP (a) and calculated thickness loss maps based on Faraday's law (b). *Source:* Adapted from Varela et al. [77], IOP Publishing.

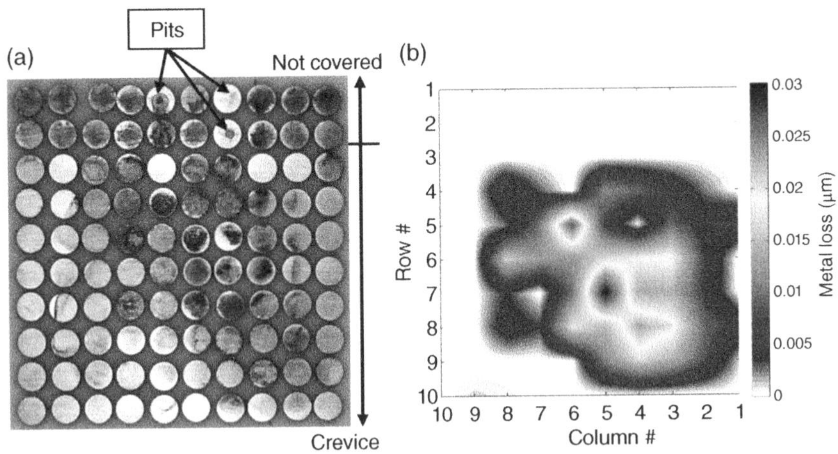

Figure 4.17 Comparison of the sensor's array surface after 23 hours immersion in 0.1 M NaCl solutions under CP (a) and calculated thickness loss maps based on Faraday's law (b). *Source:* Adapted from Varela et al [77], IOP Publishing.

of more than 2 μm outside the crevice and of the order of 0.1 μm inside it. In the cathodically protected −760 $mV_{Ag/AgCl}$ case (Figure 4.17), a less intense attack (yet darker corrosion products are visible) was observed. An acceptable correlation is observed for the electrodes inside the crevice. However, Faraday's law-based calculations would suggest zero corrosion outside the crevice because the measured currents at this location were only cathodic, suggesting a limitation on the direct application of Faraday's law to current density maps recorded by the DAS under CP. This limitation of using the DAS array can be addressed through data analysis described in reference [77]. Overall the 3D DAS array facilitated the simulation and monitoring of corrosion under disbonded coatings, and the DAS's ability to detect anodic currents under CP was explained based on the oxygen depletion expected under simulated disbonded coatings [77].

4.3.2 Probing CUDC of Varying Disbondment Geometries in Soil

The multi-electrode array probe described in references [11, 77, 89] was further utilized to visualize dynamic steel corrosion under disbonded coatings in soil. It is known that the behavior of CUDC in soil is distinctively different from that in solution, although there are reports that in saturated soil CUDC can behave similarly as it does in electrolyte solutions [11, 88, 89]. It was found that in non-saturated soil, CUDC behavior changed significantly under the effect of a number of environmental factors and coting disbondment parameters, in particular with coating disbondment gap size [87–89]. The geometry of the disbondment such as the length and gap size was reported to vary depending on the coating types, chemicals existing in the soil environment, and level of cathodic potential supplied [65]. Coating disbondment geometry is believed to affect the ability of CP current to flow through trapped solutions in a long crevice to reach the deep areas of the coating disbondment (cathodic shielding) [61, 64]. There are observations reported in the industry that if a disbonded coating is tightly bound to a steel pipe, CUDC would not normally be an issue, while corrosion issues have been observed on disbonded coatings loosely bound on pipes [66], indicating that CUDC is critically affected by the disbondment gap size. These observations have not been sufficiently understood or explained due to the lack of direct experimental evidences. The effect of coating disbondment geometry on electrode processes and mechanisms causing CUDC in soil has never been quantified primarily due to technological difficulties in probing dynamic and localized CUDC processes occurring within a narrow and long coating disbondment crevice.

Soil is a heterogeneous gas/liquid/solid multiphase and dynamically changing medium that is much more complex than aqueous solutions [92–94], and therefore the simulation and monitoring of CUDC in soil is challenging. Varela et al. [95] firstly used the electrode array probes to study CUDC in soil conditions and proved that serious CUDC could be developed in nonsaturated soil due to the formation of an air gap in the coating disbondment crevices, preventing the penetration of CP currents into the crevice and leaving an isolated section for active corrosion to occur. This suggests a critical effect of varying disbondment geometry on CUDC processes and mechanisms in soil under the effect of the wet/dry cycles, which could be induced by fluctuations of water table, rainfall, and humidity of surrounding environment [68]. Wang et al. further used the electrode array probe in an electrochemical cell shown in Figure 4.13b to simulate and probe varying CUDC processes and mechanisms under the effect of different influence factors, in particular coating disbondment geometry [89]. Experiments were designed using the soil cell setup shown in Figure 4.13b to simulate and monitor complex CUDC conditions under coating disbondment of different crevice gap sizes (0.25, 1, and 2 mm) in soil environment of changing moisture contents. This was to capture and understand the phenomena empirically observed in the industry, for instance CUDC would not normally occur if a disbonded coating is tightly bound to a steel pipe, while it occurs if a disbonded coating is loosely bound on a pipe [66], as well as wet/dry seasonal changes affect CUDC in the soil.

Figure 4.18 shows the total CP currents supplied to the 3D probes with different crevice gap sizes over their exposure to soil with varying soil moisture contents [89]. When soil was saturated, as shown in Figure 4.18, the total CP currents required for maintaining a CP potential of $-850\,mV_{CSE}$ on the probes with different crevice gap sizes reached a similar and stable value of approximately $5\,\mu A$. The draining of water from saturated sand by opening the valve at 24 hours, simulating the transition of wet to dry season when the soil gradually turned nonsaturated, caused CP current values increased quickly. This is understandable

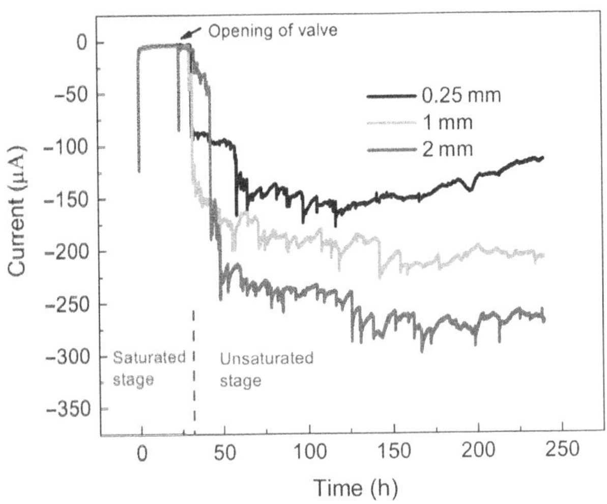

Figure 4.18 The total CP current supplied to the 3D probe surfaces with different gap size for maintaining a CP potential of $-850\,mV_{CSE}$ on the probes. *Source:* Wang et al. [89]/ With permission of Elsevier.

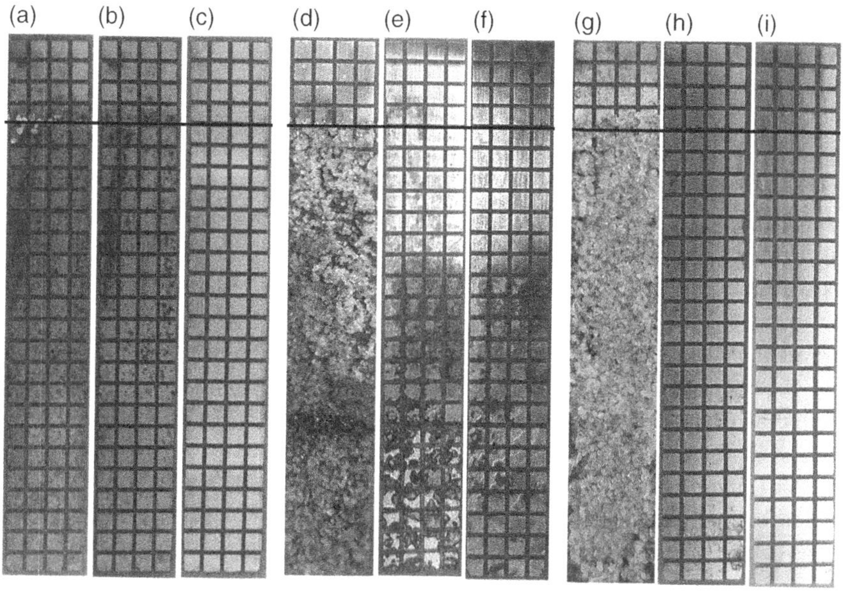

Figure 4.19 Surface photo of the WBE with gap size of 0.25 mm (a) after exposure, (b) after removing soil, (c) after removing rust; surface photo of the WBE with gap size of 1 mm (d) after exposure, (e) after removing soil, (f) after removing rust; surface photo of the WBE with gap size of 2 mm (g) after exposure, (h) after removing soil, (i) after removing rust. *Source:* Wang et al. [89], ELSEVIER.

since oxygen supply become more sufficient due to oxygen transportation through pores in soil with the soil becoming increasingly non-saturated. Another phenomenon that is observable in Figure 4.18 is that the CP current became more negative in value with the increase in gap size, indicating more CP current penetration into wider crevices. After the whole experiment was completed (10 days), as shown in Figure 4.19, the extents of corrosion damage on each of the WBE surface were significantly different. The most serious localized corrosion occurred on the probe surface under the crevice of 1 mm gap size, while less corrosion occurred on probes under either smaller (0.25 mm) or larger (2 mm) gap size crevices. This suggests that CP current monitoring alone is insufficient for understanding the actual rates or kinetics of CUDC in complex soil environments, and that in-situ corrosion monitoring is needed in this case. These results were not in agreement with findings from experiments in electrolyte solutions, where steel under disbondment was found to be protected from corrosion, independent of the crevice geometry, due to less CP current shielding in continuous and more conductive electrolytes [11]. These results clearly show that electrochemical corrosion processes and mechanisms occurring on the probe surfaces were changing dynamically and are dependent upon the coating disbondment parameters and environmental conditions [89].

Figure 4.20 shows current density maps registered over the probe surface with the crevice gap size of 0.25 mm at a CP potential of −850 mV_{CSE} during the simulated wet/dry cycle. At the beginning of exposure to the electrochemical cell filled with saturated soil, as shown in Figure 4.20a, a combination of cathodic and anodic current was observed in the crevice area, while only cathodic current was recorded outside the crevice. This suggests that conventional corrosion cells existed in the coating disbondment crevice at the beginning of experiment, although minor CP current penetration into the entrance area of the disbondment crevice is noticeable in Figure 4.20a. The anodic currents distributed along the crevice and mainly accumulated on the middle rows of the WBE array. After 24 hours the current density decreased gradually and reached a very small value of almost 0 and 5 $\mu A/cm^2$, for most of the crevice area and the opening mouth of the crevice (Figure 4.20b). This is similar to CUDC behavior obtained in dilute sodium sulfate electrolyte solution [11]. When the soil became non-saturated, as shown in Figure 4.20c, major anodic area appeared within the crevice areas of the probe surface, suggesting the initiation of localized corrosion. The CP current distribution on the exposed surface was not uniform either, indicating the soil on the surface was heterogeneous. An interesting observation in Figure 4.20c is that there was a zero-current sum zone inside the crevice, suggesting that this corrosion cell was isolated from the CP functioning outside the crevice. This type of current distribution has been found in other studies performed in non-saturated soils [95] and it was attributed to the formation of air gap within the coating disbondment. This isolated corrosion cell moved gradually to the bottom part of the crevice, as shown in Figures 4.20c–f, until no significant current was recorded inside the crevice, leaving an area of zero current density (Figure 4.20f). Interestingly, after this period, the zero current density region decreased and cathodic currents progressively appeared on the WBE maps, until the whole surface inside the crevice was covered with cathodic currents (Figure 4.20g–k). At the end of the experiment, both anodic and cathodic currents (Figure 4.20l) were registered on the probe surface immediately after CP was disrupted, suggesting the re-starting of general corrosion mechanism when CP was stopped [89]. Figure 4.20 clearly shows several changes in corrosion processes and mechanisms with environmental and CP condition changes.

Figure 4.21 shows the current distribution maps registered over the probe surface with the crevice gap of 1 mm at a CP potential of −850 mV_{CSE} during wet dry cycle. When the soil was saturated over the first 24 hours, as is shown in Figure 4.21a and b), the current density distribution maps were very similar to that shown in Figure 4.20a and b), suggesting that in saturated soil the current density distribution was nearly independent of the gap size. When the soil became

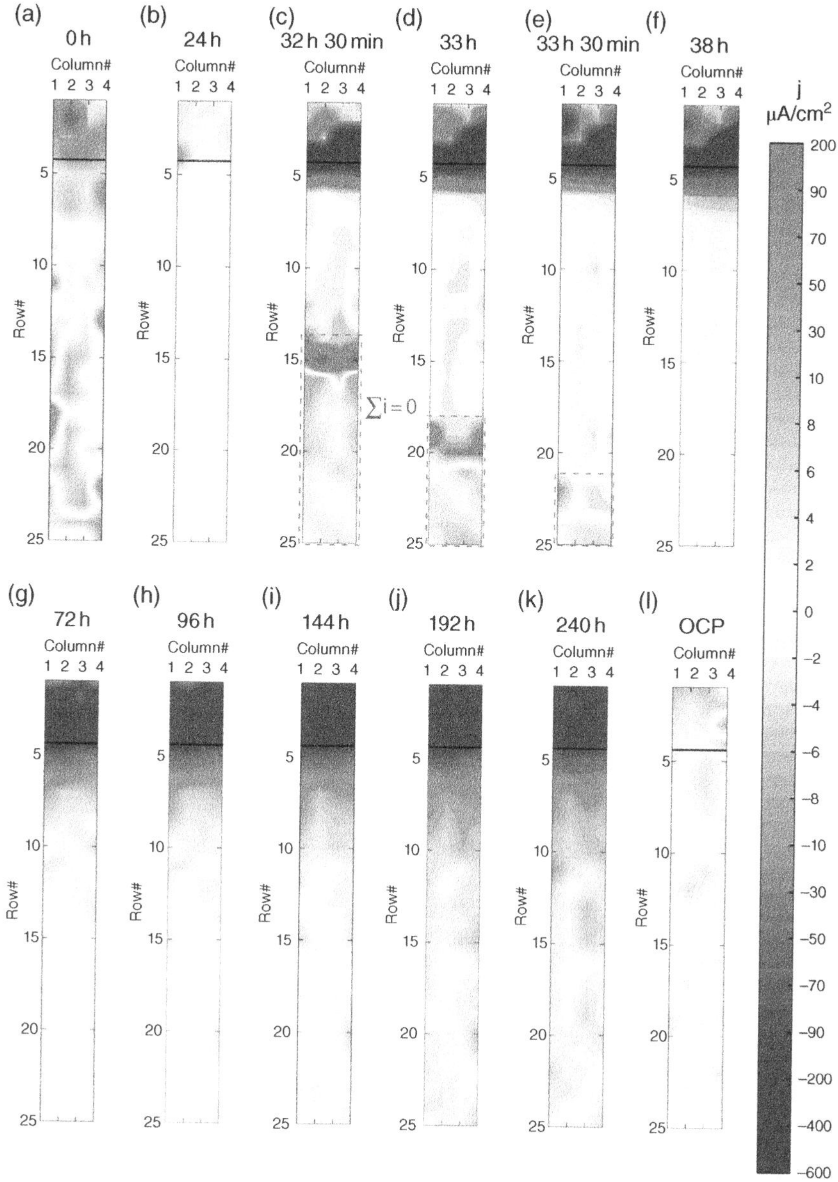

Figure 4.20 Current density distribution maps registered over the array electrode surface with the crevice gap of 0.25 mm at a CP potential of $-850\,mV_{CSE}$. *Source:* Wang et al. [89]/ With permission of Elsevier.

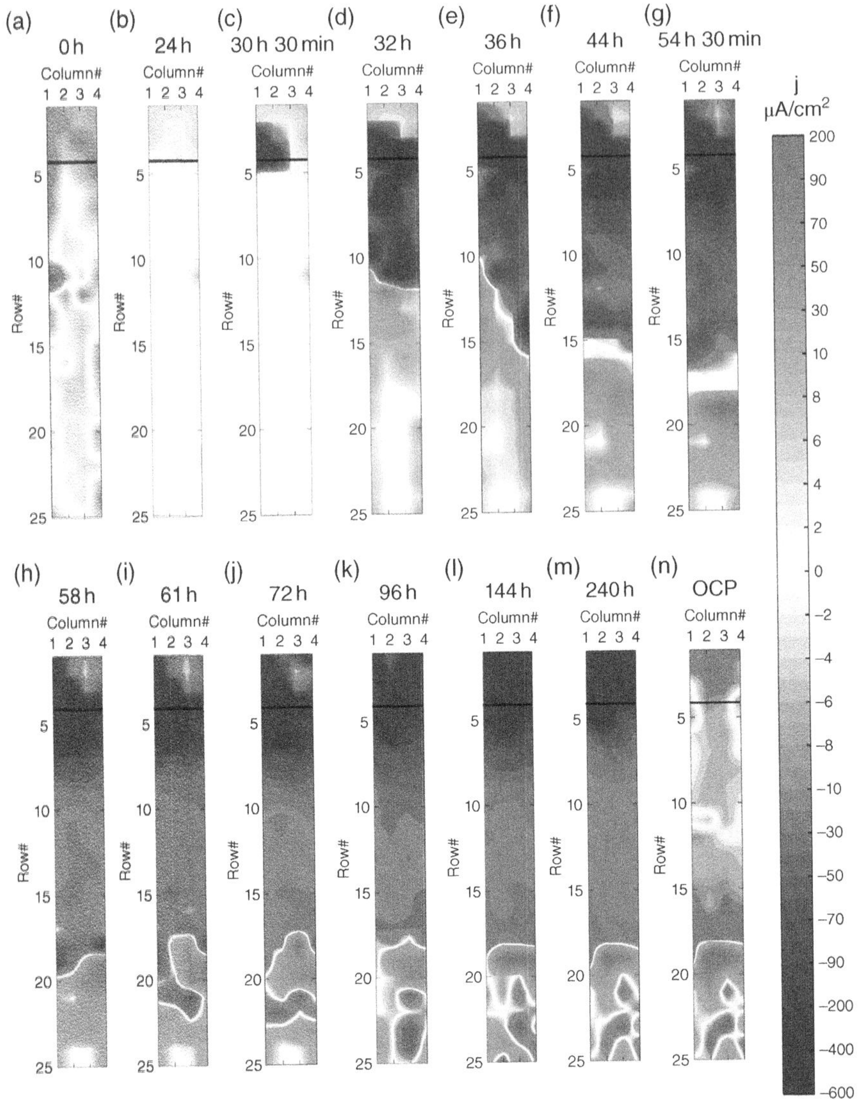

Figure 4.21 Current density distribution maps registered over the array electrode surface with the crevice gap of 1 mm at a CP potential of −850 mV_{CSE}. *Source:* Wang et al. [89]/ With permission of Elsevier.

non-saturated, the current outside the opening mouth of the crevice firstly turned into much more negative values (Figure 4.21c), corresponding to the dramatic increase of the total CP current applied to the WBE surface. After about 32 hours, as shown in Figure 4.21d, the current distribution inside the crevice became similar with that in Figure 4.20c, with both anodic and cathodic areas in the crevice area. However, in this case, there was no zero-current area inside the crevice area (Figure 4.21d), suggesting that the anodic current area inside the crevice was not isolated from the cathodic current outside of the crevice. In addition, the cathodic currents inside the crevice decreased gradually with the crevice length. As shown in Figure 4.21d–h, the anodic current area moved gradually to the bottom of the crevice. Interestingly, with the moving of cathodic current distribution, the anodic current distribution stopped moving, leaving an area of anodes at row#20, as shown in Figure 4.21i and j. With the extension of the exposure, the horizontal boundary between the anodic and cathodic current zones disappeared (Figure 4.21k), and stable anodic zones formed on the probe surface after approximately 96 hours (Figure 4.21l and m). This suggests that CP was insufficient for protecting the bottom area of the crevice from corrosion. At the end of the experiment (after 10 days), anodic currents (Figure 4.21n) were still registered on the probe surface after CP was disrupted. These results further show that electrochemical corrosion processes and mechanisms occurring on the probe surfaces were changing dynamically and are dependent upon the coating disbondment and environmental conditions [89].

Figure 4.22 shows the current distribution maps registered over the WBE surface with the crevice gap of 2 mm at a CP potential of $-850\,mV_{CSE}$ during wet dry cycles. A common behavior observable in Figures 4.20–4.22 is that in saturated soil the cathodic current inside the crevice decreased during the first 24 hours and the current distribution was independent with the gap sizes of the crevices (see Figures 4.22a, 22b). When the soil became non-saturated, the cathodic current at the opening mouth of the crevice firstly turned into much more negative values (Figure 4.22c), corresponding to the dramatic increase of the total CP current applied to the probe. After 43 hours, as shown in Figure 4.22d, large cathodic current appeared inside the crevice. Next to the cathodic region, there was a zero-current density area, with anodic current sites distributed. With the extension of the experiment (Figure 4.22e), the cathodic current moves gradually toward the crevice tip. After about 48.5 hours, the anodic current zone moved to the crevice tip (Figure 4.22f and g). The anodic current region inside the crevice shrink gradually until the crevice was covered almost fully with cathodic currents (Figure 4.22h–k) and remained almost unchanged. This suggests that CP current was able to penetrate into and protect the bottom area of crevice area from corrosion, and this result confirms that CP current was indeed could penetrate to the whole WBE array surface within the crevice and protect the bottom area of crevice area from corrosion.

Figures 4.20–4.22 clearly visualize dynamic processes and mechanisms occurring within simulated coating disbondment of different gap sizes. The anodic currents in 3D electrode array probe maps are good indicators of corrosion location and electrochemical corrosion kinetics. Figure 4.23 presents the accumulated metal loss electrochemically calculated based on Faraday's law and optically measured by the

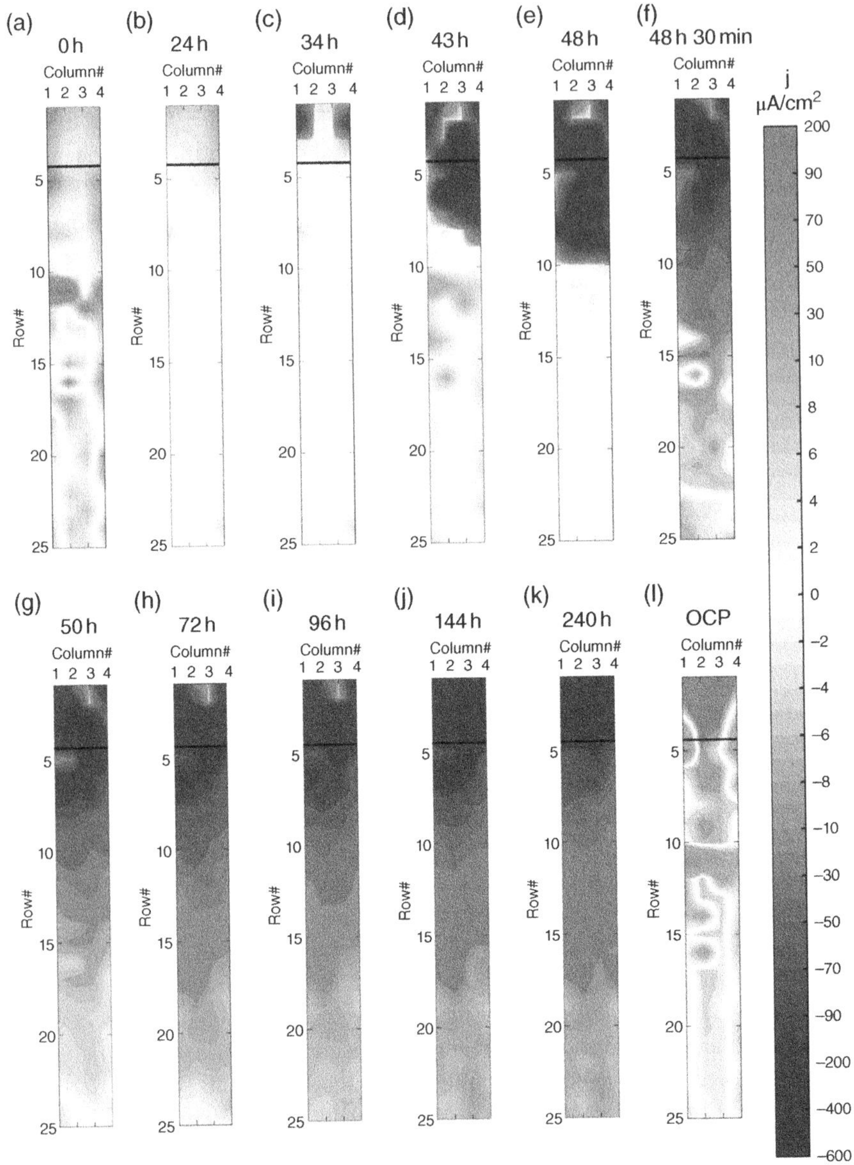

Figure 4.22 Current density distribution maps registered over the array electrode surface with the crevice gap of 2 mm at a CP potential of $-850\,mV_{CSE}$. *Source:* Wang et al. [89]/ With permission of Elsevier.

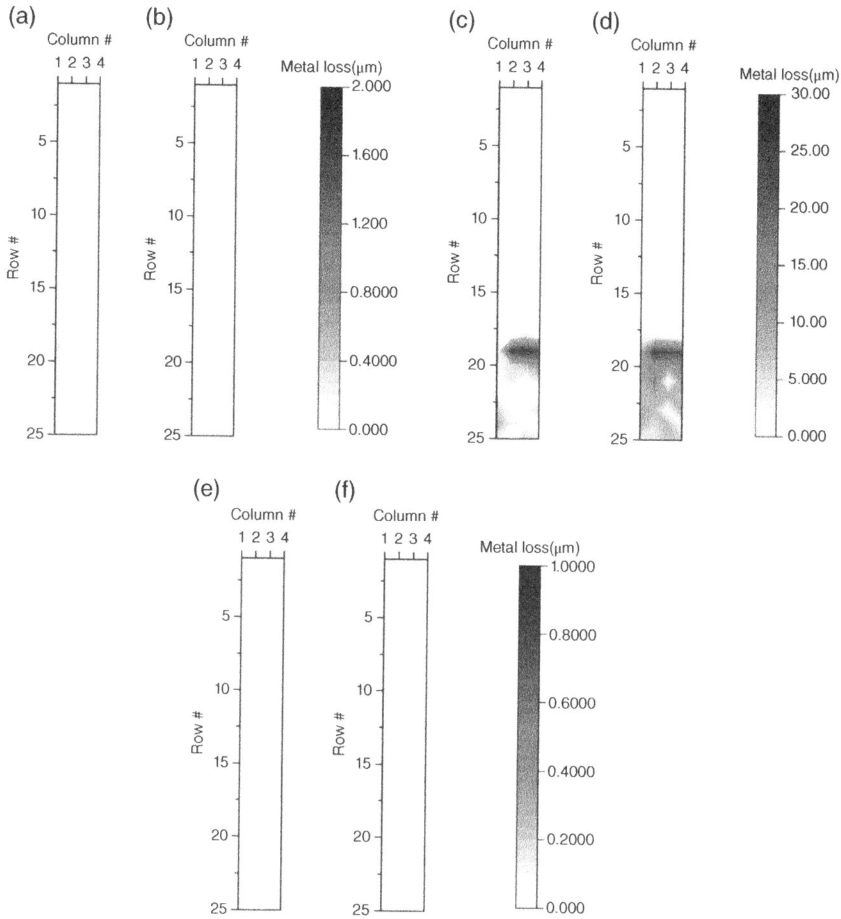

Figure 4.23 Metal loss maps calculated based on Faraday's law with the crevice gap size of 0.25 mm (a), 1 mm (c), and 2 mm (e) respectively; surface profilometry results with the crevice gap size of 0.25 mm (b), 1 mm (d), and 2 mm (f), respectively. *Source:* Wang et al. [89]/ With permission of Elsevier.

surface profilometry on the same probe surface with different crevice gap size. When the crevice gap size was 0.25 and 2 mm, although some oxide film was found on the WBE surface, no obvious corrosion damage was observed after the removal of corrosion products and the corrosion depth was too low to measure (see Figure 4.23b and f), and the maximum accumulated corrosion depth calculated using Faraday's law was less than 0.3 μm (Figure 4.23a and e). When the gap size was 1 mm, more corrosion occurred and the corrosion depths were measurable. The localized corrosion damages and their distributions along the crevice determined based on Faraday's law (Figure 4.23c) and by surface profilometer (Figure 4.23d), especially the extremely corroded sites, matched very well [89].

These results clearly demonstrate that the disbondment geometry can significantly influence the corrosion process within the crevice formed by disbonded coatings in soil with varying wet/dry conditions, which is in agreement with observations in the field – CUDC was not normally observed if the coating-disbonded region is tightly bound to a steel pipe (It could be similar with the case that the gap size was 0.25 mm), while localized corrosion under disbonded coating was observed if the disbonded coating is loosely bound on a pipe (it could be similar with the case that the gap size was 1 mm) [66]. Moreover, the results here indicated CUDC would not occur when the gap size of disbondment region increased to 2 mm. This may be similar to the case that when coating is totally loose when disbonded, e.g. coal tar epoxy coatings, corrosion may not occur. It should be noted, however, that these industrial experiences are not quantitative and therefore they could not be used for direct comparison with results reported here. These results also have implications for designing corrosion probes for monitoring corrosion in soil. It is clear that the crevice geometry should be considered to let the corrosion probes experience the maximum corrosion rate in order to provide early warning of the worst case scenario corrosion of a underground infrastructure system [89].

Overall, the spatial and temporal resolution of the electrode array has enabled the visualization of dynamic electrode processes and mechanisms occurring under simulated coating disbondments in soils [87–89] and electrolyte solutions [11, 76] under the effect of variables such as applied CP potential, crevice gap size, and solution conductivity on CP effectiveness under disbonded coatings. A model [11] was proposed to explain the effect of each of these variables on the current distribution maps, suggesting that in electrolyte the $-850\,mV_{CSE}$ CP potential criterion can be satisfied, independently of the crevice geometry, by means of concentration polarization instead of direct electrochemical polarization. This model suggests that despite the detrimental effect of cathodic shielding, the steel under disbonded areas could still be protected, independent of the crevice geometry and solution resistivity, by means of concentration polarization instead of direct electrochemical polarization [11]. Another model is illustrated in Figure 4.24 that was proposed to explain the significant effects of the crevice gap size on the behavior of CP and corrosion current density distributions on 3D electrode surfaces shown in Figures 4.18–4.23, based on different corrosion mechanisms under changing environment conditions [89]. As illustrated in Figure 4.24a and b, when the crevice is small (e.g. 0.25 mm), there would be little sandy soil getting into the crevice area. When the soil outside the narrow crevice became non-saturated, the solution trapped inside the crevice was absorbed due to capillary effect of the soil, which left an air gap within the crevice area (see Figure 4.24a). This explains the existence of a zero-current zone in the current density

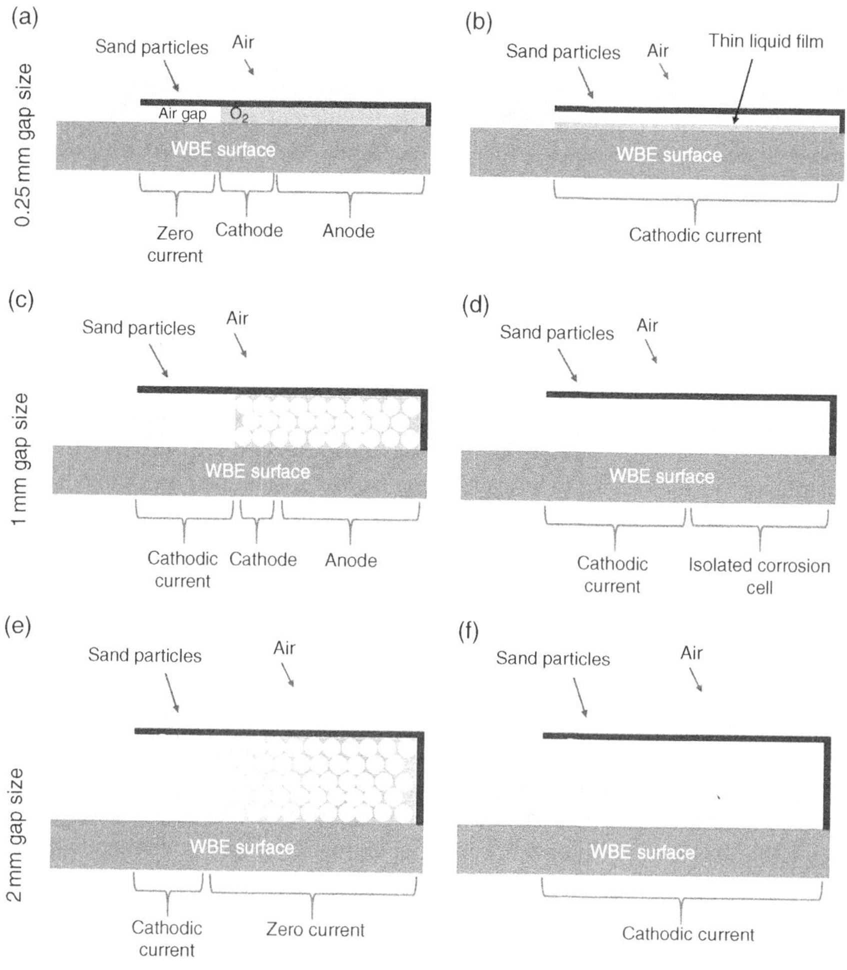

Figure 4.24 Schematic diagram of the corrosion processes and mechanisms occurring over 3D electrode array surfaces under disbonded coatings with the crevice gap size of 0.25 mm (a and b), 1 mm (c and d), and 2 mm (e and f) in non-saturated stage. *Source:* Wang et al. [89]/ With permission of Elsevier.

distribution maps shown in Figure 4.20c–e between the opening mouth of the crevice and the remained solution inside the crevice. The corrosion cathode tended to locate close to the air gap, where oxygen could be constantly replenished and the anode tended to be stable at the crevice tip. This anode/cathode distribution pattern was similar with waterline corrosion reported in the literatures [96]. With the progress of the experiment, the solution occluded inside the crevice evaporated gradually. This explains why the corrosion cell moved gradually to the crevice tip until no obvious corrosion cell existing inside the crevice (Figure 4.20f).

When the gap size of the crevice was increased to 1 mm, as illustrated in Figure 4.24c and d, sand particles could easily enter and be trapped inside the crevice. Due to the high resistance of non-saturated sand section and geometrical limitation of the crevice, a cathodic current gradient inside the crevice was observed and CP current was difficult to penetrate to the crevice to mitigate the corrosion inside the crevice. A corrosion cell was developed within the crevice, with negative cathodic currents tended to locate close to the air interface and the anode tended to accumulate at the crevice tip (see Figure 4.21d). This corrosion cell moves gradually toward to the crevice tip (see Figure 4.21(e)-6(h)), which could be attributed to the reduction of solution trapped inside the crevice and the gradual decrease of soil resistance due to the CP-induced chemical changes. The results of Dang et al. have also evidenced the decrease of soil electrolyte resistance under CP by the means of electrochemical impedance spectroscopy (EIS) [97]. Figure 4.21n indicates that at the end of the experiment, CP current was still insufficient in the crevice to mitigate the corrosion inside the crevice due to the high resistance of sand inside the crevice and geometrical limitation of the crevice, which is demonstrated in Figure 4.24d, explaining why the isolated corrosion cell evolved at the tip of the crevice (Figure 4.21m). When the gap size of the crevice was further increased to 2 mm, as illustrated in Figure 4.24e and f, more sandy soil particles would be trapped inside the crevice and the soil inside the crevice was continuous. Moreover, the large crevice gap would allow CP current to penetrate into the crevice. The continuous path of sand favors the solution within the crevice being absorbed outside of the crevice. With the decreasing of solution occluded inside the crevice and gradual decreasing of soil resistivity due to the CP-induced chemical change [97], the cathodic current area moved gradually to the crevice tip (Figure 4.22f and g). Figure 4.22l indicated that enough CP current could penetrate to the crevice, which is illustrated in Figure 4.24f. It explained why the whole surface within the crevice was covered with cathodic current in Figure 4.22h–k [89]. More detailed on this and related studies can be found in references [87–89].

4.3.3 Field Monitoring of Localized Corrosion on Buried Pipelines

As mentioned previously the lack of corrosion data significantly impedes the maintenance of "invisible" and "hidden" aging steel structures such as underground pipelines [98–102]. The development of accurate and robust probes and sensors for acquiring accurate and timely corrosion data under filed conditions has been listed as a critical need for infrastructure transformational research [101] and a grand challenge for corrosion research [103]. Traditional approaches to collecting field inspection data through field excavation and inline inspection "pigging" have limitations because historical corrosion data often do not have high enough spatial and temporal resolution required for accurately predicting the

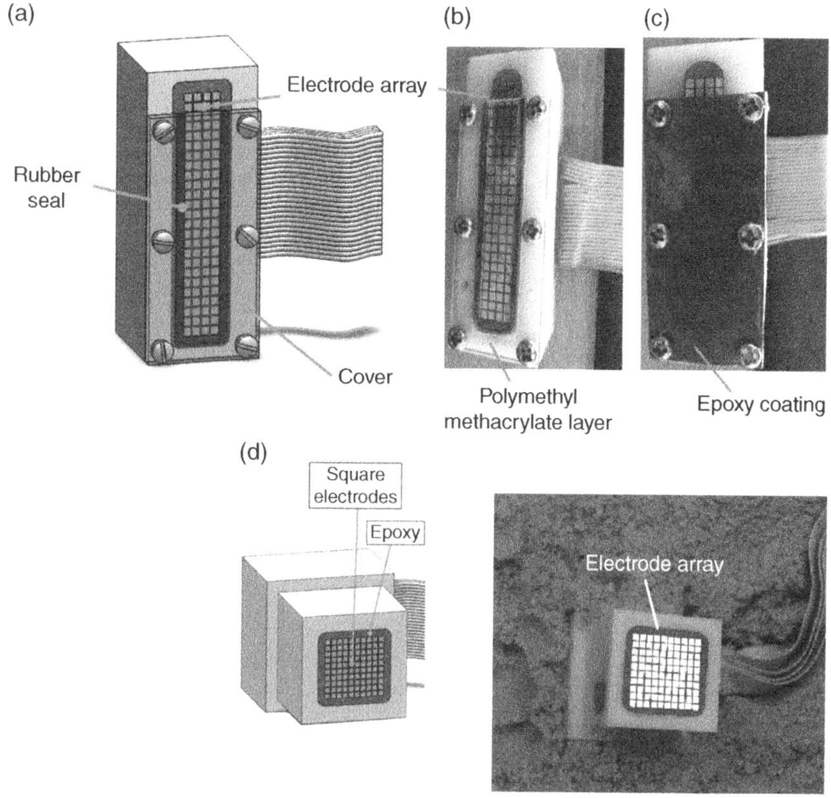

Figure 4.25 Ilustration of (a) the CUDC probe desings for simulating corrosion under disbonded coating by partially covering most of the electrode array with (b) an acrylic or (c) an epoxy pipeline coating layer that behaves as a shielding coating, and (d) Bare probe for measuring the effects of stray current-induced cathodic protection (CP) deficiency on corrosion.

failure of infrastructure under dynamic localized corrosion attack [104]. The development of corrosion probes for detecting CUDC [76, 77, 87] and stray current corrosion [21, 25] has been an industry need because these forms of localized corrosion are considered to be the worst-case scenario corrosion problems found on buried pipelines that is not detectable by existing corrosion probe [3, 85, 86, 104]. Over the past decade, corrosion probes designed based on electrochemically integrated multi-electrode array have been evaluated in an extensive field and semi-field sandbox evaluation programs in order to assess their capabilities in visualizing, probing, and monitoring critical forms of buried steel corrosion [6].

Figure 4.25a–c show the electrode array probes (using electrode arrays) designed for field monitoring of CUDC, while Figure 4.25d shows a bare electrode array probe (using normal electrode arrays) for measuring the effects of stray

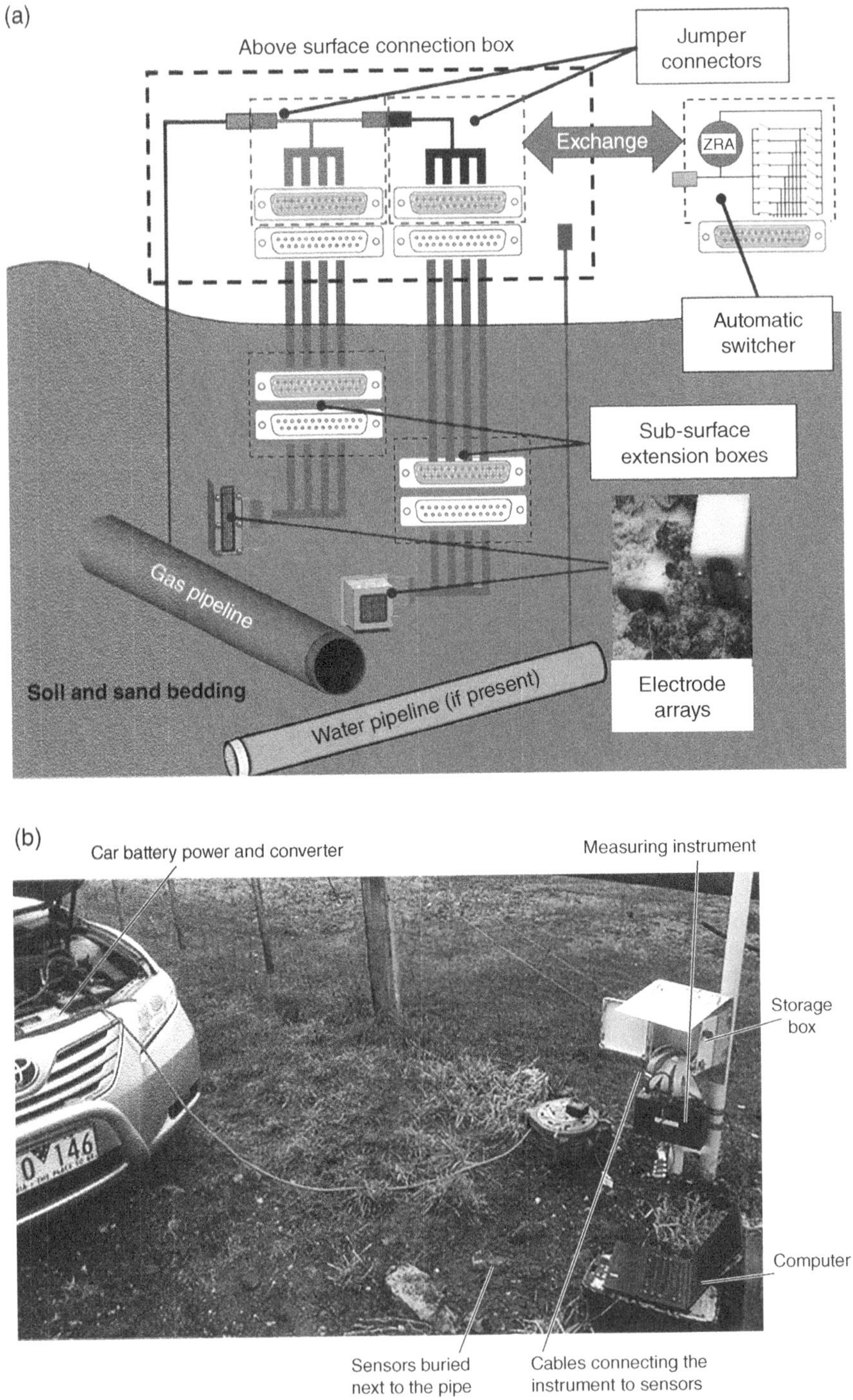

Figure 4.26 (a) Schematic representation of a probe installation site and basic electrical connection performed at each installation site. (b) Initial testing of a completed probe installation at a North Melbourne gas pipeline site. *Source:* Tan et al. [6], ELSEVIER.

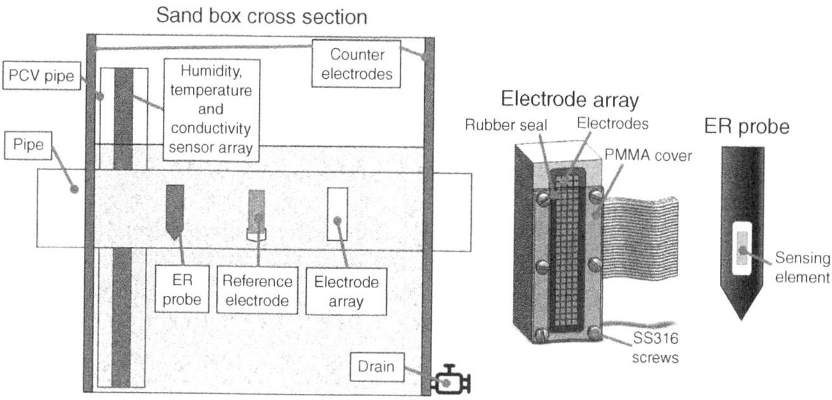

Figure 4.27 A schematic diagram of the experimental arrangement and corrosion probes [6].

current-induced cathodic protection (CP) deficiency on corrosion. Probes for simulating and measuring CUDC and stray current corrosion were deployed at underground pipeline sites and laboratory sandboxes, as illustrated in Figures 4.26 and 4.27, and measurements were performed regularly to provide in-situ and site-specific view over corrosion processes affecting buried steel pipelines. Localized corrosion rates were measured with a sensitivity in the order of 10 μm/year [6]. These 3D probes are made to create the conditions produced under disbonded coatings, incorporating a polymethyl methacrylate (PMMA) cover that behaves as a perfectly shielding coating (Figure 4.25b) or an epoxy pipeline coating film (Figure 4.25c). This cover or film was placed at a constant distance of 1 mm from the electrode surface to create a crevice that was considered to create the "worst-case scenario" CUDC condition [89]. In order to maximize the length of the crevice, the electrode array consisted of 4 row of 25 square electrodes of the same dimensions described for laboratory tests before [89]. The first two columns of electrodes directly were left to be exposed to the corrosive environment, simulating the coating holiday usually found next to disbonded areas. More details on the CUDC probe tests in the laboratory have been included references [76, 77, 87–89]. The bare electrode array probe (Figure 4.25d) is made for measuring the effects of stray current-induced CP deficiency on corrosion that has been demonstrated in laboratory to possess the necessary temporal and spatial resolutions required for the visualization of dynamically changing electrode processes and the effective quantification of the effects of dynamically changing potential excursions on buried steel corrosion [6]. Field tests were conducted by deploying eight CUDC probes in several gas pipeline sites in Melbourne and Geelong regions of Australia. Table 4.1 presents details about each installation site environmental conditions where steel pipelines covered with different coating materials existed. Table 4.2 shows details of the probes' CP and open-circuit potential recordings.

Table 4.1 Details of the probe's installation site conditions.

Tag	Site	Pipelines present	Bedding material	Observations during the installation
A	North of Melbourne	Cross-bonded water and gas pipelines	Sand	The pipe was located beneath a concrete slab
B	West of Melbourne	Cross-bonded water and gas pipelines	Sand	Independent connection to both pipelines
C	North of Melbourne	Gas pipeline	Sand	Underground running water present
D	Geelong	Gas pipeline	Sand	Clay rich soil

Source: Tan et al. [6].

Table 4.2 Probe field installation site CP and open-circuit potentials recordings.

		Potential vs. $Cu/CuSO_4$ reference electrode	
Tag	**Connected to**	**Min (mV)**	**Max (mV)**
A	Gas pipeline	−1020	−980
	Open circuit	−877	−680
B	Gas pipeline	−1050	−916
	Water and gas pipelines	−791	−769
	Water pipeline	−530	−501
	Open circuit	−780	−752
C	Gas pipeline	−1850	−380
	Open circuit	−1060	−980
D	Gas pipeline	−1426	−1320
	Open circuit	−920	−830

Source: Tan et al. [6].

Accelerated semi-field tests were also performed in sandboxes of one meter in size with an epoxy-coated pipe, installed across 220 mm in diameter. As shown in Figure 4.27, the arrangement consisted of titanium strips 2.5 cm wide installed at the four vertical edges of the box as anodes for the CP system. A copper/copper sulfate reference electrode was installed next to the center of the front wall of the sandbox. This site was selected for the reference electrode because it is a low-current density area and therefore small IR drop effects. A humidity, conductivity, and temperature measurement array was installed vertically next to one of the corners of the box. Two corrosion probes were installed at approximately the same depth as the center line of the simulated pipeline at a distance of 20 cm from it. One of the probes was a commercially available electrical resistance (ER) probe specifically designed for cathodic protection monitoring applications. Details of the ER probe application and comparison with other techniques are contained in reference [85]. At the beginning of the test, once all probes were installed and the box was filled with pre-washed sand up to a total depth of 60 cm. Tap water was added to the box until a 10 cm supernatant solution was stable above the sand. After the drainage valve was open, the system was left to evolve for approximately two weeks before the wet/dry cycles were reinitiated by closing the drainage valve and adding tap water once again. This simulates the dry/wet cycles in industry pipeline field that are due to weather/rainfall changes, commonly forming air gaps under disbonded coatings. At all times the simulated pipeline section and the probes were kept at a CP potential of $-850\,mV_{CSE}$.

Cumulative corrosion distribution maps for the bare metal probe installed in Site A are presented in Figure 4.28. It can be seen that corrosion patterns for both gas pipeline CP and open-circuit conditions are reasonably uniform indicating a low risk for localized corrosion at this pipeline site. The cumulative metal loss values after 249 days were approximately 21 and 81 μm under CP and at open circuit, respectively. Although these metal losses are low, the rate of metal loss has accelerated over the last 100 days for both conditions. Figure 4.29 presents the cumulative metal loss distribution maps for the bare metal probe installed in Site

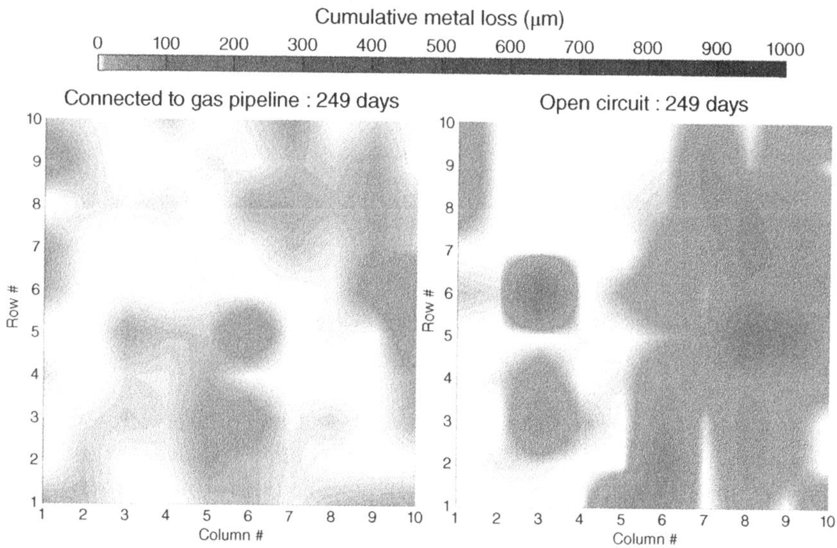

Figure 4.28 Cumulative corrosion maps for the bare metal probe installed in Site A. *Source:* Tan et al. [6].

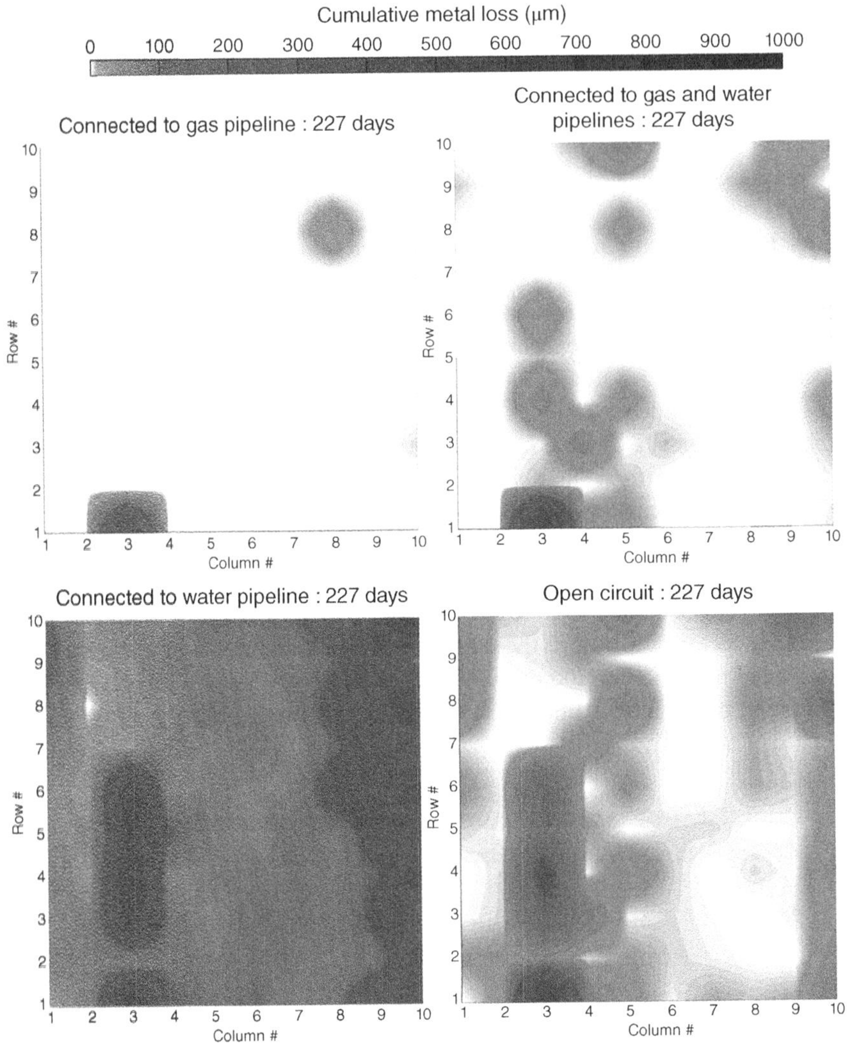

Figure 4.29 Cumulative corrosion maps for the bare metal probe installed in Site B. *Source:* Tan et al. [6].

B after 227 days for all the CP conditions. As expected, metal losses were lower for the conditions when the applied CP potential was more negative. For the case where the probe was connected to the gas pipeline, almost no corrosion was detected on the probe's surface except for one intensely corroding localized area where metal losses were calculated to be over 300 μm. This localized corrosion could be the result of stray currents or interference with the nearby water pipeline and a nearby train line (located within 50 m). The overall results provided by this

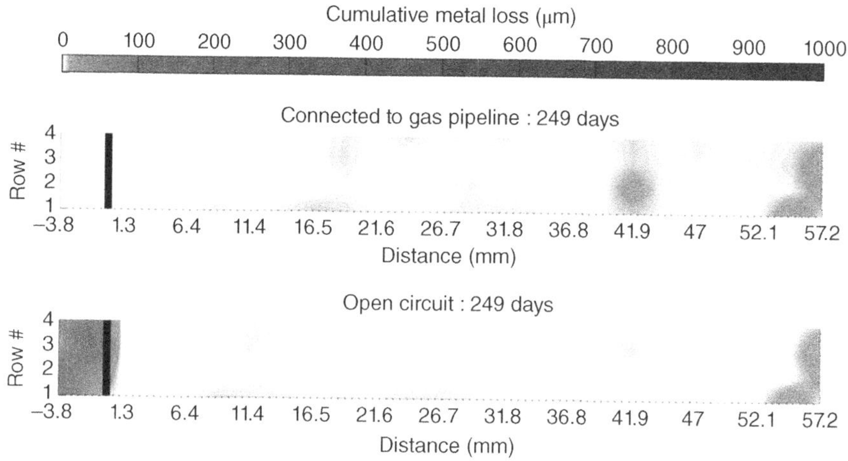

Figure 4.30 Final cumulative corrosion maps for the disbonded coating probe installed at site A. *Source:* Tan et al. [6].

probe suggest that the water pipeline is at severe risk of general corrosion at coating defect areas, while the gas pipeline is adequately protected although with some risk of localized corrosion. In a general note, the results obtained from bare metal probes in all four sites have shown similar behaviors with almost all probes indicated negligible metal losses when connected to their cathodically protected gas pipeline. In contrast, when the measurements were taken at open circuit, the final metal losses were significantly larger. The clear differences in the behavior observed under CP and at open circuit are in agreement with known pipeline protection under CP conditions and confirm corrosion monitoring consistency [6].

Cumulative metal losses maps for the 3D electrode array coating disbondment probe installed in Site A are presented in Figure 4.30 where the black vertical line represents the division between the holiday and the disbonded coating area. It can be seen in Figure 4.30 that corrosion losses occurred under the coating disbondment area although the corrosion was only minor (the maximum metal losses were less than 30 μm), while bare metal on the probe connected to gas pipeline CP did not show metal losses. This suggests that CP was ineffective due to CP shielding by the disbonded coating layer. Figure 4.31 shows photographs of corrosion under a coating disbondment found on the same gas pipeline (under similar conditions, about 2 km away from site A) after 40 years of service. When extrapolating the probes measurements for a period of 40 years, despite of the margin of error, the metal losses would only be approximately 1.6 mm, which is consistent with the small corrosion damage observed on the nearby pipe section. The fact that similar metal losses were recorded within the crevice areas of corrosion probes with and without CP (Figure 4.30) can be explained by analyzing the current distribution maps. Figure 4.32 presents characteristic current distribution maps obtained from

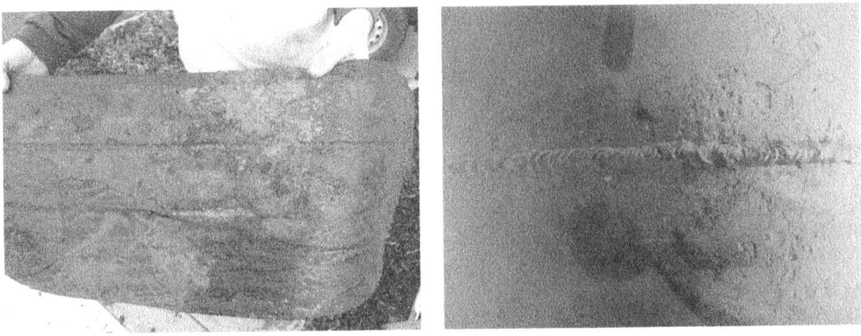

Figure 4.31 Corrosion under disbondment found on the same gas pipeline about 1.25 miles (2 km) from site A. *Source:* Tan et al. [6], ELSEVIER.

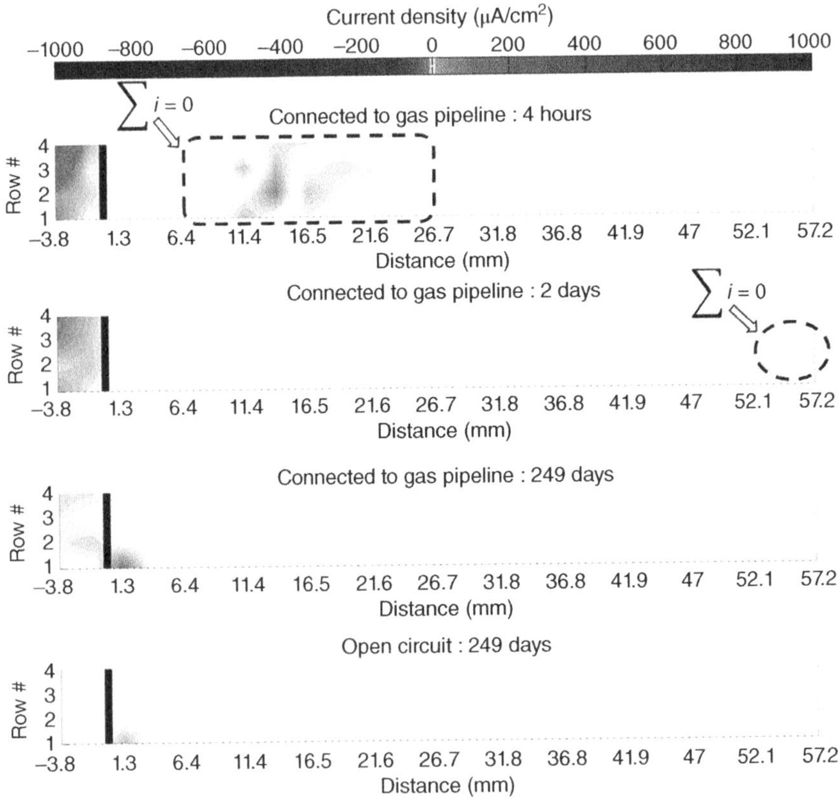

Figure 4.32 Evolution of the current density distribution maps on the disbonded coating probe installed at site A. *Source:* Tan et al. [6].

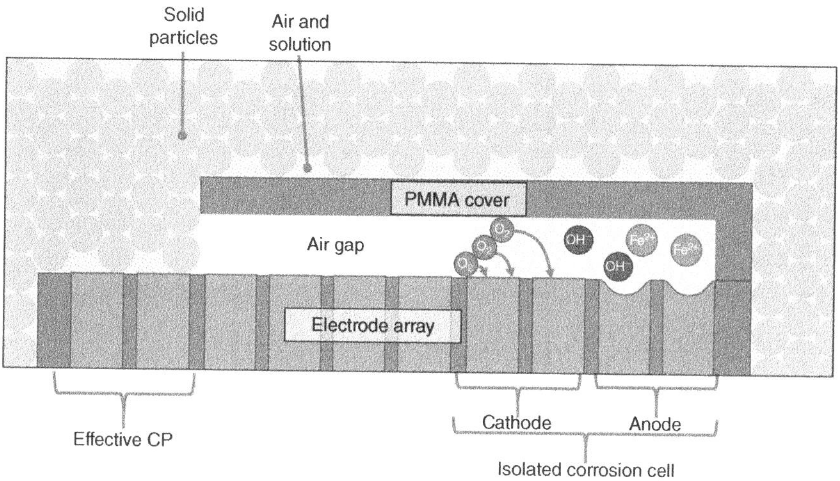

Figure 4.33 Schematic representation of air gap corrosion model under disbonded coating. *Source:* Tan et al. [6].

probes under both gas pipeline CP and water pipeline open-circuit conditions. The current distribution map obtained four hours after the probe's installation under gas pipeline CP shows large cathodic currents outside the crevice, while an area with cathodic and anodic currents was found over a region close to the crevice opening. Interestingly, the summation of the current densities found in this region was zero, indicating that this corrosion cell was isolated from CP outside the crevice. This type of current distribution has been found in previous studies performed using the same type of probes in non-saturated soils, and it is attributed to the formation of air gaps within the coating disbondment [11, 76, 77]. Figure 4.33 illustrates this concept: the non-saturated soil surrounding the probe draws part of the solution initially placed within the crevice leaving an air gap that influences the corrosion process in two ways. One, it act as a disruption in the electrolytic path required to supply CP to the further end of the crevice, and two, it also acts as a renewable source of oxygen for the isolated corrosion cell formed within the crevice. Depending on the passive state of the metal and on how long the solution within the crevice can remain, this condition can lead to severe localized corrosion. In the particular case of the probe installed in Site A, the crevice seems to dry up by the second day of exposure when only a small corrosion cell was found at the crevice tip. After this period no significant currents were detected inside the crevice, independently of the CP condition. This suggests that for site A the wet and dry cycles that trigger this corrosion mechanism are infrequent [6].

Figure 4.34 shows representative current density maps obtained from CUDC probes located at the pipeline site B (a Melbourne water and gas pipelines crossing

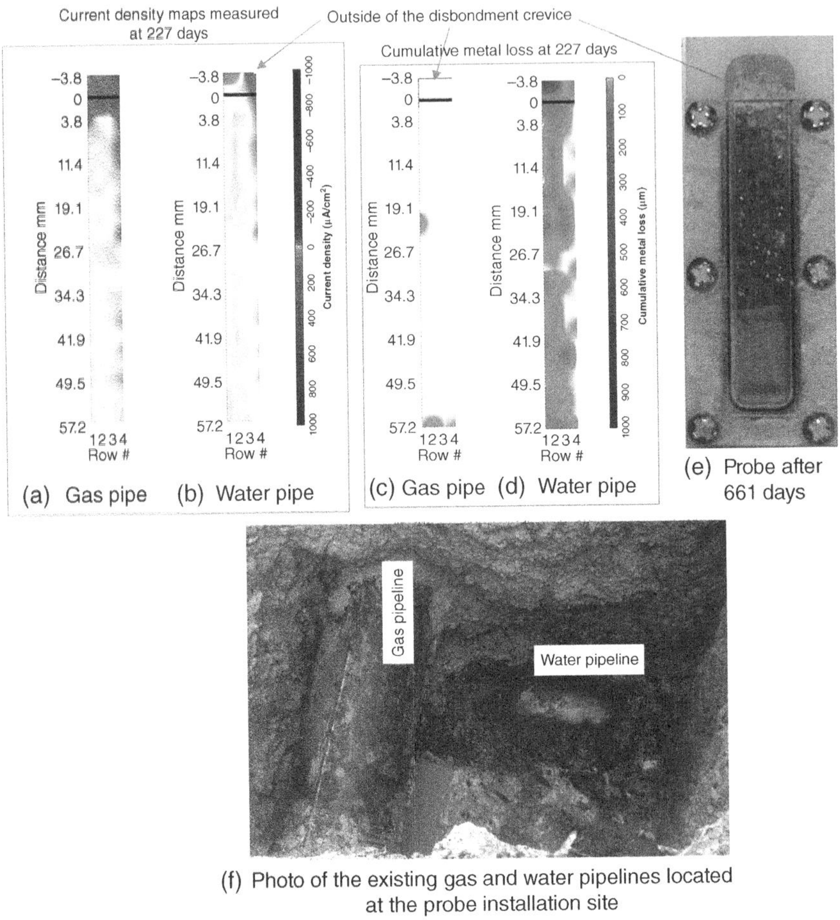

Figure 4.34 (a)–(b) Representative current density distribution maps over the CUDC probes installed at test site B (in the Altona North testing site, Melbourne) at gas pipeline (a) and water pipeline (b) potentials. (c)–(d) Final cumulative corrosion maps for the CUDC probes at gas (c) and water (d) pipeline potentials. (e) Image after the probe was withdrawn from the installation site after 661 days of testing. (f) Photo of the existing gas and water pipelines located at the probe installation site. (g) A summary of typical data collected from the bare steel probe simulating and monitoring bare steel corrosion at the Altona North Melbourne gas/water pipeline intersection site. (h) Illustrating the overall corrosion at site B after 661 days of field testing, from the disbonded coating probes. *Source:* Tan et al. [6], ELSEVIER.

site). At the site, as detailed in Table 4.1, the CP potential of the gas pipeline was fluctuating between −916 and −1050 mV and the potential of the water pipeline was fluctuating between −530 and −510 mV. A major feature in current density maps shown in Figure 4.34 is that at the gas pipeline CP potential, as shown in Figure 4.34a, the coating disbondment area was well protected with only a few

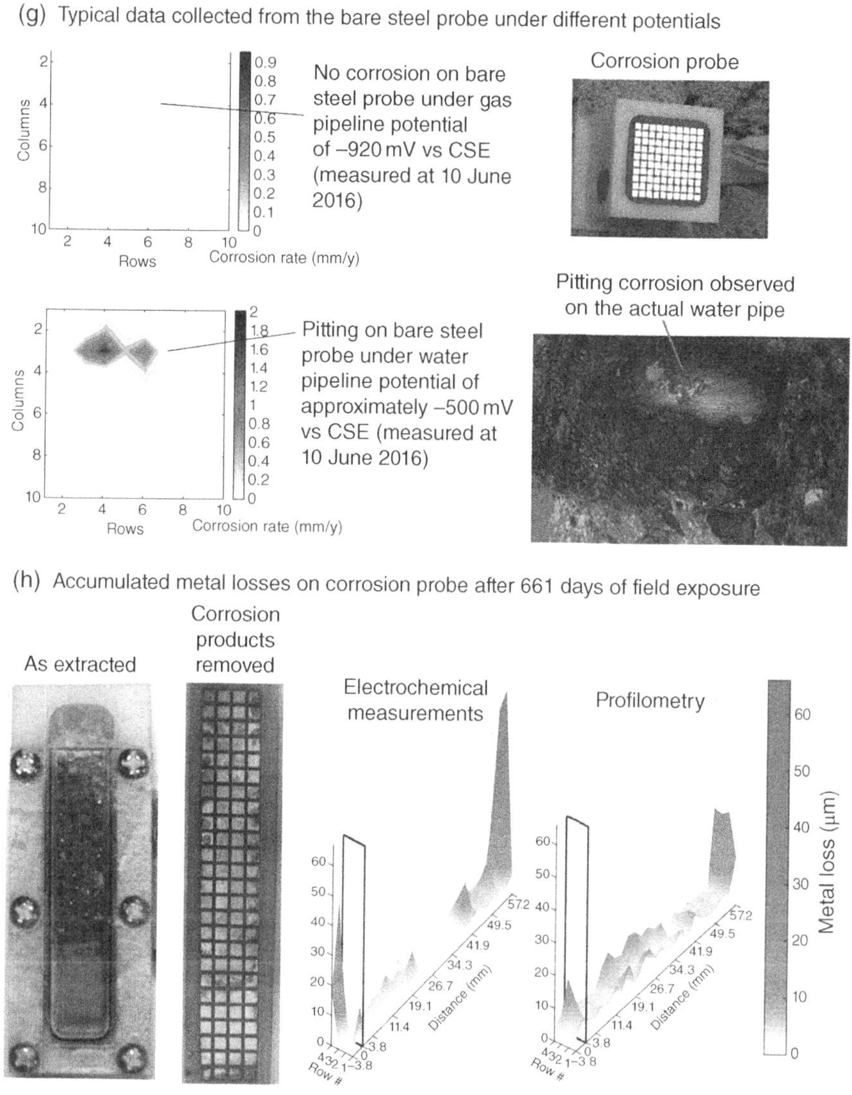

Figure 4.34 (Continued)

anodic densities registered, while the coating disbondment area at water pipeline potential was under insufficient CP with large anodic areas clearly visible in Figure 4.34b. Figure 4.34c and d show accumulated corrosion on the CUDC probes, which indicates little corrosion on the probe under the gas pipeline CP potential but significant corrosion on the probe under water pipeline potential. This is in good agreement with the corroded surface image after the probe was withdrawn from the installation site after 661 days of testing, as shown in

Figure 4.34e and the water pipeline corrosion condition shown in Figure 4.34f. In this case, the metal losses did change as a function of the CP condition and CP was working well for the gas pipeline at the site but not for the water pipeline. These results illustrate CP potential as a critical factor affecting buried pipeline corrosion exposed to unsaturated soil. Figure 4.34g and h show more results from different corrosion probes during and after completing 661 days of testing. As shown in Figure 4.34g, corrosion maps from the bare probe are in generally good agreement with the pitting behavior observed by visual inspection of the pipeline after 661 days of testing. Figure 4.34h compares the measured corrosion losses and their distribution over a probe surface after 661 days of testing, together with photos of the corroded probe surfaces. There is a good correlation between monitored corrosion depths and profilometry measurement results. This result demonstrates the accuracy of corrosion monitoring by the probes [6].

Figure 4.35 presents the final metal losses distribution maps for the coating disbondment probe installed in Site B for all four CP conditions tested by connecting the probe to the CP-protected gas pipe, to both gas and water pipes, to the unprotected water pipe, and at open-circuit condition. Similarly to the results from site A, when connected to the gas pipeline, the highest metal losses were less than 30 μm at the crevice tip location. At this site B, however, the metal losses did change as a function of the CP condition. Metal losses within the crevice were increasingly more generalized and intense as the probe was connected to the gas and water pipelines, to the water pipeline, and left at open circuit. This suggests CP was partially available at the crevice area. The results obtained from the disbonded coating probe installed in Site C and D presented similar behavior to that observed from the probe in Site B [6].

An important finding from the field testing using the CUDC probes is that corrosion appears significantly affected by the wet and dry cycles of the environmental changes within the crevice. If the environment within the crevice is kept dry, no corrosion would occur. Also if the environment is wet enough to form a continuous electrolytic path, CP can be effective under the disbonded area and corrosion can be prevented. However if the environment allows the formation of isolated corrosion cells, corrosion tends to occur locally. In order to verify the relationship between corrosion susceptibility under disbonded coatings and the hydrogeological conditions around the pipe and the type of bedding soil particles, tests were conducted in sandboxes where soil moisture conditions could be better controlled. Therefore laboratory testing in sandbox was carried out by installing CUDC and ER probes in sandboxes that simulate wet–dry cycles, as shown in Figure 4.27. The performance of the ER probes and the CUDC probes are also directly compared in these tests. Figure 4.36 presents the current density maps obtained across the surface of the electrode array soon after starting the test. The zero days map was obtained approximately two hours after the addition of water to the sandbox. In this map, relatively large cathodic current densities, driven by the CP system, were

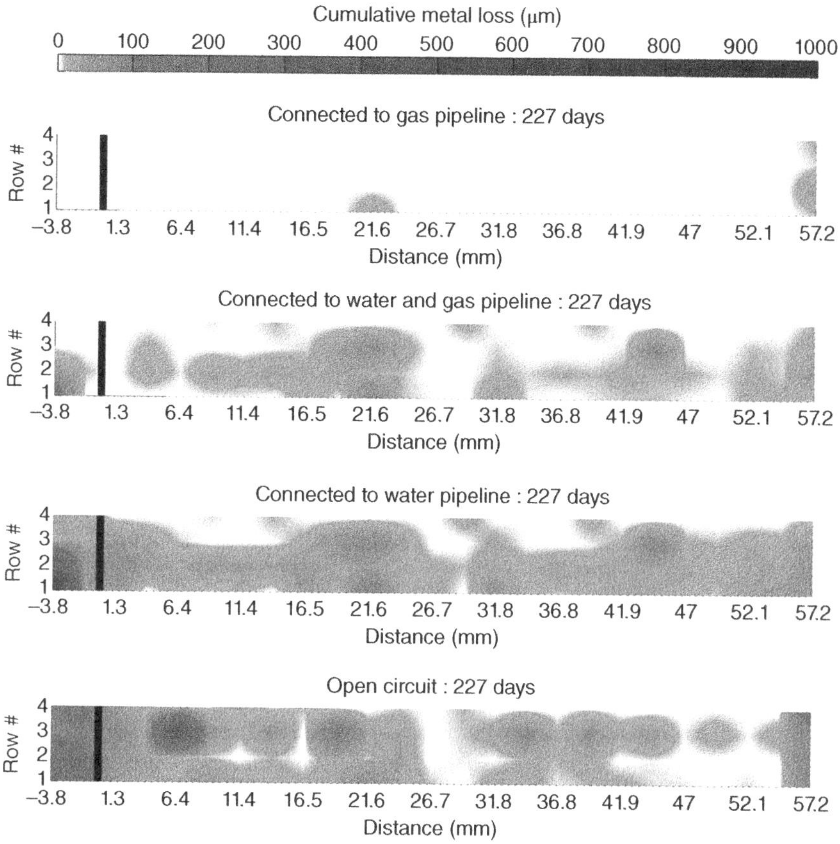

Figure 4.35 Final cumulative corrosion maps for the disbonded coating probe installed in Site B.

found outside the crevice. Inside the crevice, a sharp decreasing gradient of cathodic current densities is found close to the crevice opening. This gradient is produced by a combination of CP shielding and by lower and lower fractions of humid sand that gets into the crevice and provides an electrolytic pathway for the CP current. For distances between 10 and 50 mm from the crevice opening, no current densities were found inside the crevice. Finally, at the very tip of the crevice a combination of anodes and cathodes was found. This current density distribution is consistent with previous laboratory results [26], where the large area of zero current density inside the crevice was related to the formation of an air gap that isolated the area where CP is effective from the solution remaining inside the crevice, allowing the formation of isolated corrosion cells. These results further suggest that the formation of "zero current" zones should be due to solution drying up to form "air gaps." This is because currents should be measurable by the array

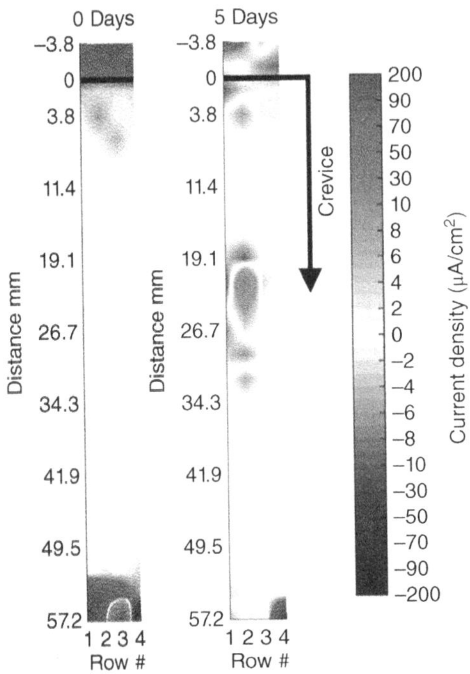

Figure 4.36 Current density distribution maps for the first wet–dry cycle. *Source:* Tan et al. [6].

electrode unless corrosion is totally uniform or the media is too resistive. After five days, the current density distribution map changed considerably. Outside the crevice and at the crevice opening, current densities were significantly less cathodic, and even some small anodic currents were measured. This is due to the lower humidity of the environment. Inside the crevice, another isolated corrosion cell was formed near the center of the crevice area and the corrosion cell at the bottom of the crevice presumably dried off. Figure 4.37 presents a series of current distribution maps that illustrate the changes observed for the second and third wet–dry cycle. At the beginning of the cycle, when water was added to the box, the isolated corrosion cells at the crevice tip and center of the array were reactivated. Then, the electrochemical activity on both corrosion cells decreased through the rest of the cycle. For the cell at the crevice tip, the decrease was abrupt, but for the cell at the center of the array the decrease was slow, taking up to 14 days to shut off the corrosion cell. From the fourth wet–dry cycle on, the behavior was different (Figure 4.38). In these cases the addition of water at the start of the cycle did not reactivate any isolated electrochemical cells inside the crevice. Instead, only an increase of cathodic current densities was found outside the crevice and at the crevice opening where CP is effective. For the rest of the cycle only an initial decrease in cathodic current density at the same area was observed. From then on, no significant changes were registered. Figure 4.39 presents the surface of the array, the profilometry results reflecting actual metal loss, and the

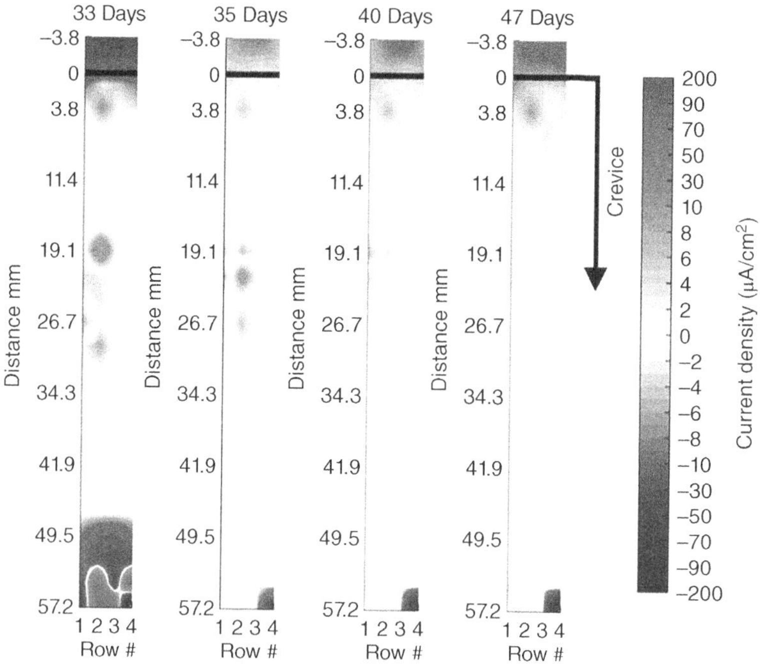

Figure 4.37 Current density distribution maps typical of the second and third wet–dry cycle. *Source:* Tan et al. [6].

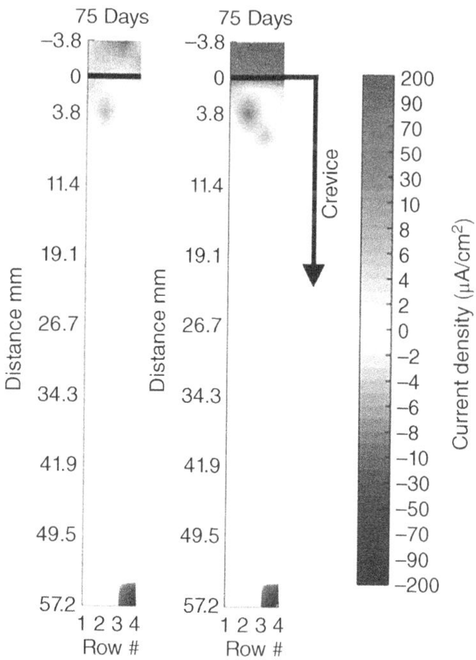

Figure 4.38 Typical current density distribution maps from the fourth wet–dry cycle. *Source:* Tan et al. [6].

electrochemically calculated metal losses based on the probes results. The surface of the sensor presented three areas with significant deposition of corrosion products. In all three cases the corrosion product mixed with sand to form a deposit that was well attached to the steel surface. It is believed that the formation of this bulky corrosion products and mix at the outside of the crevice gradually prevented water ingress. This could explain the differences observed between the current densities maps obtained during the first three wet–dry cycles and from the fourth cycle onward (Figure 4.39). The profilometry results indicate that metal losses outside the crevice were 27 μm at the most, while at the center and bottom of the array, maximum values were closer to 50 μm. The electrochemically calculated metal losses show an extremely good correlation not only in the pattern that corrosion presented inside the crevice but on the magnitude of the maximum metal losses [6].

It can be seen from results described above that current distribution maps from electrode arrays provided sufficient information to identify the site of isolated

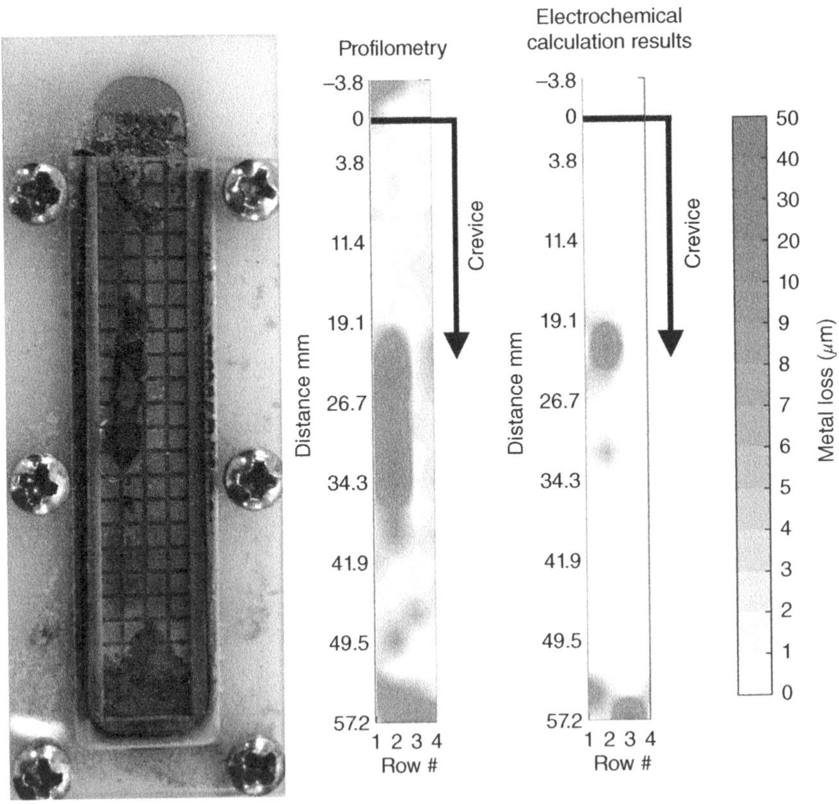

Figure 4.39 Appearance of the electrode after testing, profilometry results after corrosion product removal and results of electrochemically calculated metal loss. *Source:* Tan et al. [6], ELSEVIER.

corrosion cells inside the crevice. The results of the field trials demonstrate the potential of the probes to be used as a practical tool to electrochemically visualizing localized corrosion on invisible underground infrastructure under field conditions. The probes do not require CP interruption to evaluate corrosion rates and CP effectiveness. The probes do not require an accurate knowledge of the potential applied to the pipe or the chemistry of the environment in which they are immersed. The electrochemical nature of the probes allows to inexpensively evaluate instantaneous corrosion rates, with a sensitivity much greater than that of current methods employed by the pipeline industry. Early warning of localized corrosion a structural health monitoring system could be achieved by installing probes at the worst-case scenario sites. Testing results correlated reasonably well with long-term pipeline corrosion behavior. These results demonstrate the practicality and some unique advantages of the electrode array probes for monitoring corrosion on buried pipelines. Although the exact location of localized corrosion can be difficult to predict along the length of a buried pipeline, the locations that are sensitive to localized corrosion is predictable in corrosion engineering. Probes can be embedded at these strategic locations in order to obtain real-time data for localized corrosion monitoring and prediction, for instance, non-piggable pipelines; stray current-affected pipeline sections; river and shore crossing pipeline sections; pipeline sites that are far away from CP units; areas with localized corrosion detected during ILI; coating disbondment and degradation prone areas; coating damaged due to dents and hydrostatic testing and pipeline sections located at public safety sensitive areas [6].

References

1 NACE International Standard SP1069-2013 (2013). Control of External Corrosion on Underground or Submerged Metallic Piping Systems. NACE International, Houston, TX, USA, 1–60.

2 Australia Standard AS 2832.1-2015 (2015). Cathodic protection of metals; Part 5: Pipes and cables. Australian Standard Committee MT-014, Sydney, Australia.

3 Tan, M.Y., Ubhayaratne, K., Huo, Y. et al. (2019). Predictive maintenance based on smart monitoring and data analytics*Australasian Corrosion Association (ACA) conference: Corrosion & Prevention 2019*, Melbourne, Australia (24–27 November 2019).

4 Mahdavi, F., Forsyth, M., and Tan, M.Y.J. (2017). Techniques for testing and monitoring the cathodic disbondment of organic coatings: an overview of major obstacles and innovations. *Progress in organic coatings* 105: 163–175.

5 Leidheiser, H. Jr., Wang, W. Jr., and Igetoft, L. (1983). The mechanism for the cathodic delamination of coatings from a metal surface. *Progress in organic coatings* 11: 19–40.

6 Tan, M.Y., Huo, Y., and Varela, F. (2021). Field and laboratory assessment of electrochemical probes for visualizing localized corrosion under buried pipeline conditions. *Journal of Pipeline Science and Engineering* 1: 88–99.

7 Thompson, I. and Saithala, J.R. (2016). Review of pipeline coating systems from an operator's perspective. *Corros. Eng. Sci. Technol.* 51 (2): 1–8.
8 Funke, W. (1981). Blistering of paint films and filiform corrosion. *Progress in organic coatings* 9: 29–46.
9 Koehler, E.L. (1984). The mechanism of cathodic disbondment of protective organic coatings-aqueous displacement at elevated pH. *Corrosion* 40: 5–8.
10 Kendig, M., Addison, R., and Jeanjaquet, S. (1990). The mechanism of cathodic disbonding of hydroxy-terminated polybutadiene on steel from acoustic microscopy and surface energy analysis. *J. Electrochem. Soc.* 137: 2690–2697.
11 Varela, F., Tan, M., and Forsyth, M. (2015). Understanding the effectiveness of cathodic protection under disbonded coatings. *Electrochim. Acta* 186: 377–390.
12 Australian Standard AS4352 (1995). Test for coating resistance to cathodic disbanding. Sydney, Australia.
13 ASTM G8-96 (1996). *Standard Test Method for Cathodic Disbonding of Pipeline Coatings*. West Conshohocken, PA: ASTM.
14 Holub, J., Wong, D.T., and Tan, M. (2007). Analysis of CD TEST methods and factors affecting cathodic disbondment. In: *NACE Corrosion Conference and Expo* (CORROSION 2007), Paper No. 07022. Nashville, TN, USA (11–15 March 2007).
15 Betz, M., Bosch, C., Gronsfeld, P.J. et al. (2012). Cathodic disbondment test: what are we testing? Paper presented at the CORROSION 2012, Paper Number: NACE-2012-1285, Salt Lake City, Utah (March 2012).
16 Le Thu, Q., Takenouti, H., and Touzain, S. (2006). EIS characterization of thick flawedorganic coatings aged under cathodic protection in seawater. *Electrochim. Acta* 51: 2491–2502.
17 Shreepathi, S. (2016). Physicochemical parameters influencing the testing of cathodic delamination resistance of high build pigmented epoxy coating. *Progress in organic coatings* 90: 438–447.
18 Sørensena, P.A., Dam-Johansena, K., Weinellb, C.E., and Kiil, S. (2010). Cathodic delamination of seawater-immersed anticorrosive coatings: mapping of parameters affecting the rate. *Progress in organic coatings* 68: 283–292.
19 Rosa Rodrigues, P.C. and Fernández, C. (2011). Influence of applied potential and temperature on the cathodic disbondment for offshore structures. Paper presented at the CORROSION 2011, Paper Number: NACE-11038, Houston, Texas (March 2011).
20 Steinsmo, F.U. and Skar, J.I. (1994). Factors influencing the rate of cathodic disbonding of coatings. *Corrosion* 50: 934–939.
21 McHattie, J.S., Perez, I.L., and Kehr, J.A. (1996). Factors affecting cathodic disbondment of epoxy coatings for steel reinforcing bars. *Cement. Concrete Comp.* 18: 93–103.
22 Mansfeld, F. and Tsai, C.H. (1991). Determination of coating deterioration with EIS I. Basic relationships. *Corrosion* 47: 958–963.

23 Mansfeld, F. (1995). Use of electrochemical impedance spectroscopy for the study of corrosion protection by polymer coatings. *J. App. Electrochem.* 25: 187–202.

24 Walter, G.W. (1986). A review of impedance plot methods used for corrosion performance analysis of painted metals. *Corros. Sci.* 26: 681–703.

25 Barsoukov, E. and Macdonald, J.R. (2005). *Impedance Spectroscopy Theory, Experiment, and Applications*, 2e. New Jersey: Wiley & Sons.

26 Kendig, M.W., Jeanjaquet, S., Lumsden, J. et al. (ed.) (1993). *Electrochemical Impedance of Coated Metal Undergoing Loss of Adhesion, Electrochemical Impedance: Analysis and Interpretation*, 407–427. Philadelphia: ASTM STP 1188, American Society for Testing and Materials.

27 Papavinasam, S., Attard, M., and Revie, R.W. (2009). Electrochemical impedance spectroscopy measurement during cathodic disbondment experiment of pipeline coatings. *J. ASTM Inter.* 6: JAI101246.

28 Mahdavi, F., Tan, M.Y.J., and Forsyth, M. (2015). Electrochemical impedance spectroscopy as a tool to measure cathodic disbondment on coated steel surfaces: capabilities and limitations. *Progress in organic coatings* 88: 23.

29 Margarit, I.C.P., Oscar, R.M., Roberto, J., and Ferreira, R.M. (2001). About coatings and cathodic protection: electrochemical features of coatings used on pipelines. *J. Coat. Technol.* 73: 61–65.

30 Mahdavian, M., Naderi, R., Peighambari, M. et al. (2015). Evaluation of cathodic disbondment of epoxy coating containing azole compounds. *J. Ind. Engin. Chem.* 25: 1167–1173.

31 Isaacs, H.S. and Kendig, M.W. (1980). Determination of surface inhomogeneities using a scanning probe impedance technique. *Corrosion* 36: 269–274.

32 Huang, V.M., Wu, S.L., Orazem, M.E. et al. (2011). Local electrochemical impedance spectroscopy: a review and some recent developments. *Electrochim. Acta* 56: 8048–8057.

33 Upadhyay, V. and Battocchi, D. (2016). Localized electrochemical characterization of organic coatings: a brief review. *Progress in organic coatings* 99: 365–377.

34 Dong, C.F., Fu, A.Q., Li, X.G., and Cheng, Y.F. (2008). Localized EIS characterization of corrosion of steel at coating defect under cathodic protection. *Electrochim.Acta* 54: 628–633.

35 Yee, S., Oriani, R.A., and Stratmann, M. (1991). Application of a Kelvin microprobe to the corrosion of metals in humid atmospheres. *J. Electrochem. Soc.* 138: 55–61.

36 Stratmann, M., Streckel, H., and Fese, R. (1999). A new technique able to measure directly the delamination of organic polymer films. *Corros. Sci.* 32: 467–470.

37 Fu, A.Q. and Cheng, Y.F. (2009). Characterization of corrosion of X65 pipeline steel under disbonded coating by scanning Kelvin probe. *Corros. Sci.* 51: 914–920.

38 Leblanc, P.P. and Frankel, G.S. (2004). Investigation of filiform corrosion of epoxy-coated carbon steel by scanning kelvin probe force microscopy. *J. Electrochem. Soc.* 151: B105–B113.

39 Bi, H. and Sykes, J. (2011). Cathodic disbonding of an unpigmented epoxy coating on mild steel under semi- and full-immersion conditions. *Corros. Sci.* 53: 3416–3425.

40 Khun, N.W. and Frankel, G.S. (2013). Effects of surface roughness, texture and polymer degradation on cathodic delamination of epoxy coated steel samples. *Corros. Sci.* 67: 152–160.

41 da Fonseca, R.J.M., Ferdj-Allah, L., Despaux, G. et al. (1993). Scanning acoustic microscopy – recent applications in materials science. *Adv. Mat.* 5: 508–519.

42 Oehler, H., Alig, I., Lellinger, D., and Bargmann, M. (2012). Failure modes in organic coating studied by scanning acoustic microscopy. *Progress in organic coatings* 64: 719–725.

43 Addison, R.C., Kendig, M.W., Jeanjaquet, S.J. (1988). Measurement of cathodic disbonding using a scanning acoustic microscope. IEEE 1988 Ultrasonics Symposium Proceedings. Chicago, IL, USA (02–05 October 1988).

44 Kendig, M., Addison, R., and Jeanjaquet, S. (1990). The mechanism of cathodic disbanding of hydroxy-terminated polybutadiene on steel from acoustic microscopy and surface energy analysis. *J. Electrochem. Soc.* 137: 2690–2697.

45 Hernandez, M.A., Galliano, F., and Landolt, D. (2004). Mechanism of cathodic delamination control of zinc–aluminum phosphate pigment in waterborne coatings. *Corros. Sci.* 46: 2281–2300.

46 Galliano, F. and Landolt, D. (2002). Evaluation of corrosion protection properties of additives for waterborne epoxy coatings on steel. *Progress in organic coatings* 44: 217–225.

47 Bi, H. and Sykes, J. (2016). Cathodic delamination of unpigmented and pigmented epoxy coatings from mild steel. *Progress in organic coatings* 90: 114–125.

48 Alig, I., Tadjbach, S., Krüger, P. et al. (2009). Characterization of coating systems by scanning acoustic microscopy: debonding, blistering and surface topology. *Prog. Org. Coat.* 64: 112–119.

49 Le Thu, Q., Bonnet, G., Compere, C. et al. (2005). Modified wire beam electrode: a useful tool to evaluate compatibility between organic coatings and cathodic protection. *Prog. Org. Coat.* 52: 118–125.

50 Kong, D., Wang, Y., Zhang, W. et al. (2012). Correlation between electrochemical impedance and current distribution of carbon steel under organic coating. *Mat. Corros.* 63: 475–480.

51 Zhang, W., Wang, J., Li, Y., and Wang, W. (2010). Evaluation of metal corrosion under defective coatings by WBE and EIS technique. *Acta Phys. Chim. Sin.* 26: 2941–2950.

52 Chen, Y., Zhang, W., Ding, K. et al. (2016). Debonding mechanism of organic coating with man-made defect in the area nearby water-line by WBE technique. *J. Chinese Soc. Corros. Protect.* 36: 67–72.

53 Mahdavi, F., Tan, M.Y., and Forsyth, M. (2016). Communication – an approach to measuring local electrochemical impedance for monitoring cathodic disbondment of coatings. *J. Electrochem. Soc.* 163: C228–C231.

54 Chen, X., Du, C., Li, X., and Huang, Y. (2009). Effects of cathodic potential on the local electrochemical environment under a disbonded coating. *J. Appl. Electrochem.* 39: 697–704.

55 Xu, J., Sun, C., Yan, M., and Wang, F. (2013). Electrochemical behavior of steel A36 under disbonded coating in the presence of sulfate-reducing bacteria. *Mater. Chem. Phys.* 142: 692–700.

56 Xu, J., Wang, K., Sun, C. et al. (2011). The effects of sulfate reducing bacteria on corrosion of carbon steel Q235 under simulated disbonded coating by using electrochemical impedance spectroscopy. *Corros. Sci.* 53: 1554–1562.

57 Eslami, A., Kania, R., Worthingham, B. et al. (2011). Effect of CO_2 and R-ratio on near-neutral pH stress corrosion cracking initiation under a disbonded coating of pipeline steel. *Corros. Sci.* 53: 2318–2327.

58 Maocheng, Y., Jin, X., Libao, Y. et al. (2016). EIS analysis on stress corrosion initiation of pipeline steel under disbonded coating in near-neutral pH simulated soil electrolyte. *Corros. Sci.* 110: 23–34.

59 Angst, U., Büchler, M., Martin, B. et al. (2016). Cathodic protection of soil buried steel pipelines–a critical discussion of protection criteria and threshold values. *Mater. Corros.* 67: 1135–1142.

60 Barbalat, M., Lanarde, L., Caron, D. et al. (2012). Electrochemical study of the corrosion rate of carbon steel in soil: Evolution with time and determination of residual corrosion rates under cathodic protection. *Corros. Sci.* 55: 246–253.

61 Gan, Z.-W.S.F., Sabde, G., and Chin, D.-T. (1994). Cathodic protection to mitigate external corrosion of underground steel pipe beneath disbonded coating. *Corrosion* 50: 13.

62 Chin, D.-T. and Sabde, G.M. (1999). Current distribution and electrochemical environment in a cathodically protected crevice. *Corrosion* 55: 9.

63 Mahdavi, F., Forsyth, M., and Tan, M.Y. (2017). Understanding the effects of applied cathodic protection potential and environmental conditions on the rate of cathodic disbondment of coatings by means of local electrochemical measurements on a multi-electrode array. *Prog. Org. Coat.* 103: 83–92.

64 Wang, W., Wang, Q., Wang, C., and Yi, J. (2014). Experimental studies of crevice corrosion for buried pipeline with disbonded coatings under cathodic protection. *Journal of Loss Prevention in the Process Industries* 29: 163–169.

65 Smith, H., Bird, M., and Penna, R. (1988). Factors affecting the cathodic disbonding of pipe coatings. *Mater. Performance (United States)* 27: 11.

66 Larsen, K.R. (2016). Using pipeline coatings with cathodic protection protective coatings need to maintain integrity under operating environments and in the presence of CP. *Mater. Performance* 55: 32–35.

67 Perdomo, J. and Song, I. (2000). Chemical and electrochemical conditions on steel under disbonded coatings: the effect of applied potential, solution resistivity, crevice thickness and holiday size. *Corros. Sci.* 42: 1389–1415.

68 Perdomo, J., Chabica, M., and Song, I. (2001). Chemical and electrochemical conditions on steel under disbonded coatings: the effect of previously corroded surfaces and wet and dry cycles. *Corros. Sci.* 43: 515–532.

69 Li, Z., Gan, F., and Mao, X. (2002). A study on cathodic protection against crevice corrosion in dilute NaCl solutions. *Corros. Sci.* 44: 689–701.

70 Yan, M., Wang, J., Han, E., and Ke, W. (2007). Electrochemical measurements using combination microelectrode in crevice simulating disbonded of pipeline coatings under cathodic protection. *Corrosion engineering, science and technology* 42: 42–49.

71 Yan, M., Wang, J., Han, E., and Ke, W. (2008). Local environment under simulated disbonded coating on steel pipelines in soil solution. *Corros. Sci.* 50: 1331–1339.

72 Chen, X., Li, X., Du, C., and Cheng, Y. (2009). Effect of cathodic protection on corrosion of pipeline steel under disbonded coating. *Corros. Sci.* 51: 2242–2245.

73 Eslami, A., Fang, B., Kania, R. et al. (2010). Stress corrosion cracking initiation under the disbonded coating of pipeline steel in near-neutral pH environment. *Corros. Sci.* 52: 3750–3756.

74 Yan, M., Sun, C., Xu, J. et al. (2015). Stress corrosion of pipeline steel under occluded coating disbondment in a red soil environment. *Corros. Sci.* 93: 27–38.

75 Kuang, D. and Cheng, Y. (2015). Probing potential and solution pH under disbonded coating on pipelines. *Mater. Performance* 54: 40–45.

76 Varela, F., Tan, M.Y., and Forsyth, M. (2015). Electrochemical method for studying localized corrosion beneath disbonded coatings under cathodic protection. *J. Electrochem. Soc.* 162: C515–C527.

77 Varela, F., Tan, M.Y.J., and Forsyth, M. (2015). An electrochemical method for measuring localized corrosion under cathodic protection. *ECS electrochemistry letters* 4 (1): C1–C4. Video in Supplemental Material: https://iopscience.iop.org/article/10.1149/2.0091501eel.

78 Australian standard AS 2832.1:2015 (2015). Cathodic protection of metals – Pipes and cables, p. 18.

79 Betz, M., Bosch, C., Gronsfeld, P. J. et al. (2012). Cathodic disbondment test: what are we testing?, In Corrosion 2012. NACE International.

80 Greenfield, D. and Scantlebury, D. (2000). The protective action of organic coatings on steel: a review. *Journal of Corrosion Science and Engineering* 3: 135–176.

81 Knaster, M.B. and Parks, J.M. (1986). Mechanism of corrosion and delamination of painted phosphated steel during accelerated corrosion testing. *SAE Transactions* 95: 578–587 (No. 860110). SAE Technical Paper.

82 Hardie, D., Charles, E.A., and Lopez, A.H. (2006). Hydrogen embrittlement of high strength pipeline steels. *Corros. Sci.* 48: 4378–4385.

83 Masilela, Z. and Pereira, J. (1998). Using the direct current voltage gradient technology as a quality control tool during construction of new pipelines. *Eng. Fail. Anal.* **5** (2): 99–104.

84 Shawki, S. (2011). Above ground coating integrity assessment: experience with SUMED pipelines. In: *Integrity of Pipelines Transporting Hydrocarbons* (ed. G. Bolzon, T. Boukharouba, G. Gabetta, et al.), 311–322. Biskra, Algeria: Springer.

85 Varela, F., Tan, M.Y., and Forsyth, M. (2015). An overview of major methods for inspecting and monitoring external corrosion of on-shore transportation pipelines. *Corrosion engineering science and technology* 50: 226–235.

86 Tan, M.Y.J. (2020). A critical overview of monitoring infrastructural health using corrosion probes. *Corrosion engineering science and technology* 55: 103–117. https://doi.org/10.1080/1478422X.2019.1695390.

87 Wang, K., Varela, F.B., and Tan, M.Y. (2019). Visualizing dynamic and localized corrosion processes on cathodically protected steel exposed to soil with different moisture contents. *Corrosion* 75 (4): 398–407.

88 Wang, K., Varela, F.B., and Tan, M.Y. (2019). The effect of electrode surface area on corrosion initiation monitoring of X65 steel in soil. *Corrosion science* 152: 218–225.

89 Wang, K., Varela, F.B., and Tan, M.Y. (2019). Probing dynamic and localised corrosion processes on buried steel under coating disbondments of various geometries. *Corrosion science* 150: 151–160.

90 Brousseau, R. and Qian, S. (1994). Distribution of steady-state cathodic currents underneath a disbonded coating. *Corrosion* 50: 907.

91 Song, F.M. and Sridhar, N. (2008). Modeling pipeline crevice corrosion under a disbonded coating with or without cathodic protection under transient and steady state conditions. *Corros. Sci.* 50: 70.

92 Jiang, J., Wang, J., Wang, W., and Zhang, W. (2009). Modeling influence of gas/liquid/solid three-phase boundary zone on cathodic process of soil corrosion. *Electrochim. Acta* 54: 3623–3629.

93 Akkouche, R., Rémazeilles, C., Jeannin, M. et al. (2016). Influence of soil moisture on the corrosion processes of carbon steel in artificial soil: active area and differential aeration cells. *Electrochim. Acta* 213: 698–708.

94 Akkouche, R., Rémazeilles, C., Barbalat, M. et al. (2017). Electrochemical monitoring of steel/soil interfaces during wet/dry cycles. *J. Electrochem. Soc.* 164: C626–C634.

95 Varela, F.N. (2015). Electrochemical methods for monitoring localized corrosion under cathodic protection. Deakin University.

96 Tan, Y.-J., Bailey, S., and Kinsella, B. (2001). Mapping non-uniform corrosion using the wire beam electrode method. III. Water-line corrosion. *Corros. Sci.* 43: 1931–1937.

97 Dang, D.N., Lanarde, L., Jeannin, M. et al. (2015). Influence of soil moisture on the residual corrosion rates of buried carbon steel structures under cathodic protection. *Electrochim. Acta* 176: 1410–1419.

98 Li, X., Zhang, D., Liu, Z. et al. (2015). Materials science: share corrosion data. *Nature* 527: 441–442.

99 G. H. Koch, M. P.H. Brongers, N. G. Thompson, Y. P. Virmani, and J. H. Payer, Corrosion cost and preventive strategies in the United State, National Technical Information Service Report No. FHWA-RD-01-156, 2001

100 Newman, R.C. and Sieradzki, K. (1994). Metallic corrosion. *Science* 263: 1708–1709.

101 National Institute of Standards and Technology (2009) Technology Innovation Program report. http://www.nist.gov/tip/prev_competitions/upload/ci_wp_031909.pdf (accessed 29 January 2019).

102 Cheng, F. (2016). Monitor safety of aged fuel pipelines. *Nature* 529: 156.

103 Committee on Research Opportunities in Corrosion Science and Engineering (2011). *Research Opportunities in Corrosion Science and Engineering.* Washington, DC: National Research Council, The National Academies Press.

104 Tan, M.Y.J. (2017). Principles and issues in structural health monitoring using corrosion sensors. Corrosion/2017, NACE International, paper no. 9240.

5

Corrosion Affected by Multiple Environments and Mechanisms

A practical engineering structure can be exposed to multiple environmental zones; and corrosion processes and mechanisms in each environmental zone can be different. Shore-crossing pipelines and offshore wind turbines are typical examples of engineering structures that pass through seabed, seawater, sea–soil–air interfaces, soil, sand, and atmospheric zones. Soil, marine, and atmospheric corrosion could occur over different zones of the structures under the effects of various corrosion mechanisms such as differential aeration corrosion at the waterline areas and biologically influenced corrosion in metabolically active locations. On the other hand, corrosion protection technologies applied on each zone can also differ and could interact with each other. For instance, an impressed cathodic protection (CP) system applied on a steel pipeline section buried in coastline soil could affect the sacrificial anode-based CP system applied on pipeline section submerged in the seawater, creating a complex corrosion and protection situation. This chapter discusses such complex forms of localized corrosion and protection cases through the analysis of typical engineering structures, with particular focus on interfaces between different environmental zones. The worst-case scenario corrosion often occurs at the interfacial areas such as at the waterline, soil/air interface and shore-crossing. It is shown that corrosion data with good spatial resolutions across multiple environmental zones are required in order to ensure effective localized corrosion prediction, prevention, and management. An approach to acquiring corrosion data with sufficient temporal and spatial resolutions is to use integrated multi-probes in order to effectively simulate and *in situ* monitor the dominating localized corrosion process and mechanism at each environmental zone, focusing on the worst-case scenario corrosion processes and mechanisms. A moving probe is also described as a simplified survey tool to survey multiple zones of a long pipeline structure.

Localized Corrosion in Complex Environments, First Edition. Mike Yongjun Tan.

5.1 Localized Corrosion Affected by Multiple Environments and Mechanisms

Many major engineering structures are exposed to multiple and changing environmental conditions. Pipelines, oil platforms, drilling rigs, bridges, wind turbines, and desalination plants are common large engineering structures that are typically exposed to multiple environmental conditions. Figure 5.1 illustrates a shore-crossing pipeline and an offshore wind turbine that pass through several environmental zones. More such cases can be found in Figure 1.17.

When an engineering structure is exposed to multiple environments, different corrosion processes would occur on different parts of the structure with not only different kinetics but also different mechanisms. The most complex forms of localized corrosion usually occur at the interface of different environmental zones. Figure 5.2 shows a practical case of localized corrosion failure of a steel lamp post exposed to multi-environmental conditions, causing a fatal accident [1]. The lamp post was manually and visually inspected only three days before the accident; however, physical inspection failed to identify serious localized corrosion on the lamp post at a location just below the damp soil surface. This accident raised major public concerns over the safety of lamp posts in Singapore where in its Tampines council alone more than 300 of total 4500 such posts were corroded and replaced every year [1].

The lamp post is a common civil structure that passes through soil/air interface, in a similar manner as the shore-crossing pipeline shown in Figure 5.1a. The fallen lamp post was found to have corroded and cracked approximately 3 cm below the damp soil that contained approximately 1% NaCl and had a pH about 7. This is a typical metal structure where multiple corrosion environments and corrosion mechanisms co-existed, causing the 'worst-case scenario' corrosion to occur just below the damp soil surface. Such complex localized corrosion could be illustrated in Figure 5.3 where several different corrosion processes and mechanisms occur simultaneously on different areas of the steel structure:

i) Soil corrosion occurs on the underground part of the lamp post, as illustrated in Figure 5.3, causing localized form of corrosion in the heterogeneous, moist, and conductive soil environment. Key environmental factors that are known to affect corrosion in soil include soil moisture, salts, temperature, electrical resistance, alkalinity, acidity, permeability of water and air, oxygen content and biological organisms, stray currents, etc. More details on soil corrosion behavior of steel can be found in Chapter 3.
ii) Differential aeration cell corrosion occurs at the interface of soil and atmosphere. As illustrated in Figure 5.3, major corrosion would occur at a poorly oxygenated area below the interface of the soil (anode) that is adjacent to an

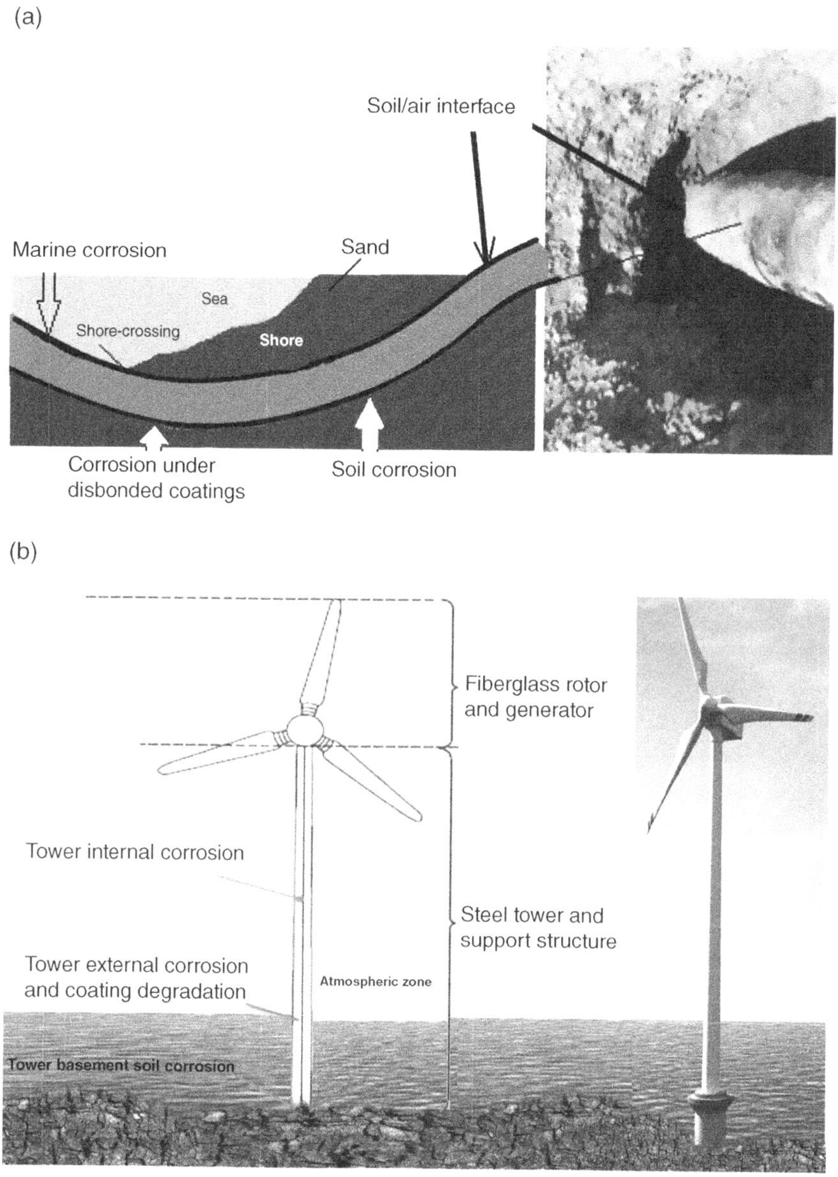

Figure 5.1 Illustration of engineering structures crossing several environmental zones. (a) A shore-crossing oil and gas pipeline; (b) an offshore wind turbine.

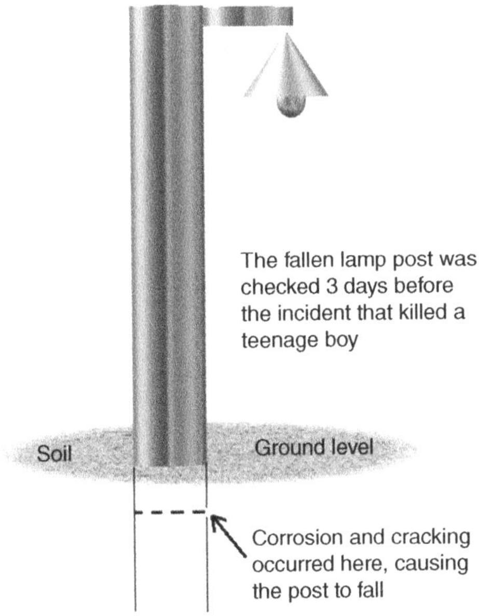

Figure 5.2 A case of localized corrosion of a lamp post due to multiple mechanisms occurring in different environments. *Source:* Ref. [1].

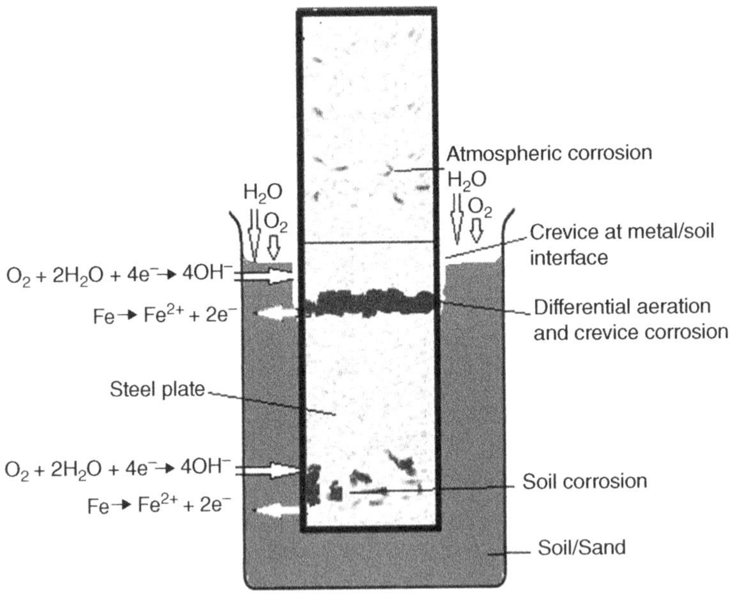

Figure 5.3 A galvanic corrosion cell formed on a steel coupon simulating a lamp post structure.

area with a good supply of oxygen (cathode). Metal oxidation and dissolution occur at the anodic areas, forming corrosion products and pits. A higher pH condition would be generated at the cathodic areas due to the oxygen reduction reaction, often leading to the formation of a passive film. In the case of the lamp post, the portion in contact with the air would act as the cathode, driving localized corrosion at the anodes that are located just below the ground level. This is similar to the waterline corrosion situation shown in Figure 1.5 that localized corrosion occurs at the anodes located just below the waterline. This differential aeration cell corrosion can occur on any metal structure where there is a difference in the concentration of dissolved oxygen between two points of the same metal surface. For instance, differential aeration cells can form at soil/water/air interfaces caused by crevices, lap joints, dirt, and debris. Under these conditions, the oxygen-starved areas are anodic, while the areas with better access to oxygen are cathodic. Such differential aeration cells are also very common on buried pipes where the portion in contact with the clay type soil usually acts as an anode, suffering preferential corrosion. The portion of the pipe in contact with the looser gravely soil is consequently the cathode due to better access to oxygen. When a pipe passes under paving, such as an airport runway or street, the portion under the paving would have less access to oxygen than does the area lying under unpaved soil and would corroded preferentially. For the same reason, a buried pipe usually corrodes preferentially at the bottom of a trench because the backfill in the trench is more permeable to oxygen (and the path is shorter) diffusing down from the surface, forming a differential aeration cell. In this case, the anode is often on the bottom surface of the pipe and the cathode is the rest of the surface.

iii) The existence of crevice between the lamp post and soil, as illustrated in Figure 5.3, would allow the initiation and propagation of crevice corrosion. The existence of crevice and oxygen concentration cell would lead to acidity changes and build-up of corrosive ions in the narrow crevice area. This could be somewhat similar to corrosion under disbonded coating described in Chapter 4.

These corrosion processes and mechanisms can interact with each other; for instance, oxygen concentration cell corrosion could lead to a relatively high pH condition at the crevice mouth area (at the interface of the air and the soil) where high oxygen content and high level of moisture existed (cathodic area), enhancing crevice corrosion further below soil surface where is obviously the anodic area of a crevice corrosion cell. The physically occluded region at the crevice area would enhance the localization of corrosion chemistry and reactions. The combination of these effects results in the most serious corrosion to occur approximately 3 cm below the damp soil surface.

For this lamp post, the main concern would be localized corrosion at the soil/air interface because it is the worst-case scenario condition affecting the integrity of the whole structure. The above ground section of the lamp post is usually protected by a galvanized layer or organic coating, and therefore atmospheric corrosion is less a concern for the structure. The main corrosion issue at the soil/air interface could be addressed in various ways. The simplest method is probably to change the design of the lamp post by extending the concrete casing above the ground level, as illustrated in Figure 5.4, in order to eliminate the multiple corrosion environments especially at the soil/air interface. This would effectively avoid the formation of a crevice and a differential aeration cell at the soil/air interface. Other methods to mitigate corrosion could include the use of a coating and waterproofing wrap to cover the corrosion sensitive soil/air interfacial section of the lamp post. The most commonly used coatings for the protection of such structure include chemically curing epoxies, polyurethanes, inorganic, and organic zinc-rich primers. Cathodic protection could also be used by placing a piece of zinc or magnesium as a sacrificial anode near the lamp post and connect to it, as

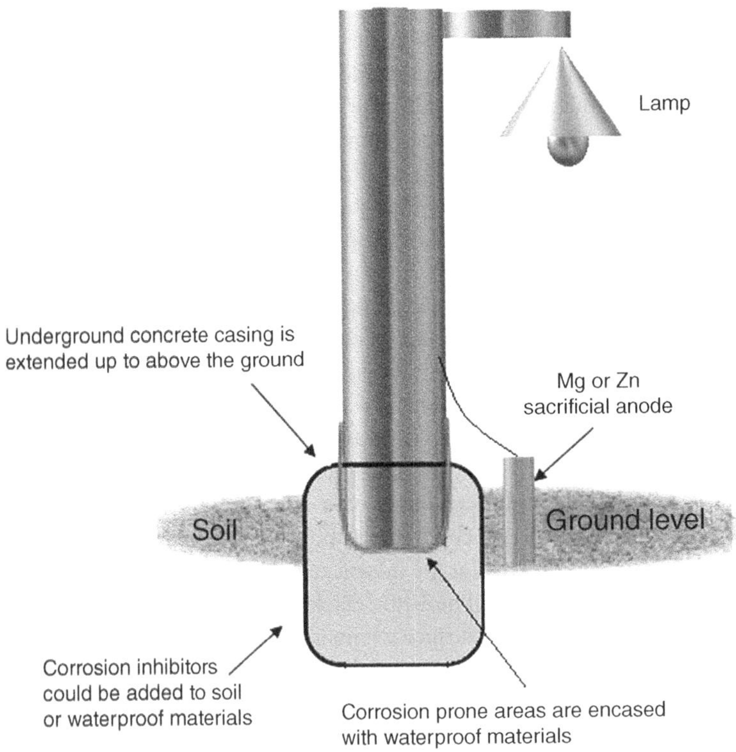

Figure 5.4 Illustrating several corrosion control methods for the lamp post.

illustrated in Figure 5.4. Adding corrosion inhibitors to the soil and coating/wrapping materials could also help mitigating corrosion. Corrosion monitoring could help verifying the efficiency of corrosion protection; however, a corrosion probe needs to be designed to effectively simulate and monitor the controlling localized corrosion process and mechanism, with particular focus on the worst-case scenario corrosion process and mechanisms at the soil/air interface.

The controlling localized corrosion process and mechanism can change when environments change and if corrosion prevention technologies and materials are applied to an engineering structure. A practical example is a steel wind turbine generator tower built on an extremely windy onshore site close to the sea, as shown in Figure 5.5, that passes through soil/air interface and is protected by surface coating and wrapping materials. In this case, the controlling localized corrosion mechanism is no longer crevice and oxygen concentration cell corrosion at the soil/air interface, instead localized corrosion and materials degradation issues are mainly associated with local coating damages and high humidity and airborne salt. The existence of a highly protective coating prevented the formation of a differential aeration cell and a crevice at the metal/coating/soil/air interface, therefore localized corrosion at the interface of soil and atmosphere becomes a less influential issue.

The wind turbine generator tower shown in Figure 5.5 is a hollow structure that constructed from a low carbon steel material and coated with a protective coating. Exposed to a marine environment, the air contains high levels of water and salt, often quickly causing general and localized forms of corrosion that could affect the structural durability of the tower. Key environmental factors that can affect corrosion in such marine environment include rainfall, moisture and salts in air, temperature, UV, fatigue and coating damage due to cyclic loading, coating's permeability of water, salt, air and biological organisms, etc. The tower shown in Figure 5.5 is exposed to three major environmental zones:

i) Marine atmospheric zone I: the external tower region that is exposed to windy and moisturized air. External surfaces of steel tower structures exposed in the marine atmospheric zone were protected by a coating of various thicknesses. Inspection of the tower generally shows that area where heavy corrosion occurred had thinner coating (260 vs 500 μm), while protected areas had thicker coating. Corrosion-resistant materials were applied for certain critical components of the tower, for example stainless steel was used for bolting and other fastening devices.
ii) Marine atmospheric zone II: the internal of the tower is indirectly exposed to moisturized marine atmosphere. There has been cases that rain water got into the internal tower from the gate. Internal surfaces of steel tower structures were also protected by a thin organic coating, as well as rustproof oils of various types.

(a)

Figure 5.5 (a) An onshore wind turbine tower that was built on an extremely windy onshore site close to the sea, and (b) a dismantled tower section.

iii) Soil zone: the buried tower region is exposed to damp soil. The buried steel surface of the tower was coated and then wrapped by a plastic layer. Inspection of the area was made by digging approximately 20 cm into the damp gravel/soil. As shown in Figure 5.5a, plastic wrapping materials and the coating were intact and no corrosion was visible from the dig-up locations. In this case differential aeration cell corrosion and crevice corrosion did not develop into corrosion controlling factors.

(b)

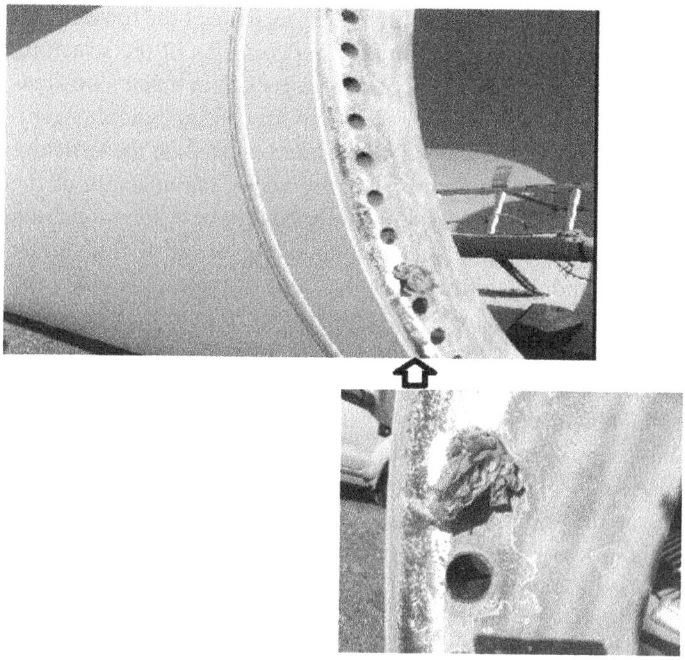

Figure 5.5 (Continued)

A major corrosion feature observed on the external surface of generator towers is that major rusting and coating damages occurred on the joint areas of tower sections, as indicated in Figure 5.5a. Significant coating damages and rusts are clearly visible on the joint areas of the tower where an uneven surface existed. More coating damages and rust spots are clearly observable on the side of the tower opposite to the sea, while much less coating damages were observed on the side facing the sea. This is believed to be due to a high salt concentration at locations opposite to the sea where condensation and evaporation of water with high airborne salt occurred, causing a highly aggressive and varying marine atmospheric corrosion environment. Please note that coating thicknesses at these locations with heavy rusts were usually low, typically less than 300 μm. More rusts and coating damages were also observed on the external surfaces of bolting, fastening devices. and other metallic components with crevices and sharp edges where the condensation and evaporation of water with high airborne salt occurred, causing an aggressive marine atmospheric corrosion environment. These areas were temporarily protected by a layer of greasy rustproofing materials or oil. Rusts were also clearly visible inside the crevice of the joint areas of the tower sections, as shown in Figure 5.5b (the tower was dismantled due to a mechanical damage), although these rusts may not cause a structural integrity issue of the tower.

Since coating was the most critical factor affecting corrosion of the structure, monitoring, repair, and maintenance of the coating, especially at major joint areas of the tower, would be necessary. Corrosion probes to simulate and monitor critical coating failure mechanisms should be designed and applied on the structure in order to provide data for informing maintenance needs. Maintenance of the tower should ensure the integrity of the coating by re-coating or by local touch-up coating. Coating thickness of 800 μm is recommended for overall coating repair areas. The performance and durability of coatings and rustproof oils may need performance verification by laboratory salt-spray testing. Due to the temporary protection nature of rustproof materials and the changing nature of corrosion processes, regular inspection and re-application of rustproofing materials could be necessary. It should be noted that when coating aging and degradation is significant, the buried section of steel tower, especially areas just below the damp soil surface with damaged coatings, would become very susceptible to corrosion in a similar manner as at the soil/air interfacial area of the lamp post shown in Figure 5.2. When the tower is aged, this soil/air interfacial area could become a critical location that should be regularly inspected or monitored. Similar to the lamp post case described above, probing different, especially the controlling, corrosion and materials degradation mechanisms over various periods would be necessary if effective structural health monitoring is desirable.

Similar analysis applies to many other engineering structures including the offshore wind turbines shown in Figure 5.1b, as well as solar farms and electrical power systems, although they may involve different materials exposed to different environmental conditions. The fact that localized corrosion is responsible for some 70–90% of all corrosion failures in industry, including types shown in Figure 5.1, suggests that it is a major issue affecting the reliability and durability of many engineering structures.

5.2 Probing Multi-Mechanism Corrosion Across Multiple Environments

The fact that the 'worst-case scenario' corrosion often occurs across different environmental zones in an engineering structure suggests the need for monitoring and management of corrosion processes and mechanisms over the whole structure exposed to different environments, especially at the interfaces of different environments. A prerequisite condition for effective management of multi-environmental corrosion is the availability of corrosion data from each zone that are required for corrosion modeling and prediction. The following sections discuss corrosion probe design and application for several practical and typical environment cases.

5.2.1 Probing Localized Corrosion Across Soil and Air Interface

An approach to avoiding unexpected failure of engineering structures is the use of corrosion probes for monitoring localized corrosion in multiple environments. This requires appropriate corrosion probe designed based on the understanding of corrosion mechanisms in each environmental zone, with special attention on the 'worst-case scenario' corrosion that often occurs in interfacial areas. Take the lamp post case shown in Figures 5.2–5.4 as an example, the fallen lamp post was checked three days before the incident but was failed to identify localized corroison a few centimeters below the gorund. On this structure, the 'worst-case scenario' corrosion occurred below the damp soil surface because this location had the most favorable corrosion conditions: differential aeration, high level of moisture and conductivity, and the existence of crevices between the lamp post and soil. Oxygen concentration cell effects and chemistry changes in the crevice area lead to significant electrochemical heterogeneity and rapid crevice corrosion at the location. This location was obviously the anodic area of a crevice corrosion cell where fastest corrosion dissolution reactions occurred. The cathodic area would be at the interface of the air and the soil where high oxygen content and high level of moisture existed. There was a significant galvanic corrosion effect between anodic and cathodic areas, leading to localized corrosion just below the damp soil surface. An effective corrosion monitoring probe would certainly need to simulate and monitor all these features causing this 'worst-case scenario' corrosion crossing the soil and atomsphere interface.

Figure 5.6a illustrates a corrosion monitoring system using metal coupons or ER probes. It is clear that metal coupons or ER probes placed above (coupon A) and below (coupon C) the ground level would only indicate the corrosivity of the atmosphere and the soil. They would not provide information about corrosion that caused the lamp post incident, because the critical crevice and galvanic effects and the differential aeration environment that are critical to the corrosion of the lamp post are not simulated on the steel coupons. Although the coupon B that is placed both in and above the ground could be able to correctly simulate the lamp post corrosion, the inspection of the coupon would need to withdraw the coupon from the soil and this could significantly change the corrosion on the coupon. Figure 5.6b illustrate a corrosion monitoring system using electrochemical probes. Various electrochemical methods including corrosion potential measurement, potentiodynamic polarization, linear polarization resistance (LPR), electrochemical impedance spectroscopy (EIS), and electrochemical noise (ENA) could be applied to acquire and analyze corrosion data from electrochemical probes. (More details on corrosion probes developed based on these electrochemical methods can be found in Section 2.3.2 of Chapter 2.) Although electrochemical techniques such as LPR, EIS, and ENA have been widely used in monitoring and estimating the rates of uniform corrosion in general corrosion systems, they have major

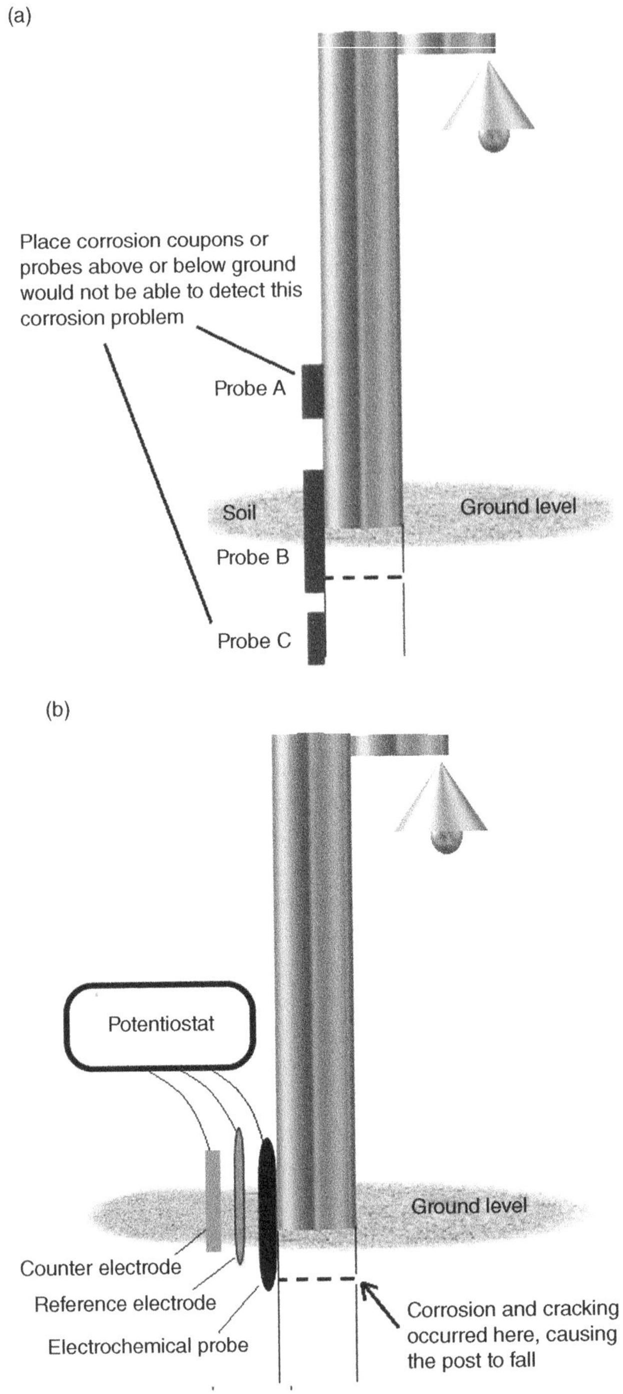

Figure 5.6 Illustrating a lamp post corrosion monitoring system using various metal coupons and probes, (a) using metal coupons or ER probes; (b) using electrochemical probes; and (c) using an integrated multi-probe system.

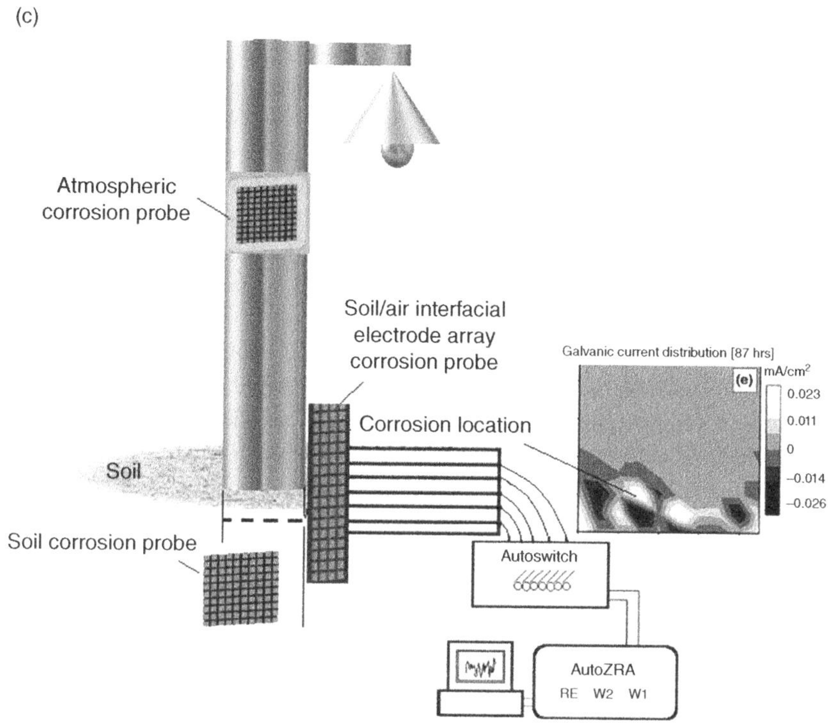

Figure 5.6 (Continued)

limitations in measuring localized corrosion kinetics and distributions [2, 3]. In the case of the three-electrode corrosion measurement system shown in Figure 5.6b, it could only be able to detect a corrosion potential or a corrosion rate that is a mixture of contributions from many local potentials or corrosion rates, none of which we can use to evaluate localized corrosion at the soil/air interface of the lamp post. Obviously, probes designed based on conventional electrochemical techniques have major technological limitations in measuring localized forms of corrosion. This is especially true for electrochemical polarization measurements in highly resistive conditions, such as in soil, that often cause a huge potential drop commonly referred to as IR drop, causing significant corrosion rate measurement errors [3, 4].

Many technological innovations have been made over the past decade to address this issue with reports scattered throughout the literature. The use of a corrosion probe designed based on an electrochemically integrated multi-electrode array [2–5], often referred to as the wire beam electrode (WBE), has been proposed as a promising solution to this challenge. Two important characteristics of the WBE method that are particularly valuable for the monitoring of corrosion in complex environmental conditions are (i) the WBE is applicable to high-resistance multi-phase environment, which has been demonstrated in previous studies, therefore it has the potential for application in different environmental conditions with varying resistivities [5]; (ii) the WBE can map corrosion processes on an instantaneous and continuous basis, therefore it has the capability of monitoring changing corrosion processes. In order to achieve the highest possible sensitivity for detecting early signs of localized corrosion, the WBE was used in conjunction with electrochemical noise analysis since ENA has the advantage of providing early warning of localized corrosion initiation [3]. As shown in Figure 5.6b, a WBE probe is placed both in and above the ground in order to simulate and probe the lamp post corrosion mechanism and processes across the soil/air interface. The ability of the WBE probe in measuring galvanic current distribution maps in soil has been demonstrated during the exposure of the WBE to dry, damp, and chlorinated sand environments [5]. The probe was able to detect characteristic corrosion at the location below the sand/air interface, similar to the corrosion behavior occurred on lamp posts, in galvanic current distribution maps of the type shown in Figure 5.6b [5].

For an engineering structure across several different environmental zones where different corrosion processes and mechanisms could occur, several corrosion probes could be needed to simulate and probe the controlling form of corrosion process in each environmental zone. As shown in Figure 5.6c, three corrosion probes are used to monitoring corrosion in atmospheric, soil, and soil/air interfacial zones, respectively. These probes can be electrically coupled together to allow and detect the interactions of corrosion over different zones over a continuous metal structural surface. It should be noted, however, in industry practice the main concern could be the monitoring and control of the 'worst-case scenario' corrosion at the soil/air interface and therefore the atmospheric corrosion and soil corrosion probes in Figure 5.6c may not be needed in practice.

5.2.2 Probing Localized Corrosion Across Water/Air and Soil/Water Interfaces

Similar approaches can be taken for measuring localized corrosion crossing the interfaces of water/air and soil/water. For instance, probes designed based on electrode arrays have also been used to measure corrosion across water/air interface, in particular to study waterline corrosion [6–9] to overcome issues with traditional methods of using a long steel coupon with part of it immersed in the seawater to simulate a steel piling [8, 10]. An electrode array similar to the corrosion probe shown in Figure 5.5c has been demonstrated to be able to measure localized corrosion at soil/water interface [11, 12]. In a typical experiment, the corrosion behavior of X70 pipeline steel in the tropical juncture area of seawater–sea mud was studied by means of the WBE and other electrochemical techniques [11]. The results showed that X70 pipeline steel formed an oxygen concentration cell in the juncture area of seawater–sea mud, with electrodes in the sea mud and in the vicinity of interface of seawater–sea mud acting as anode and those in the seawater acting as cathode. In the last stage of corrosion monitoring, the electrodes in the bottom of sea mud changed into cathodes where the main cathode reaction occurred. The corrosion rates of the electrodes in the seawater were greater than those of the electrodes in the sea mud, and there was a peak of corrosion current in the vicinity of interface of seawater–sea mud. Sulfate-reducing bacteria at the bottom of the sea mud was considered to participate in the reaction, producing iron sulfide [11]. Further test found that serious corrosion of carbon steel in the vicinity of the interface sea-water/sea-mud was controlled by dissolved oxygen concentration and the sea-mud impedance [12]. Similar tests were conducted by burying electrodes in the tropical coastal red soil to study the macro nonuniform corrosion behavior of carbon steel [13]. The nonuniform corrosion of metal was found to be easy to occur in soils due to their heterogeneity. The occurrence of corrosion of carbon steel was found to be dependent upon the length of buried time and the corresponding depth of the spot on the buried samples. The behavior was explained by the change of soil moisture with the changing rainfall within the buried testing period, and the variation of the moisture content, oxygen content, and composition of soil with the depth of soil [13].

5.2.3 Probing Localized Corrosion of Steel in Concrete

Nonuniform environments can also develop in concrete exposed to corrosive media such as seawater because of differences in chloride concentrations and nonuniform distributions. Concrete is somewhat similar to soil in resistivity and moisture conditions, but dissimilar to soil in chemistry, especially the pH. Significant work has been carried out to use electrode arrays for measuring and understanding steel rebar corrosion in heterogeneous concrete. Several corrosion probes designs for simulating and detecting localized corrosion of steel rebar in concrete structures have been described previously in Figures 2.5 and 3.2 of this book. In an experiment reported in reference [14] for understanding the effects of thickness and compactness of concrete on rebar corrosion, it was found that the thicker and denser the protective layer of the concrete, the higher the natural corrosion potential of the electrode array probe (i.e. the WBE probe), and the higher the pH value and the lower the concentration of chloride [14]. The corrosion potential shifts toward negative with the increase of immersion time, while the polarization resistance reduced and the difference of corrosion potential increased. With the increase in the embedded depth of iron electrode wires, the corrosion rates reduced and polarization resistance increased [15]. All these results are in good agreement with known behavior of steel rebar in concrete. This suggests the usefulness of the WBE probe in monitoring corrosion in concrete structures.

The electrode array method was later used to perform more detailed studies of steel rebar corrosion mechanisms and influence factors. An interesting application of the electrode array was to use the electrode array for studying the interaction between macrocell and microcell in the early corrosion process of reinforcing steel in cement mortar [16]. The corrosion potential and galvanic current of macrocell corrosion of the reinforcing steel in cement mortar were visualized by the electrode array technique during the corrosion initiation and propagation processes. It was found that the corrosion macrocell current was closely related with the difference of corrosion potential between the anodic and cathodic areas. New insights into the corrosion mechanisms of steel in concrete were obtained from the characterization of steel corrosion in concrete at the macro- and micro- levels using the electrode array in mortar exposed to periodic wetting and drying, in conjunction with steel corrosion damage assessment by micro X-ray computer tomography (CT) and scanning electron microscopy (SEM) [17]. Uneven distributions of galvanic current and open-circuit potential (OCP) on the WBE-embedded mortar sample immersed in NaCl solution was observed, and oxygen depletion in mortar was found to facilitate the negative shift of the OCP and the smoothing of the current and potential distributions [18].

Electrode arrays with different geometries and dimensions were designed for simulating and studying steel rebar corrosion in different reinforced concrete (RC) structures exposed to a wide variety of environmental conditions that may cause various damages to the RC structures. A concrete embedding rebar electrode array was designed and made based on the WBE concept to simulate the submerged, tidal, and splash marine environments and to monitor the carbonation and chloride ion penetration of concrete, the pH and chloride ion profiles of the concretes [19]. The vertical non-uniform corrosion of steel rebar was found to be caused by the difference of oxygen concentration, moisture content, free chloride content, and pH value in the vertical direction. The macroscopic galvanic corrosion between the splash zone rebars and the tidal zone rebars was inhibited and the galvanic corrosion between the adjacent rebars was promoted by the high resistivity of the splash concrete [19]. The electrode array has been a part of integrated sensor unit for assessing the corrosion status of reinforcing steel rebar in concrete, including an Ag/AgCl reference electrode probe, a metal oxide probe, a multi-electrode array sensor, and a four-pin (Wenner) array stainless steel electrode for chloride content, pH, microcell corrosion current, and localized concrete resistivity measurements, respectively [20].

Electrode arrays were further used to evaluate inhibitors and coatings for protecting steel in concrete. A WBE was used to assess the inhibition of reinforcing bar corrosion in concrete with the addition of different inhibitors such as $NaNO_2$, $NaNO_2+Na_2CO_3$, and $NaNO_2$+aniline [21]. By measuring the open-circuit potential of electrode wires, the impact of different inhibitors on the corrosion of rebar in the concretes was studied after immersion in 3% NaCl solution. It was found that best protection was achieved by adding $NaNO_2$ and aniline as combined inhibitors [21]. Electrode arrays were also used to evaluate the migration and inhibitive effect on rebar corrosion of migrating corrosion inhibitors (MCIs) in mortar. The MCIs were found to readily diffused through the mortar and increased the resistance to general corrosion much more than resistance to localized corrosion. This increase was mainly attributed to the sealing effect on mortar pores (pore-blocking effect) rather than on chemical absorption [22]. Multi-electrode array sensors fabricated from carbon steel, 304 stainless steel, and 1100 aluminum were utilized to assess the effectiveness of two corrosion inhibitors and one coating under cyclic deicer exposure conditions [23]. Both the sensor's corrosion rates and mass loss results indicated that the inhibitors had a much greater inhibiting effect in protecting the metallic substrates during the cyclic exposure to $MgCl_2$ than to NaCl containing corrosive solutions [23]. The effect of concrete coating on corrosion of reinforced rebar in concrete was also studied by the electrode array probe [24]. It was found that coating the surface of concrete could keep the open-circuit potential of steel rebar within the passivation range, ensuring the integrity of the passivation membrane [24]. All these reports suggest the capabilities of the electrode arrays for probing dynamic and localized electrode processes occurring on steel buried in highly inhomogeneous and resistive soils and concrete structures.

5.2.4 Probing Localized Corrosion in Atmospheric Conditions

Although the electrode array has been demonstrated to be suitable for heterogeneous and highly resistive soil and concrete environments, another challenge is to design corrosion probes for monitoring atmospheric corrosion over the atmospheric zones such as that in Figure 5.6c. Atmospheric corrosion is probably the most common highly resistive and inhomogeneous corrosion that is often difficult to measure by conventional electrochemical means. Atmospheric corrosion often occurs when moisture is absorbed or trapped along with airborne contaminants such as salts, ionic gas (exhaust gases, SO_2, NO etc.) and organic matters, resulting in corrosion damage to structures exposed to atmosphere such as aircrafts, occasionally in a catastrophic manner. In the aircraft industry, for instance, most corrosion damages in aircraft structures are hidden and concentrate on difficult-to-access areas such as avionics bay and boxes, fastener holes, butt-joints, leading and trailing edges of wings, configurational and structural overlaps, structural voids, access panels, fill holes, and hidden spaces. These 'hidden' areas can develop different environmental zones where corrosion problems are hard to detect through the conventional method of corrosion management (i.e. periodic time based routine inspections) since inspection of these hidden areas requires major and costly tear-down of aircraft, which is not practical in most cases. In order to avoid undetected extensive corrosion damage and to save unnecessary tear-down of aircraft, simple, reliable, and cost-effective corrosion detection and monitoring techniques are needed. Indeed, suitable corrosion detection and monitoring technique could provide aircraft 'system health' information and forecast maintenance requirement, and thus to serve as a major corrosion management tool. Sensors such as electric sensor have also been developed to monitor aircraft corrosion. For instance, a bimetallic sensor was designed using the principle of galvanic corrosion [25]. This sensor employs two dissimilar metals to expose to the environment but kept isolated from each other. When the two metals are short-circuited through a zero resistance ammeter (ZRA), they become galvanically coupled and a current is developed. This galvanic current is measured by the ZRA and is used as an indicator of environmental corrosivity. A major limitation for this method is apparent considering the fact that this method is based only on an empirical correlation between the galvanic current and the conductivity of condensed electrolyte film and its thickness. In general, this sensor can only provide a general estimation of electrolyte conductivity, it does not detect localized corrosion. Organic paint has also been used as a corrosion sensor to detect corrosion as an indicative coating system that changes color when corrosion occurs in an aircraft [26]. This method may only detect corrosion on accessible and visible areas and may not be able to detect localized aircraft 'hidden' corrosion over difficult-to-access aircraft structures.

Aircraft corrosion is the result of electrochemical activities occurring at the interface between aircraft structure and a thin electrolyte layer absorbed or trapped on aircraft surface. Although this aqueous layer is usually very thin (on the order of tens of angstroms), it can form on metals at less than 100% relative humidity and it can be quite corrosive, especially if the solution is contaminated with atmospheric pollutants. In principle, electrochemical methods should be suitable methods for measuring aircraft corrosion considering the electrochemical characteristics of corrosion. Indeed, electrochemical corrosion sensors have been attempted using methods such as the linear polarization method [27] and electrochemical impedance spectroscope [28]. However as described before [2–4], these conventional electrochemical techniques have major difficulties in measuring localized corrosion – the main problem in the aircraft industry. A corrosion probe developed based on the electrode array WBE probe has been developed for measuring and monitoring corrosion under such thin aqueous layer [4]. To demonstrate this corrosion probe designed based on the WBE concept, experiments were carried out for studying atmospheric corrosion under a small water drop [29] or a thin layer of electrolyte [30]. Figure 5.7 shows a typical experiment that was designed to study the effects of the movement of corrosive species on the process, rates, and pattern of corrosion under a thin layer of water, simulating atmospheric corrosion in moist air [30]. The WBE probe working surface was

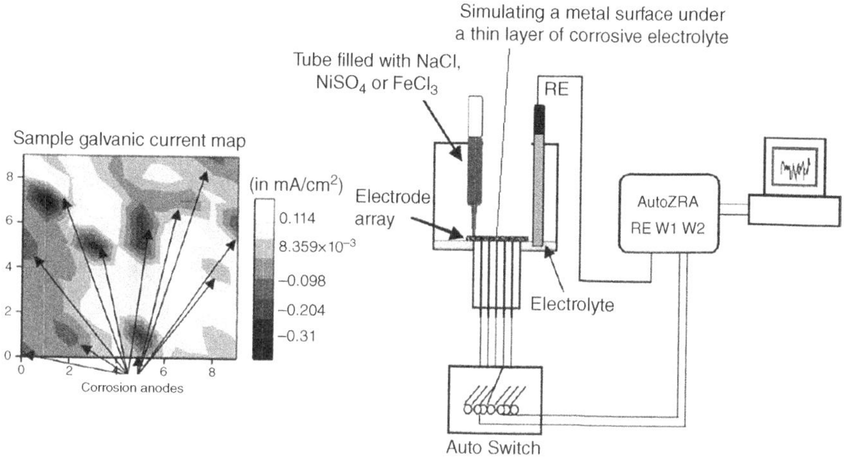

Figure 5.7 Experimental design for measuring electrochemical parameters from the WBE surface under a simulated atmospheric corrosion environment. *Source:* Aung et al. [30]/ With permission of Elsevier.

exposed to simulated atmospheric thin water layer where there were diffusion induced ion concentration gradients. Corrosion potential and current distribution maps were measured from WBE surfaces on a continuous basis. Typical patterns have been identified from these maps and the characteristics of these patterns have been found to depend heavily upon the type of electrode materials and the type of corrosive ions. For mild steel WBE surface exposed to a diffusion-corrosion environment containing $NiSO_4$ or $FeCl_3$, the characteristic pattern in maps shown in Figure 5.8a is found to emulate $NiSO_4$ or $FeCl_3$ concentration gradients, suggesting an ion-concentration-controlled corrosion behavior. When a stainless steel (SS316L) WBE surface was exposed to a diffusion-corrosion environment containing $NiSO_4$ or NaCl, significantly different features were observable. As shown in Figure 5.8b, corrosion cathodes and anodes appeared to be distributed randomly and, more importantly, these sites remained almost unchanged with the extension of diffusion-corrosion experiments. The characteristic concentration-controlled corrosion patterns shown in Figure 5.8a are totally absent in Figure 5.8b. The most probable explanation to this phenomenon is that SS316L corrosion was controlled by the nonuniformity of the passive film. It is well known that the corrosion resistance of stainless steel depends on the integrity and durability of its passive film. If passivity is not maintained and the passive film has localized weak sites or damages, these locations would become anodes where highly localized corrosion attack could occur [30]. This demonstrates the recognition and analysis of characteristic maps from WBE measurements for probing corrosion processes and mechanisms occurring under atmospheric thin water layers.

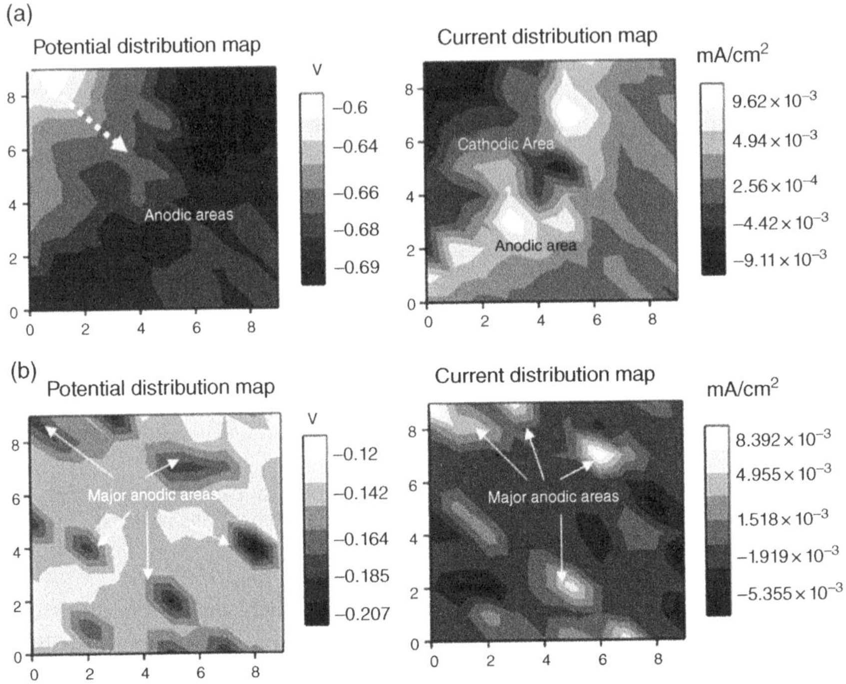

Figure 5.8 Corrosion potential (in V) and galvanic current (in mA/cm^2) distribution maps from (a) mild steel WBE and (b) SS 316L WBE exposed to a thin layer of distilled water after $NiSO_4$ diffusion for one hour. *Source:* Aung et al. [30]/ With permission of Elsevier.

In similar approaches, the corrosion behavior of mild steel and copper in thin salt solution film was studied by WBEs made of mild steel and copper wires that allowed measurements of corrosion potentials and their heterogeneous distribution in the thin salt solution film [31]. Localized corrosion of 304 stainless steel under droplets of 1 M sodium chloride solution was investigated by the WBE method, showing the stochastic characteristics of the corrosion process with the disappearance of some anodic sites. Stainless steel suffered more serious localized corrosion with the increase of the droplet size probably due to increase in the cathodic area and the three-phase boundary length [32]. Similar investigation was performed to measure the potential/current distributions of carbon steel under 0.001 M sodium chloride droplets of various sizes using a WBE [33]. During the corrosion process, carbon steel exhibited heterogeneous electrochemical characteristics with the major anodes located near the edge of the droplets. The three-phase boundary was considered to play a vital role in determining the electrochemical distribution and the degree of carbon steel corrosion. With the increase of droplet size, the average anodic current density and the corrosion degree decreased, which is believed to be due to the decrease of the three-phase boundary length per unit area [33]. The corrosion behavior of copper beneath a droplet of 0.6 M NaCl solution was studied by means of the WBE in combination with electrochemical impedance spectroscopy [34]. The results show that the distribution of galvanic current remained negative at the center and positive at the periphery of copper surface covered by the droplet at the initial stage, and the polarity reversed with the extension of time. The change of the local impedances was found to be consistent with that of the current distributions. It was found that the formation and transformation of the corrosion products dominated the changes of the distribution of the corrosion current [34].

Electrode array measurements were also made to investigate steel corrosion beneath electrolyte thin films of 0.01 M NaCl solution that was used to simulate marine atmospheric corrosion conditions [35]. The results show that atmospheric corrosion of steel was initiated around local pearlites before becoming 'uniform' corrosion afterward. According to the WBE measurements, the mean corrosion potential and the standard deviation of corrosion potential decreased, and the anode area enlarged with the exposure time [35]. The effects of water layer thickness on corrosion rates was quantified by exposing carbon steel WBEs under seawater film with various thickness [36]. It was found that the corrosion rate of carbon steel increased significantly under thin seawater film than that immersed in seawater. The current variation under seawater film indicated that the thickness of diffusion layer of oxygen was about 500 μm, and the maximal current appeared around 40 μm at which corrosion rate transited from cathodic control to anodic control [36]. The corrosion and protection of iron exposed to a wet and dry cycling corrosion test with a drop of sodium chloride solution was studied using an iron WBE by measuring the spatial distribution of coupling current and anodic/cathodic coupling charge during the progress of corrosion on an iron surface [37]. A bare WBE specimen showed transition of a coupling current map depending on the relative humidity in a short time span and transition from the induction, progress, and inhibition of corrosion due to rust accumulation on the surface in a long time span. The specimen coated with acrylic paint showed formation of coupling between electrodes under the paint before paint breaching and successive progress of the coupling distribution around the breaching point. When a scratch was made in the paint, progress of the coupling distribution along the scratch was observed [37]. The effects of short-term storage on atmospheric corrosion of copper was similarly studied using the WBE and other electrode methods [38]. It was found that after storage, the corrosion activity as well as the overall average corrosion intensity of copper WBE were reduced, but the local corrosion intensity enhanced, and the corrosion product film inhibited both the cathodic and anodic process [38].

The WBE method was also used to evaluate the corrosion behavior of a steel in acidic atmosphere with different partial pressures of CO_2. The results showed that corrosion began from the edge of droplet in dew point corrosion and gradually spread to the center of it. Corrosion current density at the edge of droplet was found to form the "cathode-anode-cathode ring" that disappeared gradually as the corrosion exposure time extended. These results indicated that the dew point corrosion reaction gradually transferred to the 'uniform' corrosion [39]. An interesting application of the electrode arrays was for probing CO_2 corrosion in the water film on the wall of large gas pipe [40]. The total anodic current density decreased slightly and then became stabilized as water film thickness increased from 648 to 2500 μm. Maximal anodic current density was observed when the water film thickness was at 350 μm. The distributions of large cathodic regions and small anodic regions were found. When the water film thickness was less than 350 μm, $FeCO_3$ was easily formed and anodic current density decreased [40]. These reports show numerous innovative applications of variously designed electrode arrays for measuring corrosion in simulated industrial and marine atmospheric conditions.

Typical experiments described above suggest that the WBE probe could be utilized to measure electrochemical corrosion in multi-environmental and difficult-to-probe corrosion conditions such as multiphase oil and gas mixtures, sand, thin water layer, under thermal insulation and under-deposit. In all these experiments, there is no need to apply an externally imposed polarization and there is no need of a counter electrode, and thus it avoids problems such as nonuniform polarization currents and IR drops associated with conventional polarization measurement. However, a conventional design of the WBE probe such as the one shown in Figure 5.7 would be difficult to be used in practice because of the difficulties of attaching it on the surface of an industrial structure. In order to address this issue, an aircraft corrosion probe has been designed, as shown in Figure 5.9, that can be attached/glued on selected surface areas of an aircraft structure, either directly exposing to corrosion environment or embedded under coatings and sealant. This probe can be a centimeter-sized device with several differently designed shapes in order to fit the uneven aircraft surface. With all the 100 terminals in the array connected together, the WBE surface forms an integrated electrochemical system that effectively should simulate the electrochemical corrosion behavior of an one-piece industrial surface. The WBE can also be integrated to the whole aircraft by coupling it to the structure. If installed properly, this sensor could simulate and monitor major types of localized corrosion including crevice corrosion, pitting corrosion, and galvanic corrosion. When embedded under coatings, the probe can measure weak areas or deteriorated areas in an organic coating film. Corrosion monitoring data obtained from such a probe would provide insight into factors controlling coating degradation on aircraft, as well as factors influencing crevice corrosion rates. Electrochemical measurements can be carried out using a portable instrument that is connectable to the sensor via a computerized auto-switch, and data could be acquired remotely and stored in a data base for analysis and modeling purposes.

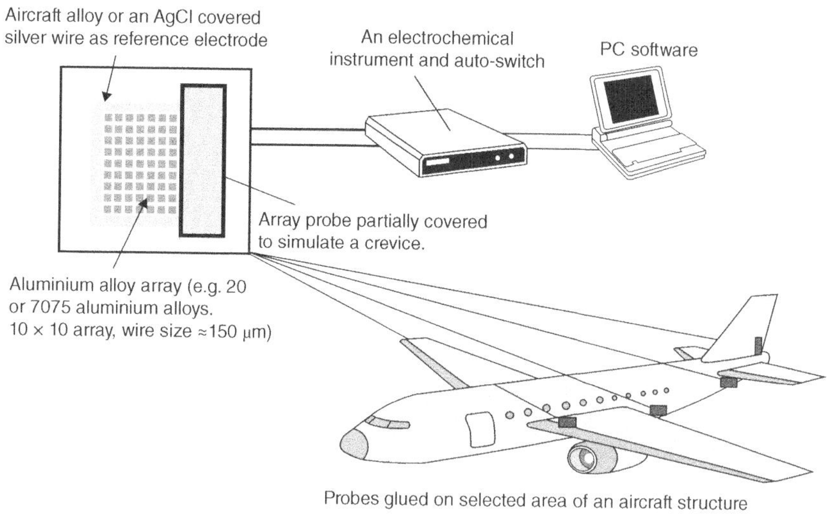

Figure 5.9 An aircraft corrosion probe that can be conveniently installed/glued on selected surface areas of an aircraft structure.

5.3 Cases of Probing Localized Corrosion Over Multiple Environments

5.3.1 Probing Localized Corrosion in Marine Environments

The ocean is a complex and harsh corrosion environment for major engineering structures such as oil and gas platforms, solar and wind farms, as well as pipelines that are partially or totally submerged in seawater. Figure 5.10a illustrates an industry case of an electrode array corrosion probe designed for simulating and monitoring the co-existence of three types of critical corrosion on a pipeline (see Figure 5.10b): pitting corrosion at welds along segments of pipes, crevice corrosion at bolts on flanges between pipe sections, and crevice corrosion between flanges bolted together. The probe contains sections simulating weldment corrosion and crevice corrosion for field testing by exposing corrosion probes in the ocean, as illustrated in Figure 5.10c. *In situ* electrochemical monitoring of the initiation, propagation, and mechanism of localized forms of metal corrosion under simulated oil and gas infrastructure conditions can be conducted to probe localized corrosion under the effect of various factors such as microbiological activities, biofouling, coating degradation, and cathodic protection. Data from both the literature and corrosion monitoring experiments can be used to fit to corrosion models, generating corrosion trend diagrams showing long-term seawater corrosion under the effect of major marine environmental conditions, producing

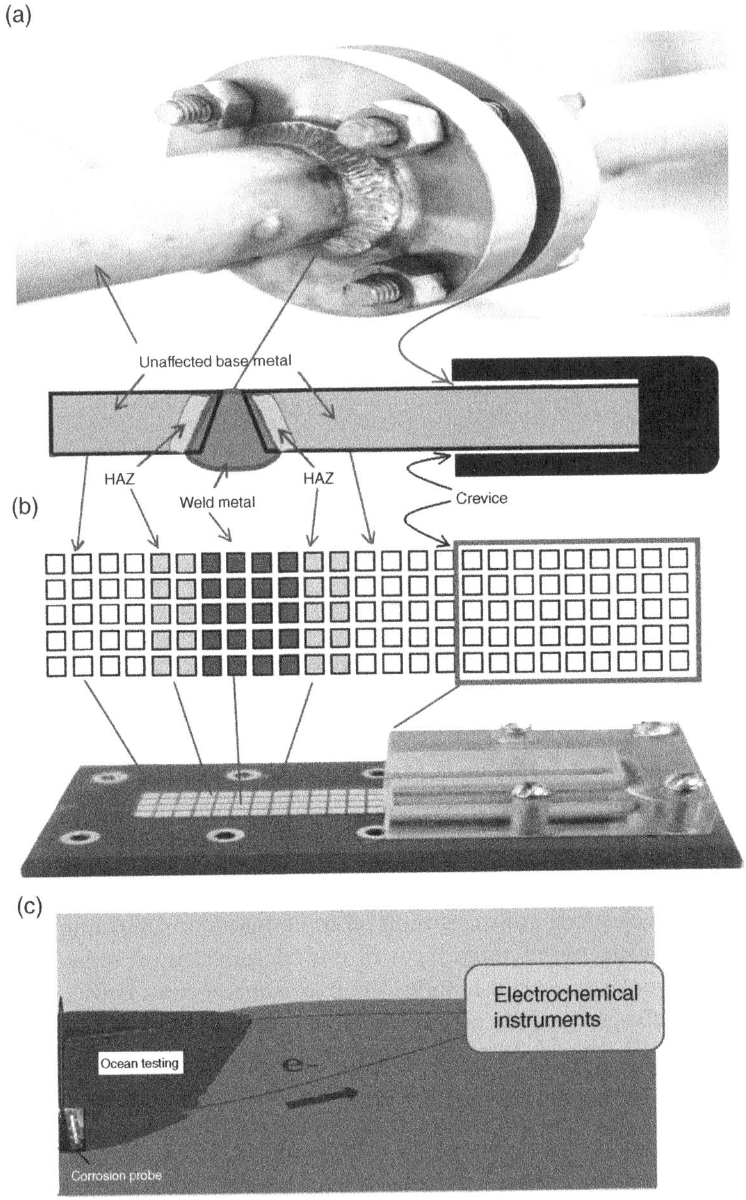

Figure 5.10 (a) Shows the corrosion at different regions of a pipeline. The region in yellow rectangle indicates the co-existence of three types of localized corrosion, i.e. pitting corrosion at welds along spools, crevice corrosion at bolts on flanges between spools, and crevice corrosion between flanges bolted together. (b) The schematic representation of the designed electrode array probe to simulate the industry conditions shown in (a). (c) Illustration of field testing by exposing corrosion probes in the ocean.

a practical tool for quantifying the long-term trends and rates of degradation of materials and corrosion of metallic infrastructural components.

Although pitting and crevice corrosion are commonly observed on marine structures, most serious corrosion often occurs around the waterline of maritime environments [6–12, 41, 42]. As ships and steel sheet piling are always partly immersed in the sea water, the monitoring and understanding of the waterline corrosion is important for managing and protecting these marine structures. Waterline corrosion has been studied for a long time [43] and is believed to be due to localized corrosion cell driven by large potential differences between areas above the waterline that are highly oxygenated (with a higher potential) and the area below the waterline, which is less oxygenated [7]. Although it has been repeatedly verified that differential aeration cell effect is the key factor affecting corrosion at the waterline area, practical corrosion behavior of the steel in the dynamically changing maritime environment could be rather different [8, 10, 44]. It is generally observed in industry that serious corrosion happens at locations approximately 250 mm above the waterline and 100–200 mm below the waterline. The high corrosion rate above the waterline is believed to be induced by the splashing of the wave [8], while the serious corrosion below the waterline is considered to be caused by both the oxygen cell and microbiologically influenced corrosion (MIC) [44]. Different corrosion behaviors of the steel piling have been reported around the various harbors due to the complex ocean conditions [44]. The multiple environmental zones existed on metal structures across waterline and complex influence factors such as bacteria and water movement suggest a need to monitor corrosion conditions and rates, especially at the 'worst-case scenario' corrosion zone at the waterline area. The conventional methods used for the study of the waterline corrosion in the sea is using a long steel coupon with part of it immersed in the sea water to simulate the steel piling [8, 10, 44]. After a long term of immersion, the coupon is taken out and the waterline corrosion behavior can be studied by visual inspecting, weight loss measurement, and microscope observation. The anodic area and the cathodic area can be roughly identified from the local weight loss of the test sample; however, the important real-time information cannot be obtained based on these traditional methods. As the transition of the anodic area and the cathodic area around the waterline is a dynamic process, there is a need to determine the anodic and cathodic current distributions in order to understand the initiation and propagation of the waterline corrosion. Probes designed based on electrode arrays have also been used to measure corrosion across water/air interface, in particular to study waterline corrosion [6–9].

Over the past two decades, the multi-electrode array probe (i.e. WBE probe) has been widely used in the study of the steel corrosion under the marine environment [6–12, 41, 42]. Using corrosion probes designed based on the WBE, the anodic area and the cathodic area can be well distinguished over the array with 100 electrodes [6, 7]. The large potential difference between the different areas was found to cause serious galvanic corrosion currents flowing between distinctly separated anodic and cathodic areas on the metal surface [6, 7], often inducing the failure of the marine structures such as piling and platforms [8]. It is generally found that in the static electrolyte, the anodic sites are concentrated at areas below the waterline after a few days of immersion, while the addition of the corrosion inhibitor such as the sodium carbonate can make the anodic area moving toward the waterline area [6–12]. In recent studies, it was also found that the anodic and cathodic distributions on the WBE have a significant influence on corrosion in the seawater containing with or without bacteria [45]. Figure 5.11 shows two probes designed for simulating and monitoring waterline corrosion. Figure 5.11a shows a sensor system developed based on an integrated 10 elements electrode array in conjunction with an electrical resistance (ER) method [41] to simulate and measure waterline corrosion in both static and flowing salt water. The combined use of the WBE and ER probes allowed simultaneous measurements of galvanic currents flowing among the multi-electrodes and the total metal loss caused ER changes, without the needs for the complication of electrochemical noise measurements [46]. It was found that in the static 3% NaCl aqueous condition, the major anodic area was located below the waterline, suggesting the waterline corrosion is controlled by the oxygen differential cell. However, in the flowing condition, the major anodic areas were found to be close to the waterline and the bottom region behaved as a cathodic area, suggesting that the flowing of the electrolyte is an important influence factor of the waterline corrosion [41].

Figure 5.12 shows typical results of galvanic currents measured during 16 days of exposure test in the static 3% NaCl aqueous solution [41]. It is seen that at the beginning, as shown in Figure 5.12a, the major cathodic area was concentrated on the waterline level. The maximum anodic current appeared below the waterline that was close to the major cathodic areas, and gradually moved to the bottom in the following test periods, as shown in Figure 5.12b. The galvanic current distributions in the flowing salt water, as shown in Figures 5.12c and d, are quite different from that in the static solution condition. The electrode located at the bottom behaved as a major cathode that could be due to flow-induced oxygen transformation to that location. The major anodic areas were concentrated immediately below the waterline and above the bottom area of the corrosion probe. This phenomenon suggests that the flowing of the solution contributed to the locations of the major anodic and cathodic sites. The corrosion depths were measured by ER method, as shown in Figure 5.13 [41]. The area with the highest corrosion depths (59 μm) was corresponding to its highest anodic current shown in Figure 5.12. It suggests that the galvanic corrosion was the main cause for the metal loss for the waterline corrosion, and corrosion within each electrode (internal corrosion) was

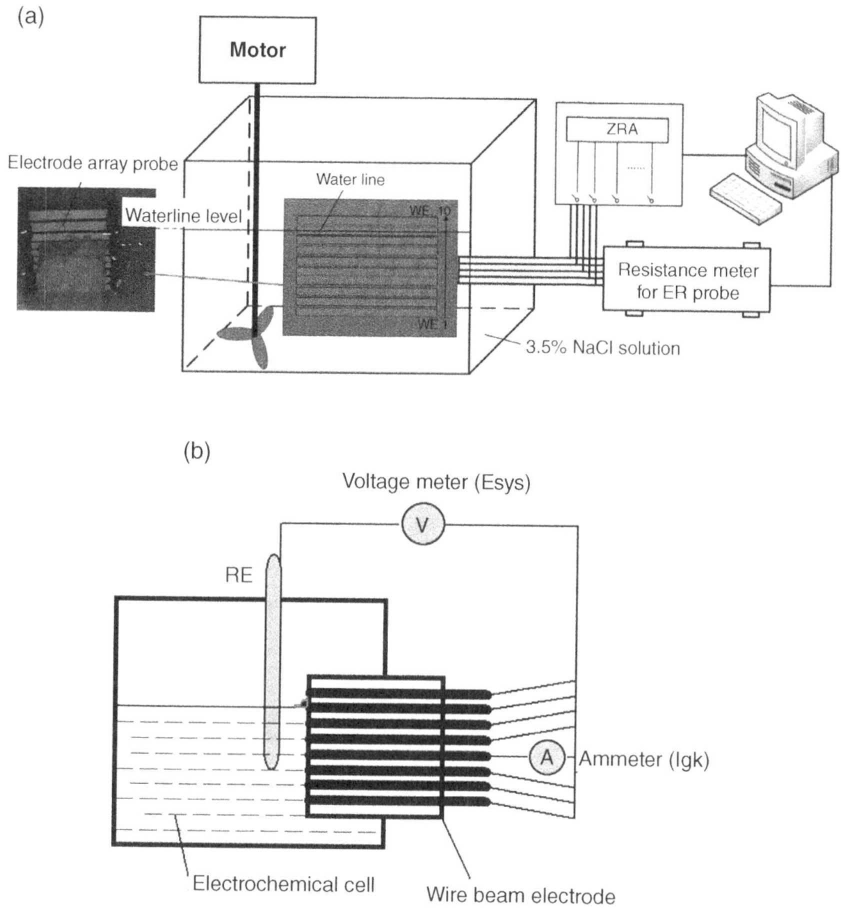

Figure 5.11 A schematic diagram showing corrosion probes designed for simulating and monitoring waterline corrosion. (a) The line sensor for ER and galvanic current measurement. *Source:* Xu et al. [41], Australasian Corrosion Association. (b) Galvanic current and the system corrosion potential measurements using a WBE (partial immersion). *Source:* Tan [7], Allen Press, CC BY.

minor in this highly localized corrosion system. The comparison of the corrosion depths calculated from the galvanic current and the ER method are presented in Figure 5.14a and 5.14b [41]. It is also clear that the corrosion depth measured by the ER method is generally larger than the corrosion depth calculated from the galvanic current. It suggests that although the galvanic corrosion was the main cause for the metal loss for the waterline corrosion, and corrosion within each electrode (internal corrosion) also contributed to corrosion. Since the 10 element electrode array used relatively large electrodes, relatively large internal currents

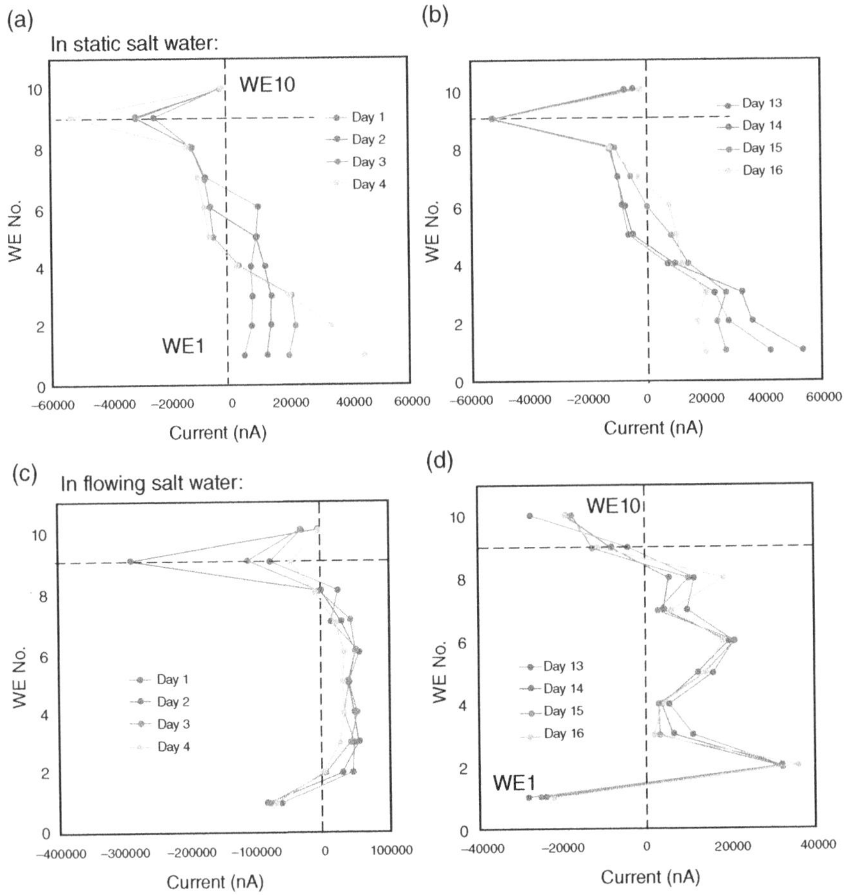

Figure 5.12 The galvanic current distributions (a, b) in the static sate water and (c, d) in flowing salt water. *Source:* Xu et al. [41]/ Australasian Corrosion Association.

existed on the electrodes, leading to the underestimate of the corrosion depth by the WBE probe. Because the corrosion depth measured by ER method is according to the geometrical information of the probe, the total metal losses included contributions form all internal and galvanic anodic corrosion currents. It is why that the corrosion depth measured by the ER method was higher than that calculated from the anodic currents. In this case, the combination of both galvanic current and ER measurements can provide more accurate information for the evaluation of the corrosion rates and patterns. Better corrosion measurement accuracy could be achieved if a WBE with larger number of electrodes is used in this experiment, as demonstrated in references [6, 7].

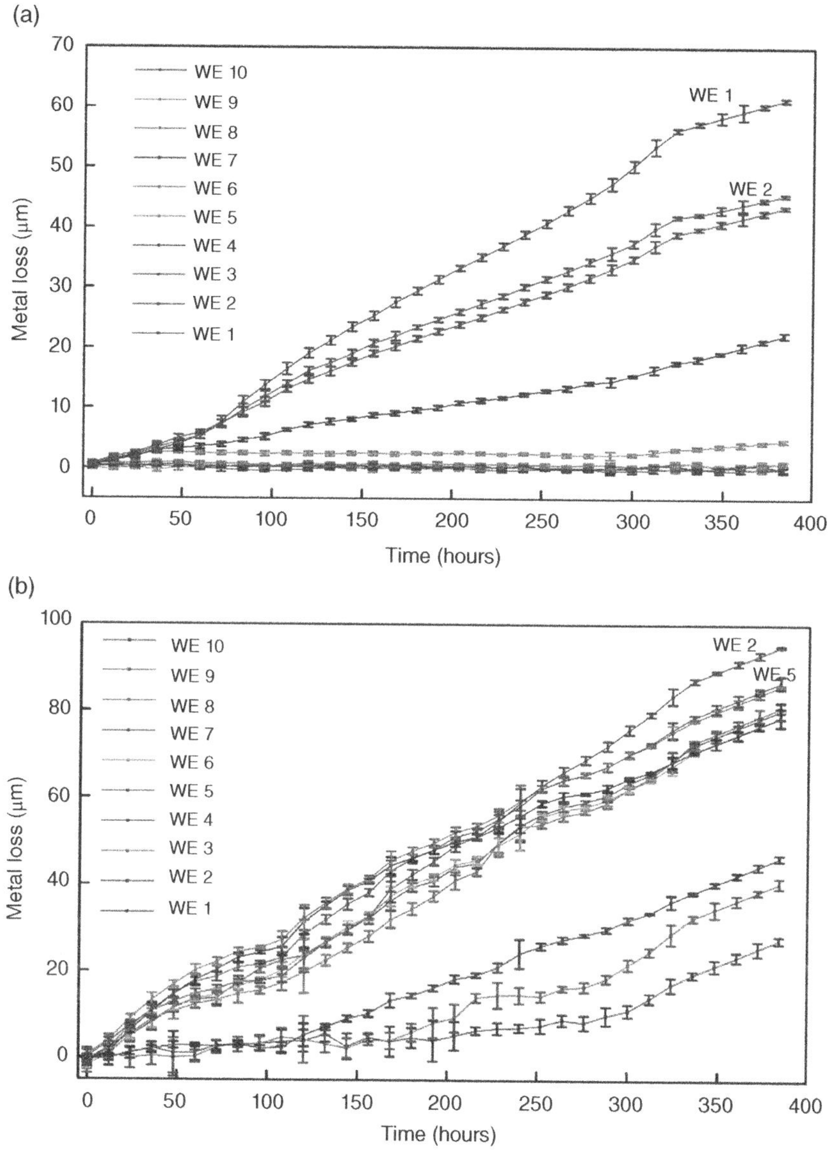

Figure 5.13 The corrosion depths of WE 1 to WE 10 measured by ER method (a) in the static condition; and (b) in the flowing condition. *Source:* Xu et al. [41]/ Australasian Corrosion Association.

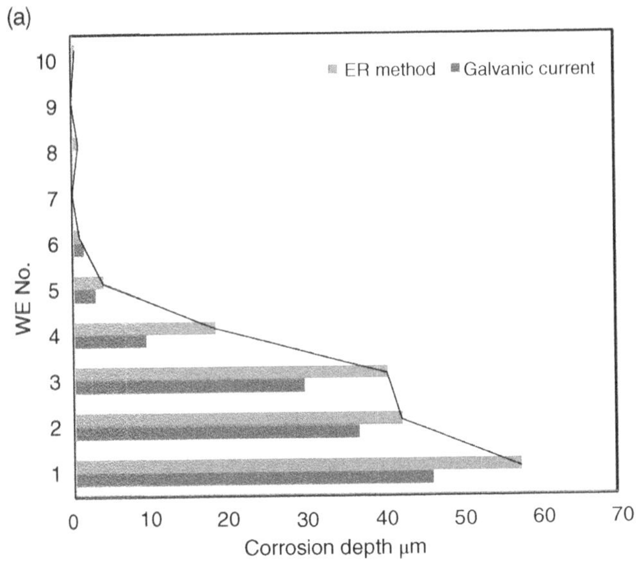

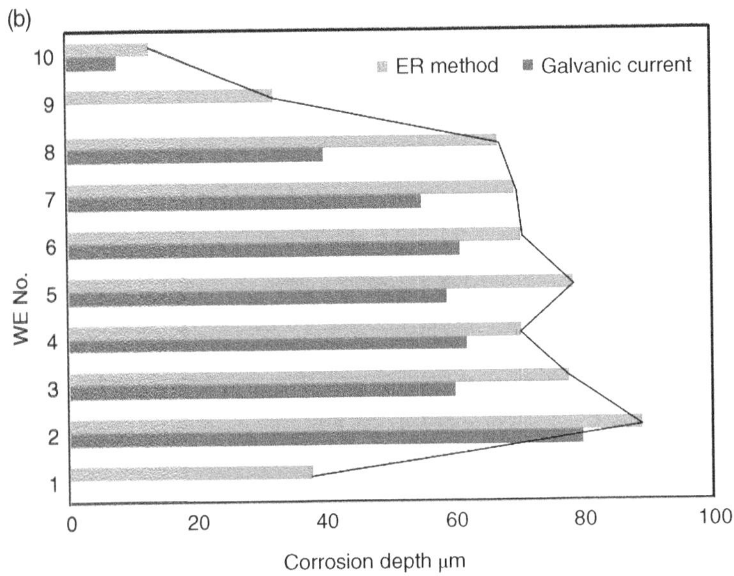

Figure 5.14 (a) The comparison of the corrosion depths measured by the ER method and the galvanic current in the static condition. (b) The comparison of the corrosion depths measured by the ER method and the galvanic current in the flowing condition. *Source:* Xu et al. [41]/ Australasian Corrosion Association.

Better resolution to localized waterline corrosion was enabled by using a 100-electrode WBE probe, as shown in Figure 5.11b, that was partially immersed in 0.017 M NaCl solution [7]. The potential and galvanic current distributions over the electrode surfaces were measured and visual observations were made at different stages of the exposure tests. As shown in Figures 5.15 and the photo in Figure 5.11b, corrosion attack occurred on almost all areas of the WBE surface except at the waterline area, similar to the corrosion pattern shown in the photo of Figure 5.11a. Figure 5.15 shows results of potential and galvanic current measurements and visual observations at different stages of the exposure test. Initially, as shown in Figure 5.15a, localized corrosion was concentrated on the areas close to waterline. It was a small anode and large cathode corrosion system. After 18 hours' exposure, as shown in Figure 5.15b, the anodic areas expanded. Finally, as shown in Figure 5.15c, the majority of the electrode surface became anode. As a result, the electrode surface became a large anode and small cathode corrosion system. The areas just below the waterline behaved as a cathode and remained unattacked. Waterline corrosion becomes more intense when a steel surface is partly immersed in an unstirred solution containing a corrosive salt (e.g. sodium chloride) with insufficient amount of inhibitor (e.g. sodium carbonate). In another experiment, highly localized corrosion behavior was observed along the waterline area on the mild steel plate and also on the surface of the WBE, which had been partially immersed in a 0.017 M NaCl and 0.008 M Na_2CO_3 solution, while there were only a few isolated anodic sites below waterline areas. Corrosion was concentrated on anodic areas and the large cathodic areas were protected. There were large potential differences between different areas of the electrode surface. Details of the experiment can be found in reference [7]. Similar results were also obtained from a long-term experiment (over one year in duration) reported in reference [6].

Compared to the relatively poor correlation between galvanically measured corrosion and ER probe measured corrosion shown in Figure 5.14 [41], much better agreement between visual observations and the corrosion potential/galvanic current distribution was obtained [7]. This can be attributed to smaller internal current component and much better measurement resolution of the 100-element electrode array shown in Figure 5.11b than that of the 10-element corrosion probe shown in Figure 5.11a.

More recently cases of array probe application to measure corrosion at the waterline zones have included the use of coated and uncoated WBE probes. For instance, Ding et al. used the WBE probe to measure coating degradation and metal corrosion beneath the coating around the waterline zone of an electrode in 3.5%NaCl solution [47]. Their results showed that at the beginning of the immersion, the distribution of cathode and anode area was mainly affected by the water penetration through coatings, probably at coating's weakness points. As the

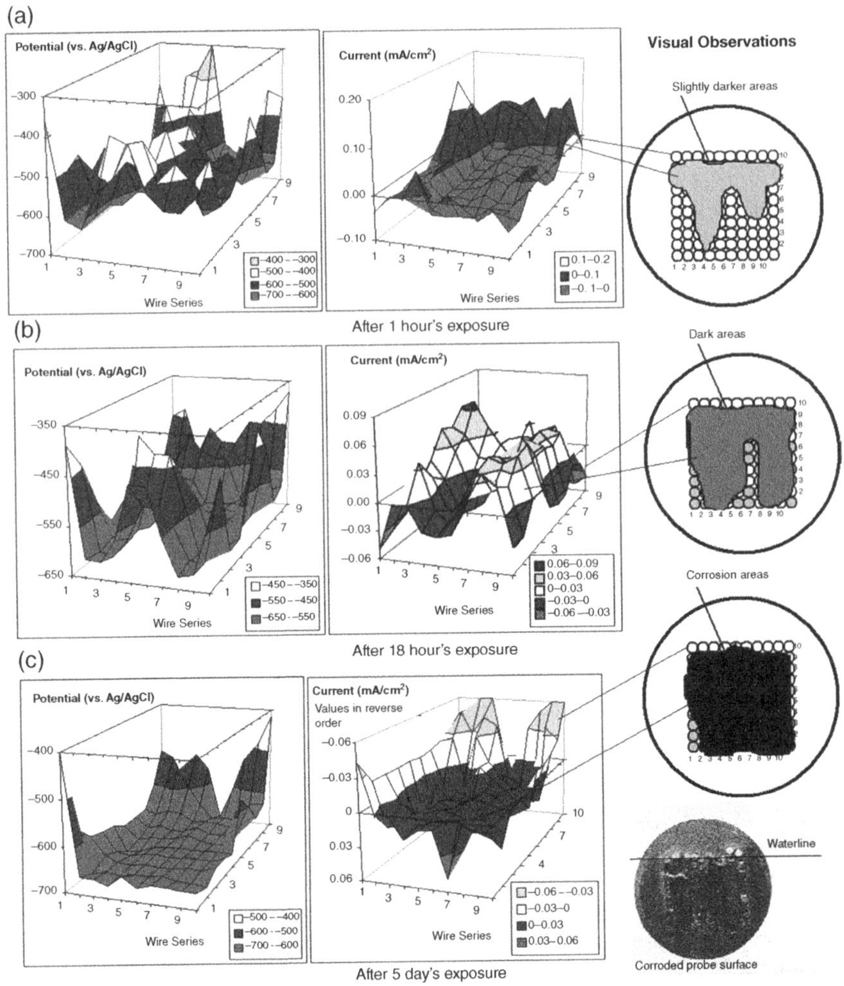

Figure 5.15 Potential and galvanic current distributions over a WBE working surface during exposure to 0.017 M NaCl solution over different periods of immersion: (a) after 1 hours of immersion; (b) after 18 hours of immersion; and (c) after 5 days of immersion. *Source:* Tan [7] / NACE International.

immersion extended, the effect of waterline on corrosion was observed, with the waterline zone acting as cathode, and the electrodes in bottom zone acting as anode. Due to the high concentrations of dissolved oxygen at the waterline zone, the cathode reaction was active at the location, leading to serious coating degradation and severe corrosion beneath the coating. Once a damage occurred on the coating, the spot would become a active and stable anode area. Similar observation was made by Chen et al. who measured coating deterioration and underlying metal corrosion processes in waterline area using a small WBE probe to monitor the current density distribution [48]. The distance between each electrode was very small (with the gap size of 0.3 mm), thus allowing it to simulate a practical metal component with a continuous metal surface. The results showed that cathodic and anodic sites tended to be weak points of the coating at the initial stage. With the continuous degradation of the coating, the cathodic zone tended to occur in the above the anodic zone due to the effect of differential aeration cells. Subsequently, the cathodic zone expanded to the waterline and the polarity reversed to the anodic zone, causing the coating to peel and blister continuously from the bottom of the probe to up parts. When the cathodic zone extended to the air phase area above the waterline, this area became the strongest cathodic zone under the thin liquid film, thus significantly accelerating the corrosion of the metal at the bottom of the corrosion cell [48]. These tests confirm the applicability of the electrode arrays for probing localized corrosion and its influence factors under complex waterline conditions.

5.3.2 Probing Localized Corrosion on Shore-Crossing Pipeline by Moving Probes

A practical industry structure that is exposed to complex waterline and soil/air interfaces is shore-crossing structures such as pipelines. Shore-crossing pipelines are commonly constructed by horizontal direction drills (HDD) for oil and gas transportation [49–51]. Corrosion is one of the main degradation factors, and external coating and cathodic protection (CP) are commonly used technologies for protecting pipelines against corrosion. Unfortunately coatings on HDD pipelines are often damaged by severe abrasion and mechanical impacts during the installation of the HDD pipes. Figure 5.16 illustrates various forms of coating damages observed on HDD pipes that could not be effectively protected by conventionally designed CP systems [50–53].

Coating damages such as those shown in Figure 5.16 could cause major pipeline integrity issues. Firstly these coating damages could not be easily assessed by conventional techniques used by the pipeline industry. This is because the environmental conditions around HDD pipelines are more complex than conventionally laid pipelines in many aspects, and therefore conventional techniques for assessing coating defects could not be used on HDD pipelines [54–60]. For instance, the most widely used potential survey method for locating defects in the pipeline coating, direct current voltage gradient (DCVG), is unable to be applied on HDD pipes because of decreasing sensitivity with the depth of a pipe. This survey could not indicate the cathodic shielding areas because these areas shield the CP current, and therefore, there is no potential gradient produced [57, 58]. Cathodic shielding is also the main reason for CP inefficiency under HDD pipeline circumstances. Another practical problem with most HDDs is that they are

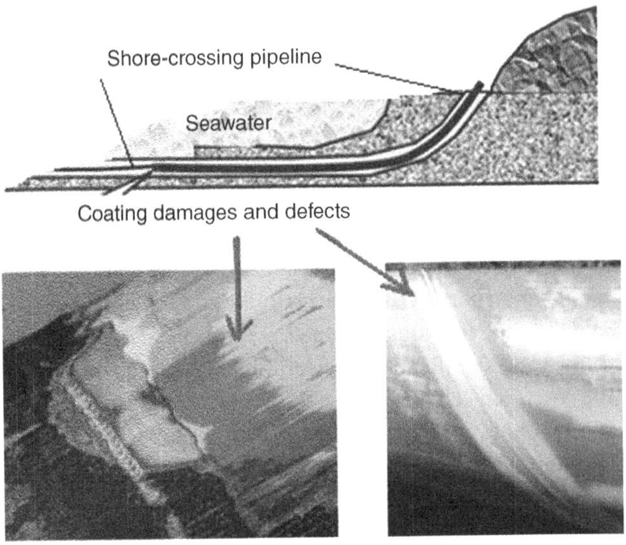

Figure 5.16 Various forms of coating damages observed on HDD pipelines.

installed in locations that are often not accessible for DCVG surveys. Another survey method, the alternating current voltage gradient (ACVG or Pearson survey) method, involves applying an AC signal to the pipe by means of a transmitter unit. At some coating damage areas, where the pipeline is in direct contact with the soil, this AC current would leak from pipe generating a zone of high current density. This in turn produces a detectable potential gradient in the soil in particular highly resistive soil [59, 60]. There are also some other methods such as the polarization change survey method for measuring and calculating 'coating resistance' in terms of 'current density' per square meter per 100 mV IR drop [60–62]. However this method has limitations in cases that the applied testing current is drawn by a small number of large defects [49–51]. There is a need for new techniques that are more reliable and accurate for the post-installation assessment of HDD pipe coating damages and CP efficiency.

A possible method to overcome limitations discussed above is electrochemical impedance spectroscopy (EIS). Due to the electrochemical nature of the pipeline surface, EIS can be a tool for measuring surface parameters to differentiate surface areas with intact and damaged coatings. By carrying out EIS analysis such as coating capacitance (C_c), double layer capacitance (C_{dl}), polarization resistance (R_p), pore resistance (R_{po}) and solution resistance ($R\Omega$) could be determined for investigating corrosion occurring on coated metal surfaces [63]. It has been reported that the coating can be categorized by measuring the impedance of coatings at 100 mHz; impedance of 1000 $M\Omega/cm^2$ indicates an excellent barrier coating, impedance of 10–1000 MΩ/cm^2 indicates the presence of capillaries in coating, while impedance from 0.1 to 10 MΩ/cm^2 suggest coating with pinholes [63, 64]. Parameters including the coating conductance, coating resistance, and polarization current density have been used to determine coating defects because a correlation was found between the measured conductance/resistance/currents and bare area of the coating under some cases. Figure 5.17 shows the relationship between the coupon surface capacitance (combination of coating capacitance (C_c) and double layer capacitance (C_{dl})) and coating defect sizes. An approximate linear relationship is clearly shown. This relationship suggests that capacitance is proportional to the coating defect surface area. These relationships provide a basis for determination of defect size based on electrochemical impedance measurements and analysis [49].

Although the relationship shown in Figure 5.17 provides a means of evaluating coating defects based on EIS measurements and capacitance values, however the measured coating defect area would be the total bare metal area without information of the distribution and size of each coating defect. There is a need to estimate the locations, numbers, and actual sizes of coating defects. This can be achieved by designing a multi-sensor system that employs multiple local probes to perform local EIS measurements. Figure 5.18 shows a concept of multi-sensor system for measuring local EIS for estimating coating damage location and size.

Although the multi-sensors are able to measure local impedance for estimating local coating damages, the system is complex to install and maintain in engineering practice. This local EIS measurement can be achieved by designing a 'moving' probe. As illustrated in Figure 5.19, local EIS measurements are carried out by 'pulling' a 'sinker' probe along a HDD pipeline, rather than installing multiple

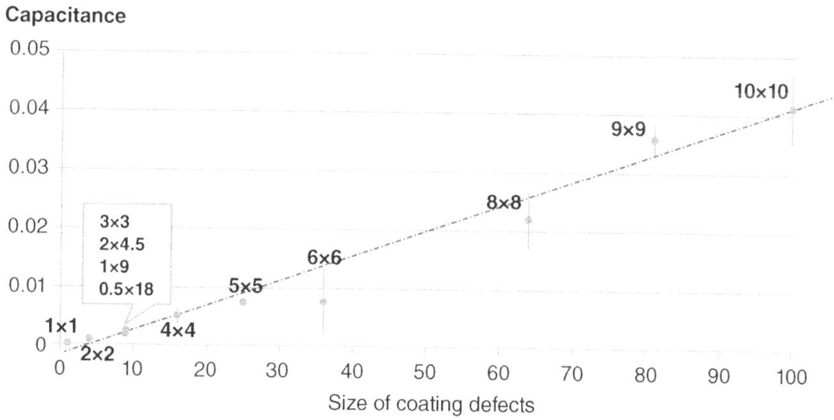

Figure 5.17 The relationship between defect size (cm^2) and test coupon surface capacitance (F). *Source:* Huo et al. [49]/ Business Cooperative research Centres Programme.

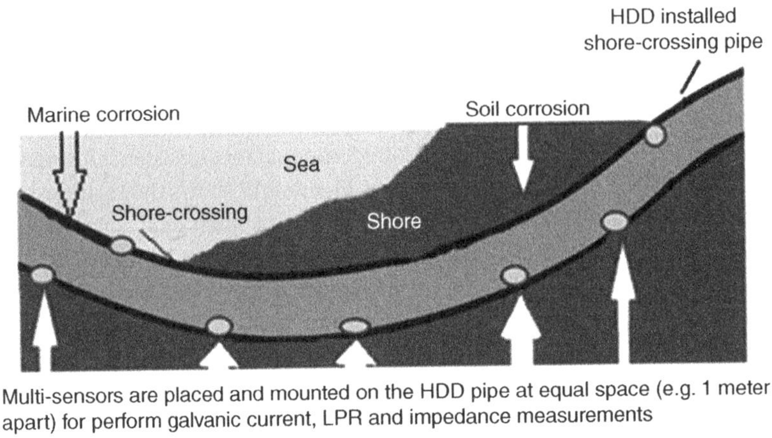

Figure 5.18 Illustrating multi-sensors mounted on a HDD pipe, allowing local EIS and polarization measurements.

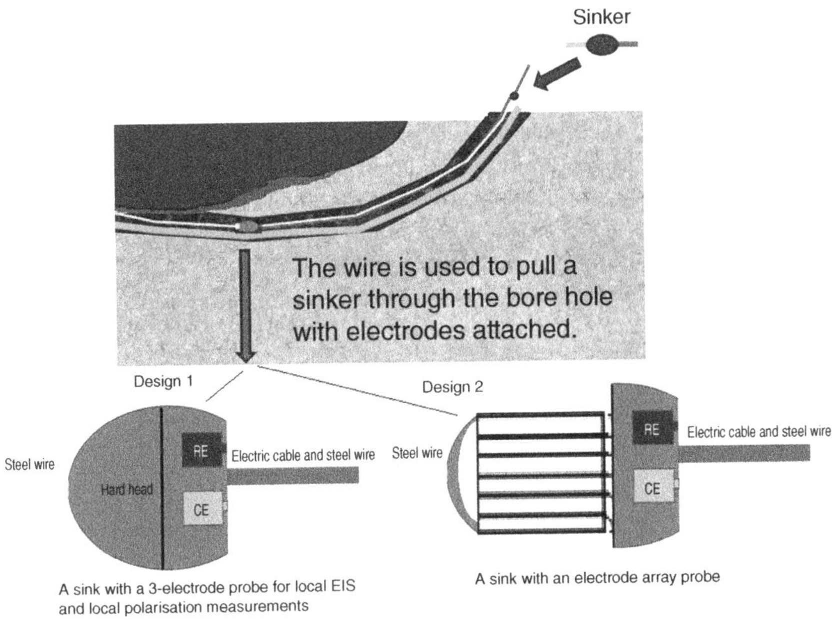

Figure 5.19 Probes that is able to be moved through the HDD borehole for local EIS measurements and corrosion mapping for detecting coating defects and localized corrosion sites on HDD pipes [49].

electrodes/sensors permanently (as shown in Figure 5.18). The moving probe can be used to help locate coating damages, as well as to estimate CP efficiency by carrying out local polarization and EIS measurements for the post-installation assessment of coating integrity. Figure 5.20 shows results from a test using a steel sample coated with a defective coating in a simulated HDD borehole. It is clearly shown in Figure 5.20a that significant changes in capacitance occurred when moving across a coating defect, and that the largest capacitance values are recorded at the defect location (i.e. just over 0.03 F at the defect compared with just over 0.01 F at non-defect locations). Capacitance results in Figure 5.20a clearly show a 2.5 cm × 44 cm long defect at the 171–215 cm location [49].

With an aim to further simplifying the local defect measurements, the possibility of measuring local CP currents as a simplified means of locating local coating defects has also been explored. The method is to apply a −1.4 V vs SCE CP potential on the steel sample in simulated HDD borehole and measure the CP current while moving the CP anode located on the sinker probe. Due to the IR drop difference in the pipe, the applied current at different locations will be different, depending upon how far it is from defect. The largest CP current is expected at

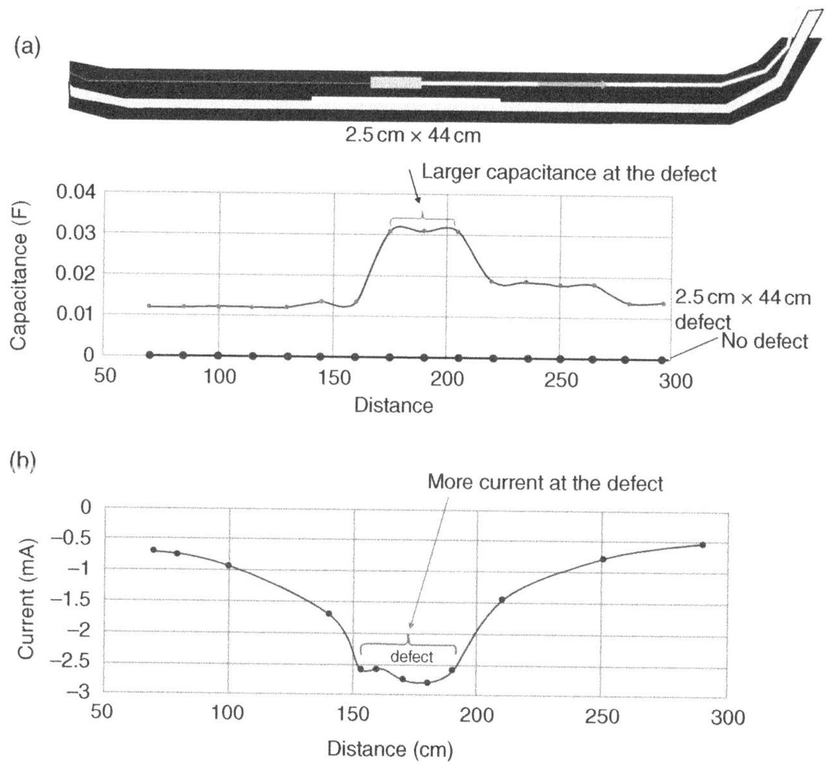

Figure 5.20 (a) Showing results from a local EIS test using a steel sample coated with (a) a defect-free coating, and (b) a steel sample coated with a 2.5 cm × 44 cm long defect at 171–215 cm location; and (b) results from a local CP current test using a steel sample coated with a 2.5 cm × 44 cm long defect at 171–215 cm location [49].

location closest to the defect. It is true, as shown in Figure 5.20b, that largest CP current was detected at the defect location [49]. The relationships shown in Figure 5.20 provide very useful means of evaluating local coating defects based on EIS measured capacitance and CP current values. Each of the local EIS test method (see Figure 5.20a) and local CP current method (see Figure 5.20b) has advantages and limitations. Local EIS method is more accurate in determining defect size; however, it takes too long for inspecting a long HDD pipeline with multiple defects. Each local CP current measurement takes less than 10 seconds;

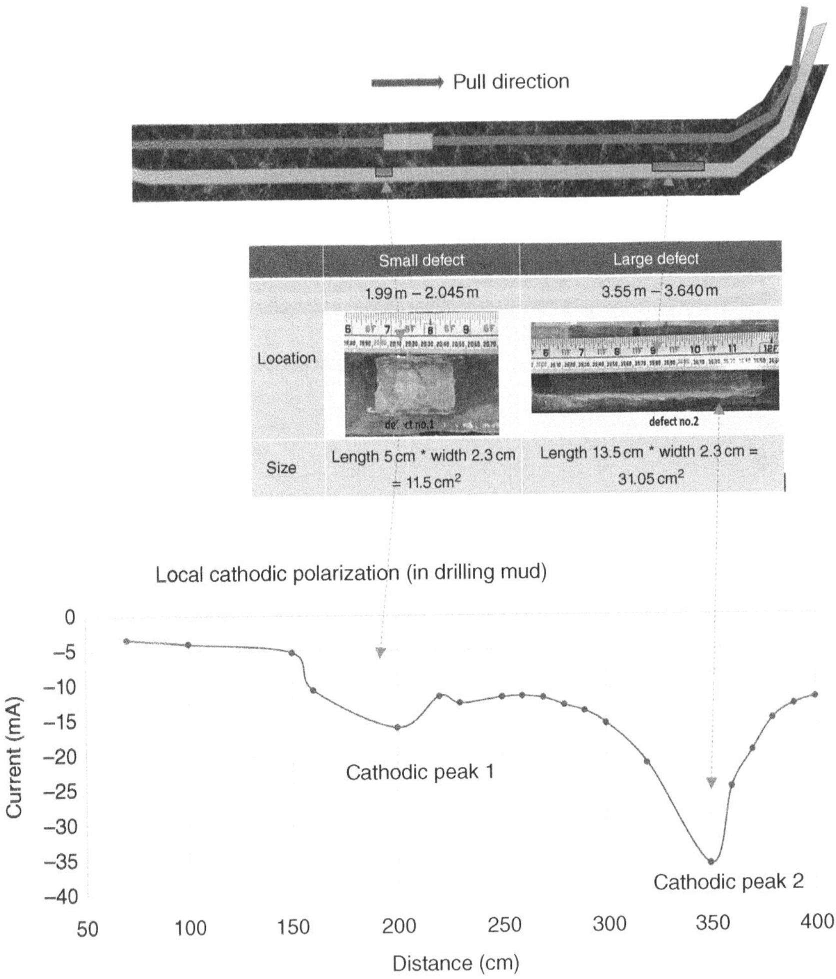

Figure 5.21 Local cathodic polarization current measurement through the testing sample surface with two artificial coating defects in a simulated HDD borehole [49].

however, local CP currents may not accurately reflect defect sizes. An idea is to use them in combination. This would enable quick survey of coating damages along a HDD pipe (using local CP method) and also measure coating defect sizes at locations of coating damage (using EIS method). Figure 5.21 shows local cathodic polarization current measurement through a testing sample surface in a simulated HDD borehole. It is clear that two coating defects at different locations are correctly detected. After detecting the coating defect location, a local CP unit can be installed close to the defect that can help enhancing local CP and avoiding cathodic shielding. Figure 5.22 illustrates a setup for detecting local CP efficiency and localized corrosion in a simulated HDD borehole using an array probe as part of the 'sinker probe'. The WBE current maps would be able to show corrosion and CP conditions at each location. Overall, it is shown that through the use of either multi-probes or a running 'sinker' probe for local EIS and CP current measurements, the location and size of coating defects as well as CP efficiency at coating defects can be quickly and reasonably accurately detected under HDD pipeline conditions [49]. These illustrate different corrosion probe design concepts for probing and controlling corrosion in complex and multiple environmental conditions.

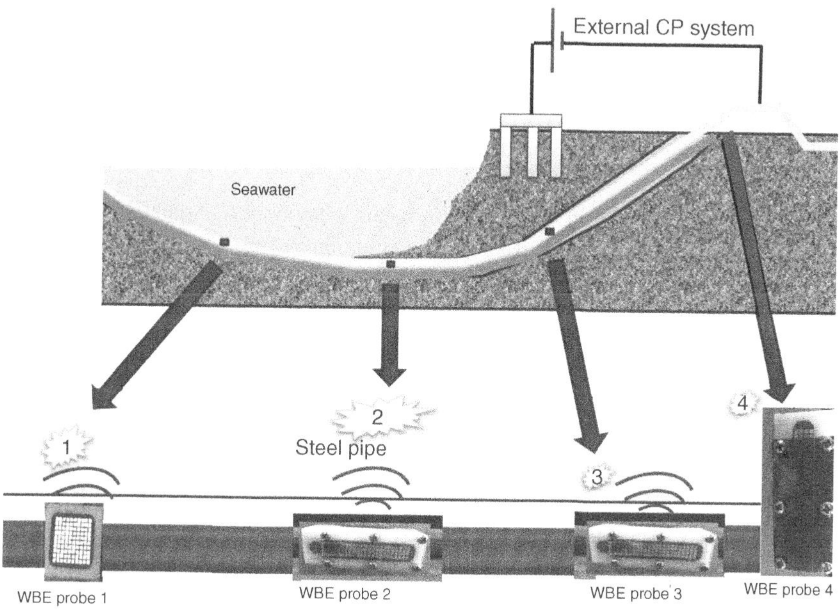

Figure 5.22 A setup for detecting CP efficiency in a simulated HDD borehole with an integrated multi-probe system.

References

1 The Straits Times (2003). Boy hit on head by lamp post dies. (Wednesday, 7 March).

2 Tan, M.Y.J. (2020). A critical overview of monitoring infrastructural health using corrosion probes. *Corrosion Engineering, Science and Technology* 55: 103–117. https://doi.org/10.1080/1478422X.2019.1695390.

3 Tan, Y. (2013). *Heterogeneous Electrode Processes and Localized Corrosion*, 1–246. Hoboken, NJ: Wiley.

4 Tan, Y.J. (2011). Experimental methods designed for measuring corrosion in highly resistive and inhomogeneous media. *Corrosion Science* 53: 1145.

5 Aung, N.N. and Tan, Y.J. (2004). A new method of studying buried steel corrosion and its inhibition using the wire beam electrode. *Corrosion Science* 46: 3057–3067.

6 Tan, Y.-J., Bailey, S., and Kinsella, B. (2001). Mapping non-uniform corrosion using the wire beam electrode method. III. Water-line corrosion. *Corrosion Science* 43: 1931–1937.

7 Tan, Y. (1998). Monitoring localized corrosion processes and estimating localized corrosion rates using a wire-beam electrode. *Corrosion* 54: 403–413.

8 Jeffrey, R. and Melchers, R.E. (2009). Corrosion of vertical mild steel strips in seawater. *Corrosion Science* 51: 2291–2297.

9 Chen, Y., Zhang, W., Ding, J. et al. (2016). Research on water-line corrosion of carbon steel by wire beam electrode technique. *MATEC Web of Conferences, EDP Sciences* 39: 01004.

10 Melchers, R.E. and Jeffrey, R. (2012). Corrosion of long vertical steel strips in the marine tidal zone and implications for ALWC. *Corrosion Science* 65: 26–36.

11 Deng, P., Liu, Q., Li, Z. et al. (2018). Corrosion behavior of X70 pipeline steel in the tropical juncture area of seawater-sea mud. *Journal of the Chinese Society of Corrosion and Protection* 38: 415–423.

12 Hu, J., Li, X., Deng, P. et al. (2015). Evaluation of carbon steel corrosion in vicinity of interface sea-water/sea-mud by techniques WBE and LP. *Corrosion Science and Protection Technology* 27: 551–558.

13 Hu, J., Deng, P., Zhang, J. et al. (2017). Macro non-uniform corrosion of Q235 steel in tropical and coastal red soils. *Corrosion Science and Protection Technology* 29 (3): 1002–6495. 03-0233-08, pp. 233–240.

14 Shi, H.-S. and Deng, K. (2005). Simulating research on the corrosion of steel bar in concrete by wire-beam electrode. *Jianzhu Cailiao Xuebao/Journal of Building Materials* 8: 682–686.

15 Hu, J., Li, G., Deng, K., and Liu, Y. (2005). Study on non-uniform corrosion of rebar embedded in concrete by wire – beam electrode. *Journal of the Chinese Society of Corrosion and Protection* 25: 88–92.

16 Li, L.Q., Dong, S.G., Wang, W. et al. (2010). Study on interaction between macrocell and microcell in the early corrosion process of reinforcing steel in concrete. *Science China Technological Sciences* 53: 1285–1289.

17 Česen, A., Kosec, T., and Legat, A. (2013). Characterization of steel corrosion in mortar by various electrochemical and physical techniques. *Corrosion Science* 75: 47–57.

18 Shi, W., Dong, Z.H., Kong, D.J., and Guo, X.P. (2013). Application of wire beam electrode technique to investigate initiation and propagation of rebar corrosion. *Cement and Concrete Research* 48: 25–33.

19 Hu, J., Deng, P., Li, X. et al. (2018). The vertical non-uniform corrosion of reinforced concrete exposed to the marine environments. *Construction and Building Materials* 183: 180–188.

20 Yu, H. and Caseres, L. (2012). An embedded multi-parameter corrosion sensor for reinforced concrete structures. *Materials and Corrosion* 63: 1011–1016.

21 Liubin, S., Daowu, Y., Sanjun, P. et al. (2011). Electrochemical study of inhibitors to improve the anti-corrosion performance of reinforced bar in the concrete. *Anti-Corrosion Methods and Materials* 58: 22–25.

22 Shi, W., Wang, T.-Z., Dong, Z.-H., and Guo, X.-P. (2017). Application of wire beam electrode technique to investigate the migrating behavior of corrosion inhibitors in mortar. *Construction and Building Materials* 34: 167–175.

23 Wu, J., Yu, H., and Shi, X. (2016). Effectiveness of products in managing metallic corrosion induced by cyclic deicer exposure: laboratory study using multielectrode array sensors, electrochemical impedance, and laser profilometer. *Journal of Materials in Civil Engineering* 28 (5): 04015186.

24 Yang, D., Song, L., Wang, C., and Xiao, Z. (2009). Electrochemistry study on epoxy resin to anti-corrosion performance of reinforced bar in the concrete. *TMS Annual Meeting* 3: 735–740.

25 Agarwala, V.S. (1982). A probe for monitoring corrosion in marine environments. In: *Atmospheric Corrosion* (ed. W.H. Ailor), 183–192. N.Y.: John Wiley & Sons, Inc.

26 Port, O. (1998). Paint that sees rust before you do. *BusinessWeek* USA (13 April).

27 Glass, R. S. and Ciarlo, D. R. (1995). Method for monitoring environment and corrosion. US Patent 5,437,773.

28 Davis, G.D. and Dacres, C.M. (1999). Electrochemical sensors for evaluating corrosion and adhesion on painted metal structures. US Patent 5,859,537.

29 Tan, Y.J. (1999). Wire beam electrode: a new tool for studying localized corrosion and other heterogeneous electrochemical processes. *Corrosion Science* 41: 229–247.

30 Aung, N.N., Tan, Y.J., and Liu, T. (2006). Novel corrosion experiments using the wire beam electrode: (II) Monitoring the effects of ions transportation on electrochemical corrosion processes. *Corrosion Science* 48: 39–52.

31 Zhong, Q. (2002). Study of corrosion behaviour of mild steel and copper in thin film salt solution using the wire beam electrode. *Corrosion Science* 44: 909–916.

32 Wang, Y., Wang, W., Liu, Y. et al. (2011). Study of localized corrosion of 304 stainless steel under chloride solution droplets using the wire beam electrode. *Corrosion Science* 53 (9): 2963–2968.

33 Wang, Y.H., Liu, Y.Y., Wang, W. et al. (2013). Influences of the three-phase boundary on the electrochemical corrosion characteristics of carbon steel under droplets. *Materials and Corrosion* 64: 309–313.

34 Zhang, P., Wang, Y., Peng, X. et al. (2014). Study of corrosion behavior of copper beneath a droplet by means of wire beam electrode technology. *Journal of the Chinese Society of Corrosion and Protection* 34: 459–464.

35 Zhang, X., Gao, Z., Hu, W. et al. (2017). Correlation between corrosion behavior and image information of Q235 steel beneath thin electrolyte film. *Journal of the Chinese Society of Corrosion and Protection* 37: 444–450.

36 Liu, Z., Wang, W., Wang, J. et al. (2014). Study of corrosion behavior of carbon steel under seawater film using the wire beam electrode method. *Corrosion Science* 80: 523–527.

37 Azumi, K., Naganuma, A., and Sato, Y. (2015). Coupling current mapping of corroding iron in a wet and dry cyclic corrosion test. *Journal of Solid State Electrochemistry* 19: 3543–3550.

38 Feng, L., Wang, Y., Zhong, L. et al. (2016). Influence of short-term storage on corrosion behavior of copper. *Journal of the Chinese Society of Corrosion and Protection* 36: 375–380.

39 Zhao, B., Yu, Y., Guo, J. et al. (2020). Evaluation of dew point corrosion in acid atmosphere by wire beam electrode. *IOP Conference Series: Materials Science and Engineering* 838 (1): 0120052019.

40 Guan, X., Zhang, D., Wang, J. et al. (2017). Numerical and electrochemical analyses on carbon dioxide corrosion of X80 pipeline steel under different water film thicknesses in NACE solution. *Journal of Natural Gas Science and Engineering* 37: 199–216.

41 Xu, Y., Tan, M.Y.J., and Huang, Y. (2018) The study of water-line corrosion in static and dynamically flowing electrolytes using multi-electrode sensor in combination with electrical resistance method. *Proceedings of the Corrosion and Prevention Conference*. Australasian Corrosion Association, Corrosion & Prevention 2018, Adelaide, Australia (11–14 November 2018).

42 Qaddoumi, N., Chen, Y., Zhang, W. et al. (2016). Research on water-line corrosion of carbon steel by wire beam electrode technique. *MATEC Web of Conferences* 39: https://doi.org/10.1051/matecconf/20163901004.

43 Evans, U.R. and Hoar, T.P. (1932). The velocity of corrosion from the electrochemical standpoint Part II. *P R Soc Lond a-Conta* 137: 343–365.

44 Melchers, R.E. and Jeffrey, R.J. (2013). Accelerated low water corrosion of steel piling in harbours. *Corrosion Engineering, Science and Technology* 48: 496–505.

45 Wu, J., Wang, P., Gao, J. et al. (2017). Comparison of water-line corrosion processes in natural and artificial seawater: the role of microbes. *Electrochemistry Communications* 80: 9–15.

46 Tan, Y.J. (2009). Sensing electrode inhomogeneity and electrochemical heterogeneity using an electrochemically integrated multielectrode array. *Journal of the Electrochemical Society* 156: C195–C208.

47 Ding, J., Zhang, W., Wang, J. et al. (2016). Evaluation of water-line zone corrosion of an electrode with coating in NaCl solution by WBE technique-I. *Journal of the Chinese Society of Corrosion and Protection* 36: 463–470.

48 Chen, Z., Zhang, W., Zhan, Y. et al. (2020). Coating deterioration and underlying metal corrosion processes in water-line area. *Role of DACs, Coatings* 10: 684.

49 Huo, Y., Tan, M.Y.J., Bryson, A. et al. (2019) New methods for assessing coating integrity and cathodic protection efficiency under horizontal directional drilling pipeline conditions. *Proceedings JTM conference 2019*, Joint Technical Meeting, Brisbane, Australia (28 April–3 May 2019).

50 Martin, B. (2005). Evaluating pipeline coating condition on horizontal directional drills. *The Australian Pipeliner* (April). p. 62.

51 Bryson, A. (2012) Horizontal Direction Drills (HDDs) – How much coating damage is acceptable? *Australian Electrolysis Committee presentation, ACA Annual Conference*, Australasian Corrosion Association 2012, Melbourne, Victoria, Australia (11–14 November 2012).

52 Australian Standard AS 2832.1:2015 (2015) Cathodic protection of metals, Part 1: pipes and cables. Appendix L, Testing of Pipelines Installed Within Horizontal Directional Drilled Crossings.

53 Hair, J. (2002). Coating requirements for pipelines installed by horizontal directional drilling. *Pipelines* 1–8. https://doi.org/10.1061/40641 (2002).

54 Segall, S., Gummow, R. A., Zahraee, M., and Fingas, D. (2015). Evaluation of Current Practices and Equipment Used for Assessing the Integrity of Coating Systems on Pipelines Installed in Trenchless Crossings (Final Report), PRCI, Contract PR-444-133602, 15 May.

55 Gummow, R.A. and Segall, S.M. (1998). In-situ evaluation of directional drill/bore coating quality. Report for PRC International Corrosion Committee, American Gas Association, by Correng Consulting Service Inc.

56 Czaszejko, T. and Edmond, L. (2008). Through thickness coating defects in pipe laid using HDD techniques. Final report to APIA, Monash University.

57 Masilela, Z. and Pereira, J. (1998). Using the direct current voltage gradient technology as a quality control tool during construction of new pipelines. *Engineering Failure Analysis* 5: 99–104.

58 Varela, F., Tan, M.Y.J., and Forsyth, M. (2015). An overview of major methods for inspecting and monitoring external corrosion of on-shore transportation pipelines. *Corrosion Engineering, Science and Technology* 50: 226–235.

59 Kowalski, A. (2014). The close interval potential survey (Cis/Cips) method for detecting corrosion in underground pipelines. In: *Underground Pipeline Corrosion* (ed. M.E. Orazem), 227–246. Cambridge: Woodhead Publishing.

60 Eyre, D. (2014). The pearson survey method for detecting corrosion in underground pipelines. In: *Underground Pipeline Corrosion* (ed. M.E. Orazem), 247–254. Woodhead Publishing.

61 Brockhaus, S., Ginten, M., Klein, S. et al. (2014). 10 – in-Line Inspection (Ili) methods for detecting corrosion in underground pipelines. In: *Underground Pipeline Corrosion* (ed. M.E. Orazem), 255–285. Woodhead Publishing.

62 Cordell, J. and Vanzant, H. (2003). *Pipeline Pigging Handbook*, 3e. CLARION Technical Publishers.

63 Mansfeld, F. (1995). Use of electrochemical impedance spectroscopy for the study of corrosion protection by polymer coatings. *Journal of Applied Electrochemistry* 25: 187.

64 Kendig, M.W., Jeanjaquet, S., Lumdsden, J. et al. (ed.) (1993). *Electrochemical Impedance – Analysis and Interpretation*, 407. Philadelphia: ASTM.

6

Localized Corrosion Impacted by Flow and Erosion

In practical engineering systems, corrosion media are often not stagnant; instead, they may be flowing rapidly and mechanically impacting metal surfaces, significantly contributing to the variability and complexity of localized corrosion processes and mechanisms. Examples of flow-affected localized corrosion include steel pipes eroded by seawater and marine debris in the splash zone of the sea, pipelines impacted internally by high-speed multiphase oil, gas and water flows, and mining pumps eroded by mineral slurries. This chapter firstly identifies issues and obstacles in managing engineering structural damages caused by flow-induced localized corrosion, followed by discussion of approaches to probing and controlling flow-affected localized corrosion. The corrosion and inhibition of internal oil and gas steel pipelines and the erosion-corrosion behavior of additively manufactured stainless steel 316 are selected as typical examples to illustrate the characteristics, prediction, mechanism, and control of localized corrosion affected by flowing liquid and solid particles. It is shown that the acquisition of corrosion data is again critical for probing, predicting, controlling, and managing flow-induced localized corrosion and erosion corrosion.

6.1 Localized Corrosion Impacted by Flowing Liquid and Solid Particles

Many industrial structures such as oil and gas pipelines, mining slurry pipes, ship hulls and impellers, water injection piping systems, nuclear power cooling systems, and hydro turbines are impacted by flowing liquid and solid particles [1–6]. Corrosion in these structures are often referred to as flow affected corrosion (FAC) and erosion corrosion although many other forms of corrosion could also occur. FAC is electrochemical corrosion in nature, complicated by flow-induced environmental changes [7–10], while erosion corrosion involves simultaneous

Localized Corrosion in Complex Environments, First Edition. Mike Yongjun Tan.

occurrence of electrochemical corrosion and mechanical erosion, impingement, abrasion, and cavitation [11–14]. FAC and erosion corrosion remain major concerns in industry, for instance, mining pumps need to be replaced frequently (once every few months in some cases) due to erosion corrosion by flowing mineral slurries. Similar situations exist in oil sand production facilities. A further complication is that metal losses caused by the coupled effect of erosion and corrosion can sometimes be dramatically higher than the sum of the metal loss caused by individual pure erosion and pure corrosion [10–19], significantly accelerating structural damage and making the prediction of structural failure much more difficult. Over the past decades, various studies have been conducted trying to understand and quantify the interaction and synergy between erosion and corrosion. Some models have been developed to predict metal losses under either erosion-corrosion-dominated or corrosion-erosion-dominated flow conditions [20], however there are many unsolved problems including those discussed in the following Sections.

6.1.1 Issues with Localized Corrosion Affected by Flow and Erosion

Figure 6.1 shows oil and gas pipeline corrosion affected by several corrosion mechanisms that are affected, more or less, by flowing liquid and solid particles. A common and typical example of FAC and erosion corrosion is the corrosion of pipeline where flowing fluids and slurries impact the internal pipeline surface, causing localized erosion corrosion due to local breakdown of protective scales and inhibitor films. Oil and gas flowlines contain multiphase flows comprising carbon dioxide (CO_2), water, oil, and gas. CO_2 corrosion can cause severe damages to a variety of downhole and surface oil and gas-producing equipment such as oil and gas wells, pipelines, pumps, condensers, gas scrubbers, storage systems. Under the impact of high-speed flow, local corrosion anodes are initiated from local damages in protective scales and inhibitor films by flow-induced mechanical impact, which is probably the most common factor causing localized corrosion in oil and gas flowlines [21–24]. For this reason, the persistency of inhibitor films is an important requirement for inhibitors used for environments affected by flowing liquid and solid particles.

The existence of flowing liquids and solid particles in corrosion environment can lead to changes in corrosion processes and mechanisms. Different metals could respond differently to flow and erosion, for instance passive metals such as stainless steel and active metals such as mild steel can respond completely differently to aeration and the stirring of solutions. In the case of stainless steel exposed to neutral brine, the corrosion controlling factor is usually the passivity of the metal surface. Stainless steel in a stagnant brine solution could have higher pitting corrosion activities than in a stirred solution because more efficient oxygen transportation

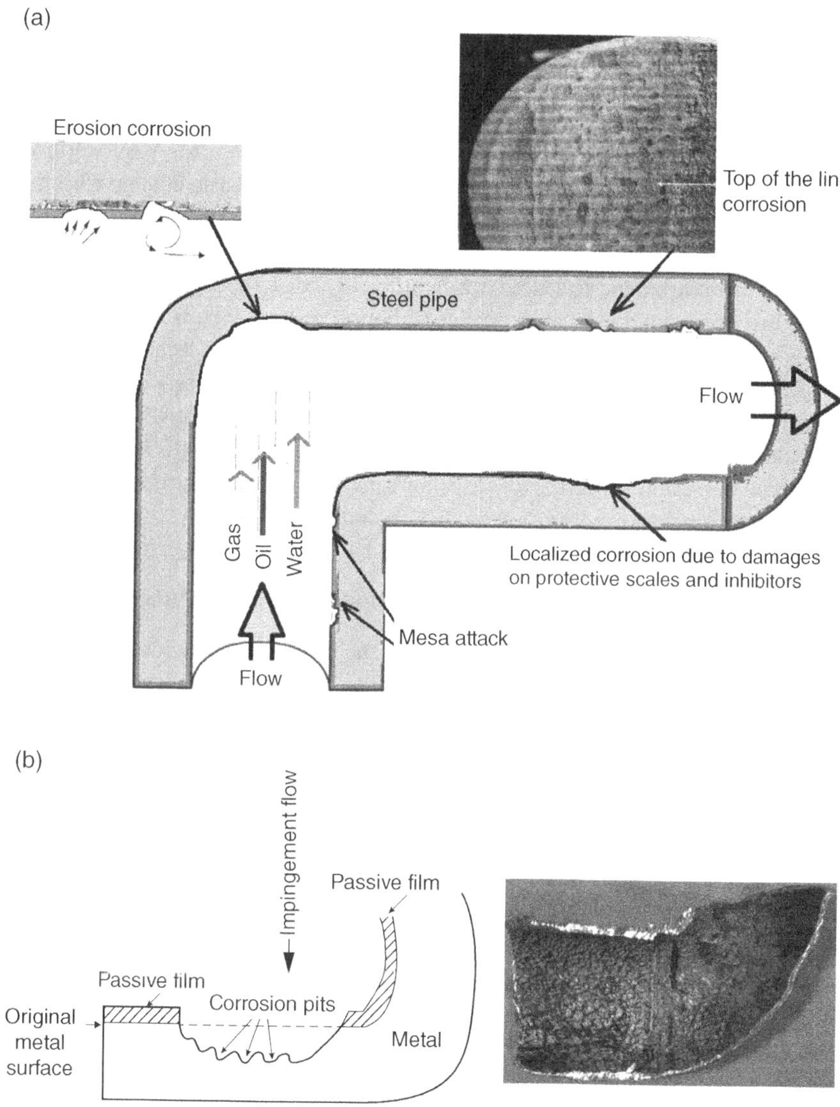

Figure 6.1 (a) Showing cases of localized corrosion in an oil and gas pipeline infrastructures that are affected, more or less, by flowing liquid and solid particles. (b) Illustrating impingement-induced localized corrosion in a multiphase flowline. The schematic on the left relates to the bottom wall of the pipe section on the right.

and electrolyte diffusion in a stirred solution could help repairing damaged passive films and maintaining the passivity of stainless steel surface. However flowing solution can cause more rapid corrosion to occur on mild steel because a higher oxygen concentration in stirred brine solutions can enhance the cathodic oxygen reduction reaction that is often the corrosion controlling reaction. A further complication is that this corrosion behavior can change again if mild steel is exposed to high pH solution (e.g. pH=11) because steel can become passive, like stainless steel, at high pH, and a higher oxygen concentration in stirred solution could help repairing passive films and maintaining the passivity of the steel surface. Furthermore the effect of flow to corrosion mechanism can also change with the variation of corrosion test temperature and duration. For instance, carbon dioxide (CO_2) corrosion of steel at a temperature above 70°C can lead to the formation of a protective iron carbonate scale that can significantly affect the rate and mechanism of pipeline corrosion [25]. Under this condition, if a short exposure test (e.g. one day) is selected for evaluating CO_2 corrosion behavior of steels, it may lead to the reporting of falsely high corrosion rates since this short-term tests can overlook corrosion mechanism change with the later formation of a protective iron carbonate scale on steel. Given the complex and changing nature of the effects of flow and erosion on corrosion processes and mechanisms, it is therefore critical to acquire timely corrosion data for corrosion prediction and control.

6.1.2 Acquiring Corrosion Data from Flowing and Erosion Systems

An approach to acquiring data for informing the corrosion behavior of materials affected by flowing liquid and solid particles is to design and perform accelerated laboratory tests that simulate real corrosion conditions and mechanisms. It is a highly challenging task to design laboratory tests that effectively and reliably simulate and evaluate corrosion behavior in actual flow environments and assess corrosion control techniques such as corrosion inhibitors. This is because FAC or erosion corrosion is well known to be determined by many factors such as the type of materials, solution pH, flow velocity and patterns, impact angle, fluid temperature, test duration and solid particle concentration, etc. The most straightforward method used to simulate corrosion in flowlines is to employ jet rigs and flowloops that could create a flow conditions similar to actual flowlines. For instance jet rig systems are commonly employed for simulating erosion corrosion under slurry impingement conditions [26–36] by providing control of the impingement parameters such as the impact velocity and angle of the sand particles. Compared to jet rigs, flowloops is a more straightforward method to simulate realistic flowline erosion–corrosion conditions. A typical small flowloop designed for laboratory FAC and erosion–corrosion testing is shown in Figure 6.2 that is composed of a water tank, a centrifugal pump, a flow meter, and a piping system with corrosion probes. Multiple probes/sensors can be placed at different areas (e.g. the elbows) of the flowloop to acquire data for monitoring, probing, and predicting localized erosion corrosion. The instantaneous corrosion rates are usually measured using electrochemical probes such as linear polarization resistance and electrochemical impedance spectroscopy based on electrochemical principles discussed before [21–25].

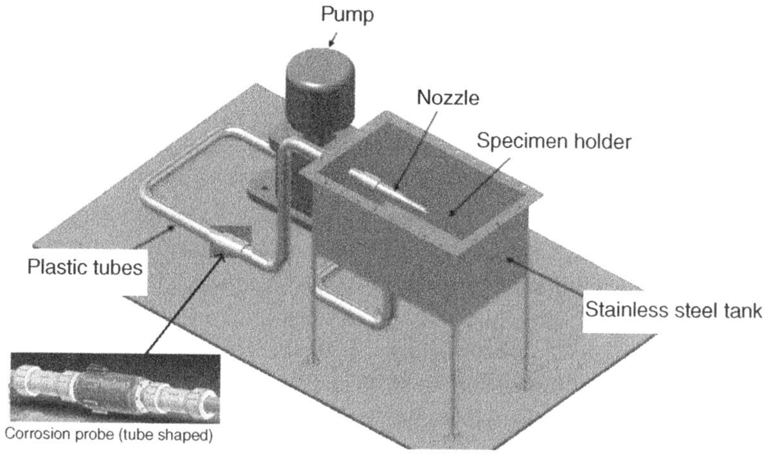

Figure 6.2 Schematic illustration of flowloop with jet impingement apparatus for simulating and monitoring flowline corrosion using a pipe corrosion probe.

Total metal losses caused by erosion corrosion including Faradaic and non-Faradaic parts of the metal losses can be measured using electrical resistance (ER) probes. The ER probe measures metal losses based on changes in the resistance of each sensing element that directly exposes to the erosion–corrosion environment [37]. The ER probes can be placed at multiple directions to ensure the sufficient impingement of sand particles on the sensing element [38, 39]. Various forms of ER sensors have also been developed for monitoring the nonuniform corrosion in pipelines including the field signature method [40] and ring pair electrical resistance sensor [41]. These sensors are designed based on dividing a pipe into small sections and measuring the resistance of each section to estimate localized erosion–corrosion depths at these sections. Unfortunately slurry jet rig and flowloop tests are time consuming, troublesome, and costly. Practical issues such as the wear of the pump in the flowloop and the cleaning of the flow system can be real issues affecting the reproducibility and reliability of these tests. In order to minimize the number of tests required for modeling and predicting erosion–corrosion behavior, computational fluid dynamics (CFD) simulation is often used in conjunction with flowloop and jet rig tests because CFD simulation can be used to predict the distributions of the flow velocity, wall shear stress, and sand concentration at a certain pipe section [1, 28, 36].

A significantly simplified experimental method commonly used to evaluate erosion corrosion is the rotating cylinder/disc electrode (RCE) system. The RCE is a tool for creating aeration and mass transportation changes in laboratory testing conditions and is commonly used to mimic turbulent corrosion conditions found in large-scale industrial structures. The RCE is described as an effective means of controlling the velocity of fluid flow over an electrode surface, creating a surface shear stress. A common application of the RCE is in corrosion inhibitor assessment tests where the diffusion of oxygen is no longer the rate controlling factor and the addition of a corrosion inhibitor could become the rate determining factor, therefore adding an inhibitor to a stirred solution may show much greater effects on corrosion rates than adding an inhibitor to a stagnant solution. On the other hand, inhibitor film formation and destruction could be assessed under the impact of fluid in a stirred solution. For these reasons RCE with solution stirring is used as a means of improving inhibitor test sensitivity in accelerated laboratory tests and for evaluating inhibitor film persistency. An electrochemical cell shown in Figure 6.3 is an example of using an RCE for studying the formation of inhibitor films in a concentrated inhibitor solution, and the destruction of inhibitor films under the effects of fluid shear stress created by rotating the electrode at selected speeds [21–24]. In these experiments, electrochemical impedance spectroscopy (EIS) was used to probe and monitor inhibitor film formation and destruction and for evaluating the film persistency under simulated CO_2 containing oil pipeline conditions [21, 24]. Electrochemical parameters such as the resistances and capacitances of inhibitor layers, charge transfer resistance, and double-layer capacitance were used to analyze the inhibitor mechanism and to calculate the corrosion rate [21, 24]. Corrosion and erosion sensors are often used in conjunction with jet rigs, flowloops, and RCE devices for probing flow-affected

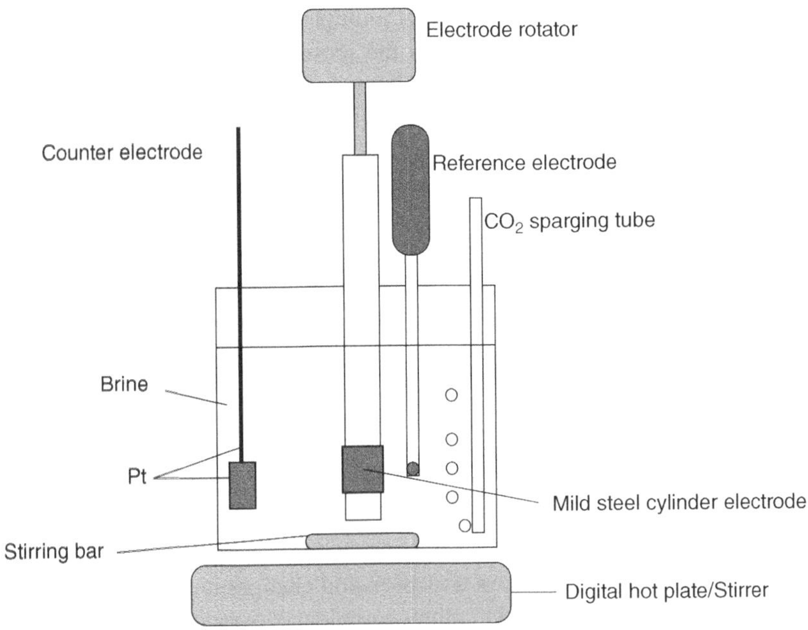

Figure 6.3 A schematic diagram of probing flow influenced corrosion using rotating disc electrodes [21]. *Source:* Tan et al. [21], Kinsella et al. [42].

corrosion and the effects of corrosion inhibitors. A limitation many corrosion sensors have is in the monitoring of rapid and localized changes in electrode processes because techniques such as EIS take quite a long time (often more than one hour) for each measurement. Electrochemical noise analysis (ENA) is a method that has been used for monitoring dynamic changes in electrode processes by measuring the electrochemical noise resistance (R_n) on a continuous basis. A typical experiment setup is illustrated in Figure 6.4 for assessing the persistency of an inhibitor imidazoline film on a dual mild steel RCE [22, 23]. More details on the experimental and test results are described in Section 6.2.2 about monitoring inhibitor film damage by flowing liquid.

Experimental methods based upon RCE and conventional electrochemical sensors have a limitation in measuring localized corrosion because data from traditional electrochemical polarization and EIS measurements only indicate overall general corrosion rates under flow dynamic conditions. In order to monitor localized corrosion under flow conditions, an electrode array-based corrosion probe, often referred to as the wire beam electrode (WBE), is used in experimental setups such as the one shown in Figure 6.5 [43]. The WBE probe has two important characteristic that make it particularly useful for studying localized corrosion and its inhibitors in complex multiphase flow conditions: (i) the WBE can map corrosion potential and galvanic current on an instantaneous and continuous basis, providing sufficient spatial and temporal information on localized corrosion processes; and (ii) the WBE is applicable to high resistance multiphase environment and is thus able to simulate corrosion under resistive and complex multiphase flow conditions. On the basis of the distributions of the potential and galvanic current, the dynamic variation of the anodic sites and cathodic sites can be captured, indicating that the propagation of localized corrosion in the multiphase flowing

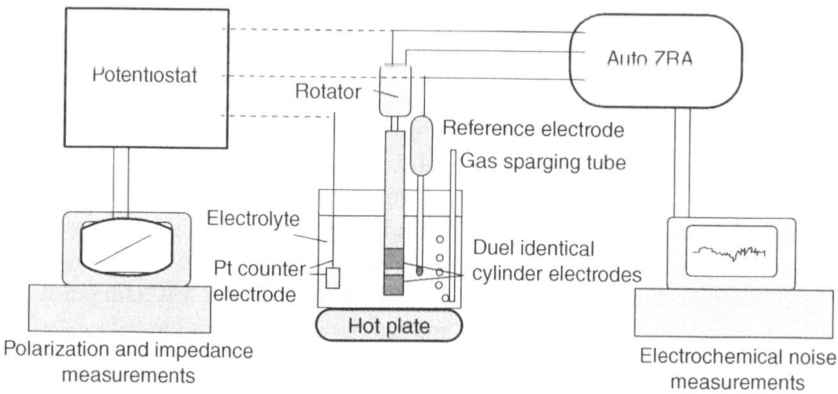

Figure 6.4 A schematic diagram of probing flow influenced corrosion using dual rotating disc electrodes for electrochemical noise recording [22]. *Source:* Tan et al. [21], Kinsella et al. [42].

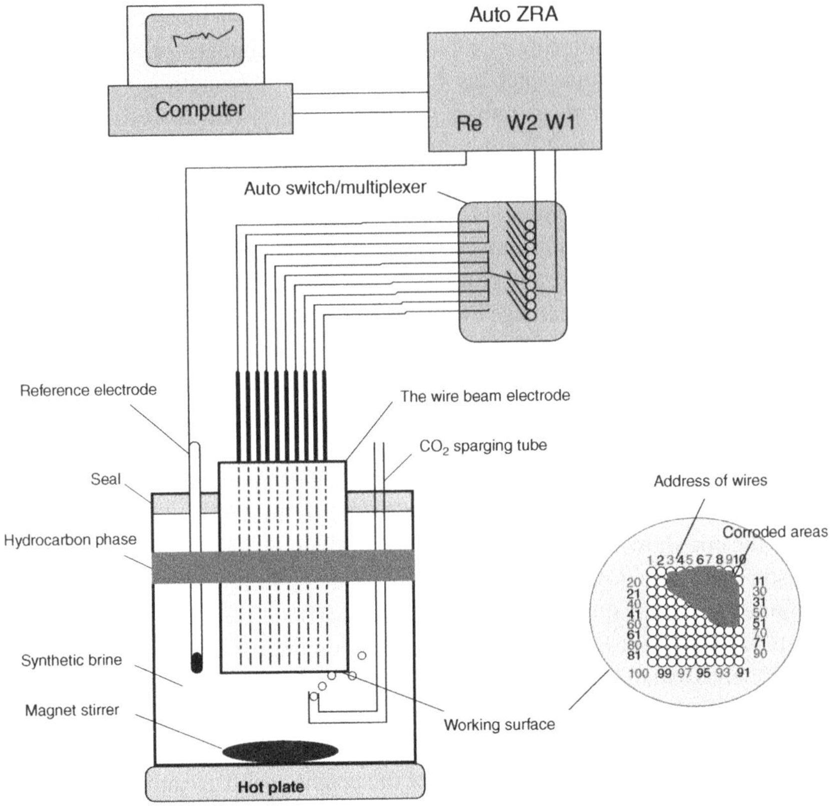

Figure 6.5 Electrochemical cell for simulating multiphase oil flowline corrosion environment (viewed with no stirring). *Source:* Tan et al. [45] / With permission of Elsevier.

conditions [43–45]. More recently, variously designed electrode arrays have been used by Xu et al. as tools for simulating, visualizing, and monitoring the dynamic processes of FAC and erosion corrosion under natural and practical corrosion conditions [46–48]. It has been demonstrated that electrode array-based corrosion probes are able to effectively mimic steel surfaces in FAC and erosion-corrosion behavior. FAC and erosion corrosion were found to be able to evolve and propagate freely on the electrode array surface and that similar corrosion patterns were observed on the array and metal coupon surfaces after being exposed to the same flow conditions. This method has enabled the visualization of FAC and erosion–corrosion processes with high spatial and temporal resolutions, providing a new means of probing, monitoring, and assessing FAC and erosion corrosion and their prevention. Results suggest that FAC and erosion are the results of chemical, galvanic, and flow interactions between different areas of a metal surface [46–48]. Two experimental setups using differently designed electrode arrays are illustrated in Figure 6.6.

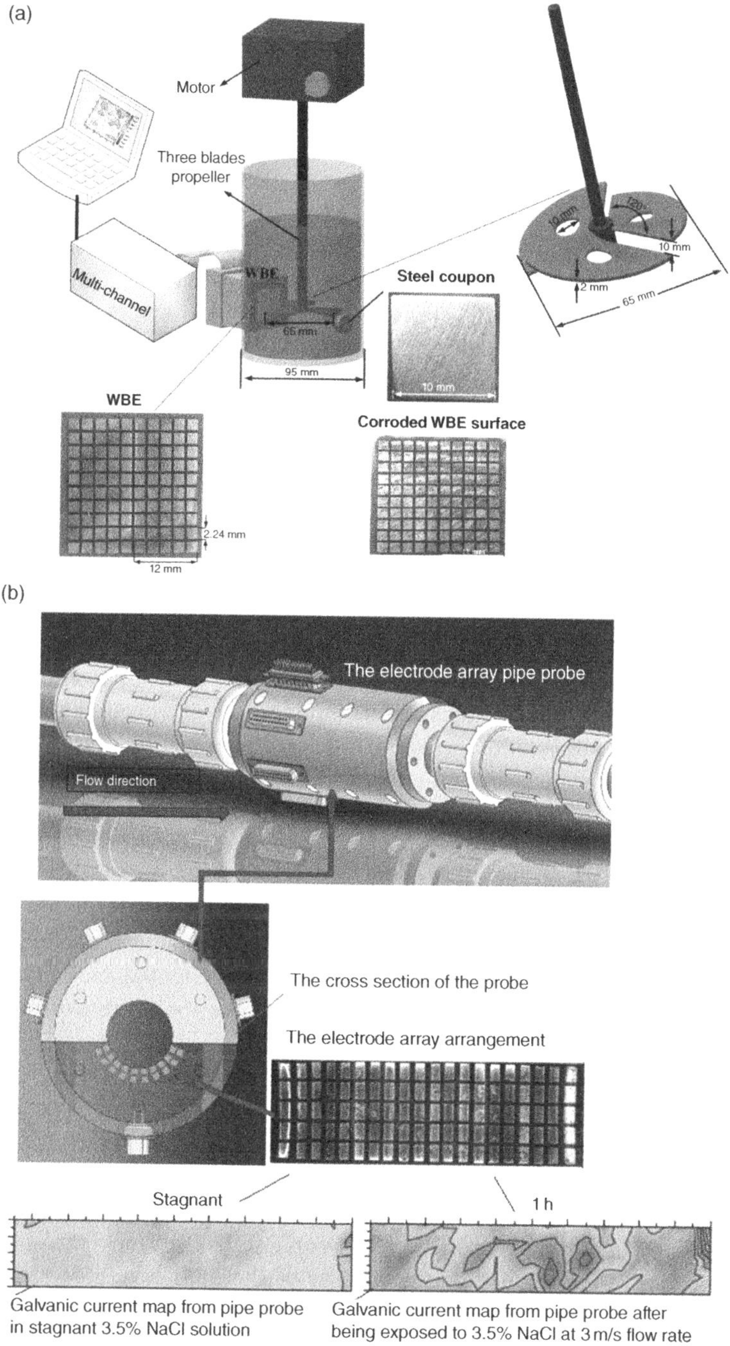

Figure 6.6 Schematic diagram of the setup for probing localized corrosion in a flowline affected by flowing liquid and solid particles. (a) A pipe wall probe for monitoring FAC and erosion corrosion. *Source:* Xu and Tan [46–48], ELSEVIER; (b) A pipe-shaped probe for simulating flowline corrosion.

The pipe-shaped WBE probe shown in Figure 6.6b is designed for fitting to a flowloop that was used to create a typical flowline corrosion condition (see Figure 6.2) to monitor internal pipe corrosion. In a typical design, as shown in Figure 6.6b, electrode array consists of 5 row of 20 square electrodes of the same dimensions. Each of the 100 electrodes is individually addressable allowing for sequential measurement of the net current flowing through it while maintaining the whole array at the open-circuit corrosion potential or a polarized potential. The gaps between electrodes were filled with a cold mounting epoxy resin and cured under pressure to minimize the size of bubbles. The epoxy resin acts as an insulator between the electrodes and also locks them in position. The probe wiring consisted of a set of four 25 ways flat ribbon cables. Laboratory tests were conducted by deploying the internal pipe corrosion probes in a small flowloop designed to simulate flowline corrosion, as shown in Figure 6.2. Representative current density maps obtained from an internal pipe corrosion probes exposed to stagnant electrolyte and to electrolyte flowing at 3 m/s are also shown in Figure 6.6b. Significant changes in corrosion pattern occurred when the solution changes from stagnate to flowing conditions. The movement of solution had significant effect on the corrosion patterns, mechanisms, and rates in the pipeline. This type of electrode array probes have also been used by Zeng et al. [6] who used array electrodes arranged in pipe elbow shapes, in conjunction with CFD simulation, to study erosion–corrosion behavior of an X65 pipeline elbow. Erosion-corrosion rates at different locations of a pipeline elbow were quantified, including pure corrosion rate, pure erosion rate, erosion–enhanced corrosion rate and corrosion-enhanced erosion rate. The relative percentage contributions of these erosion–corrosion rates were determined to identify the dominant influence factor at different locations of the elbow. The distribution of erosion corrosion rates were found to be in good agreement with the hydrodynamics and sand particles distribution at the elbow [6]. This was extended to the probing of the erosion-corrosion behavior of stainless steel pipe elbows. It was demonstrated that more negative corrosion potential and pitting potential, higher passive current density, and higher erosion corrosion rate would occur at the outer wall of elbow outlet. The electrodes at the outer wall of elbow inlet and at the inner wall act as cathodes, while the electrodes at the outer wall of elbow outlet as anodes after coupling. The passivation behavior and galvanic effect of stainless steel elbow were found to be associated with the hydrodynamics and sand particles distribution at the elbow [6]. These reports confirm the usefulness of the electrode array probes as a unique approach to probing the complex and changing effects of flow on FAC and erosion–corrosion processes and mechanisms, pushing the frontier of the multi-electrode array method for addressing the difficult task of simulating and visualizing the dynamic and complex FAC and erosion corrosion.

6.2 Cases of Probing Corrosion and Inhibition in Flowing Liquids

6.2.1 Probing Erosion–Corrosion Behavior of Stainless Steels

Experimental devices such as those shown in Figures 6.2–6.6 have been used in various tests for evaluating and predicting FAC and erosion corrosion of different engineering materials in different environmental conditions. In a typical study, the erosion–corrosion behavior of additively manufactured stainless steel 316 (316L SS) was investigated by Laleh et al. using a flowloop and jet impingement apparatus illustrated in Figure 6.2 that consisted of a high-pressure pump, a 20 l stainless steel tank, a specimen holder, a nozzle, a flow rate gauge, and valves [49]. A 0.6 M NaCl solution containing 2.5 wt% sand particles was driven through a nozzle impinging on the specimen surface at velocities up to 17 m/s. Electrochemical measurements and post-test specimen surface profilometry and SEM-based analyses were performed to understand the erosion–corrosion behavior, as well as possible factors contributing to the erosion–corrosion performance [49].

316L SS has a naturally formed passive film on the surface, which usually provides an excellent corrosion resistance; however, in the presence of abrasive particulates in a fluid stream, the impact of the erosive particles can cause local disruption of the passive film and subsequent corrosion attack on the exposed fresh surface. Figure 6.7 shows potentiostatic polarization behavior recorded from both commercial and SLM-produced 316L SS specimens under erosion–corrosion test conditions and under polarization to an anodic potential of +100 mV (vs. Ag/AgCl, to keep both specimens in the stable passive range). The potentiostatic polarization method has been commonly used as an effective technique to monitor the current fluctuations during the erosion–corrosion test to evaluate the critical flow speed for passive metals and alloys. As shown in Figure 6.7, the general trend recorded from both the commercial and SLM-produced specimens are rather similar. Under static conditions, the current density kept decreasing to a very small value (less than 1 $\mu A/cm^2$) and then kept steady until the pump was switched on and fluid started flowing. The flowing of the fluid caused the anodic current density an abrupt increase, showing effects of flowing on corrosion processes. With every increase in flow speed (1.5 m/s each 5 minutes), the current density rose step by step, suggesting that the moving sand impacted and caused breakdown of the passive film on stainless steel surfaces. The current fluctuations could be due to passive film breakdown (depassivation) and repair (repassivation) processes on the surface. By comparing the average current density at each flow speed (Figure 6.7c), it can be seen that the current densities recorded for the SLM-produced 316L SS are higher (~20–100%) than those obtained for the commercial specimen, suggesting a significantly higher metal dissolution rate for the

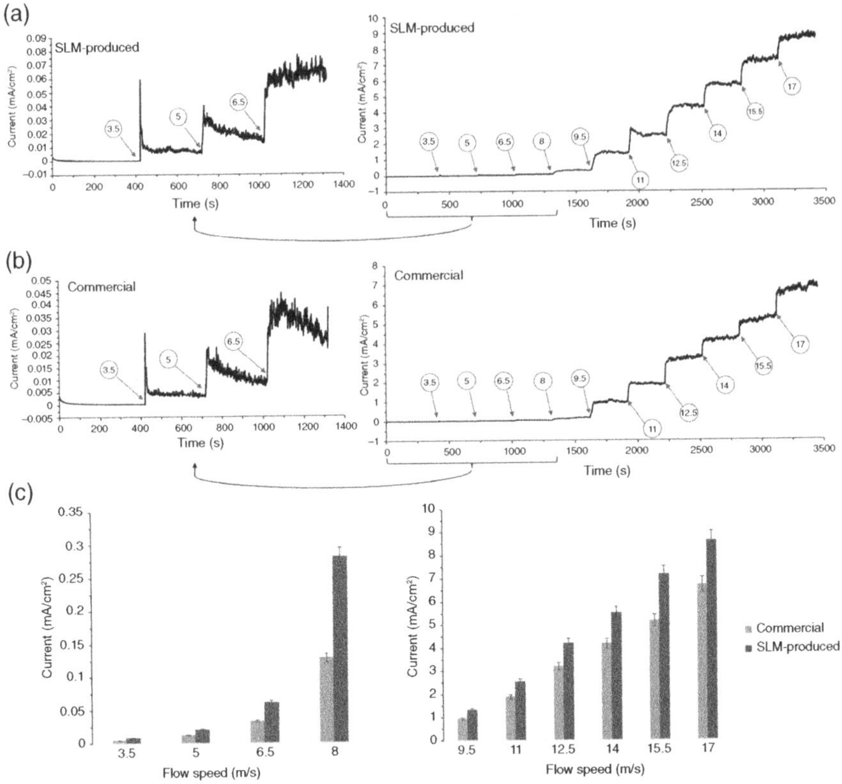

Figure 6.7 Potentiostatic polarization measurements for SLM-produced (a) and commercial (b) 316L SS specimens recorded during erosion–corrosion test. The left-hand graphs are the high magnification of the measurements at the early stages of the erosion–corrosion tests, which are shown with the brace in the figure. The average current density of the specimens at each flow speed was depicted in (c). Error bars represent the standard deviation obtained from at least five measurements under identical conditions. A 0.6 M NaCl slurry solution containing 2.5 wt% sand particles was used as the test solution. The specimens were anodically polarized to +100 mV during the test to keep them in the passive state. Every five minutes, the flow speed was increased by 1.5 m/s to reach a final speed of 17 m/s [49]. *Source:* Laleh et al. [49] / With permission of Elsevier.

SLM-produced 316L SS. These results indicate very different effects of the flowing fluid on SLM-produced and conventional 316L SS specimens [49].

In order to explain this unusual erosion–corrosion behavior of the SLM-produced 316L SS specimen, surface examination of post-corrosion specimen and potentiostatic polarization measurements were carried out. Figure 6.8 shows for the current-time response of both commercial and SLM-produced 316L SS specimens under potentiostatic polarization. It can be seen by examining the stage 3 results in Figure 6.8 that for the SLM-produced 316L SS the final current density

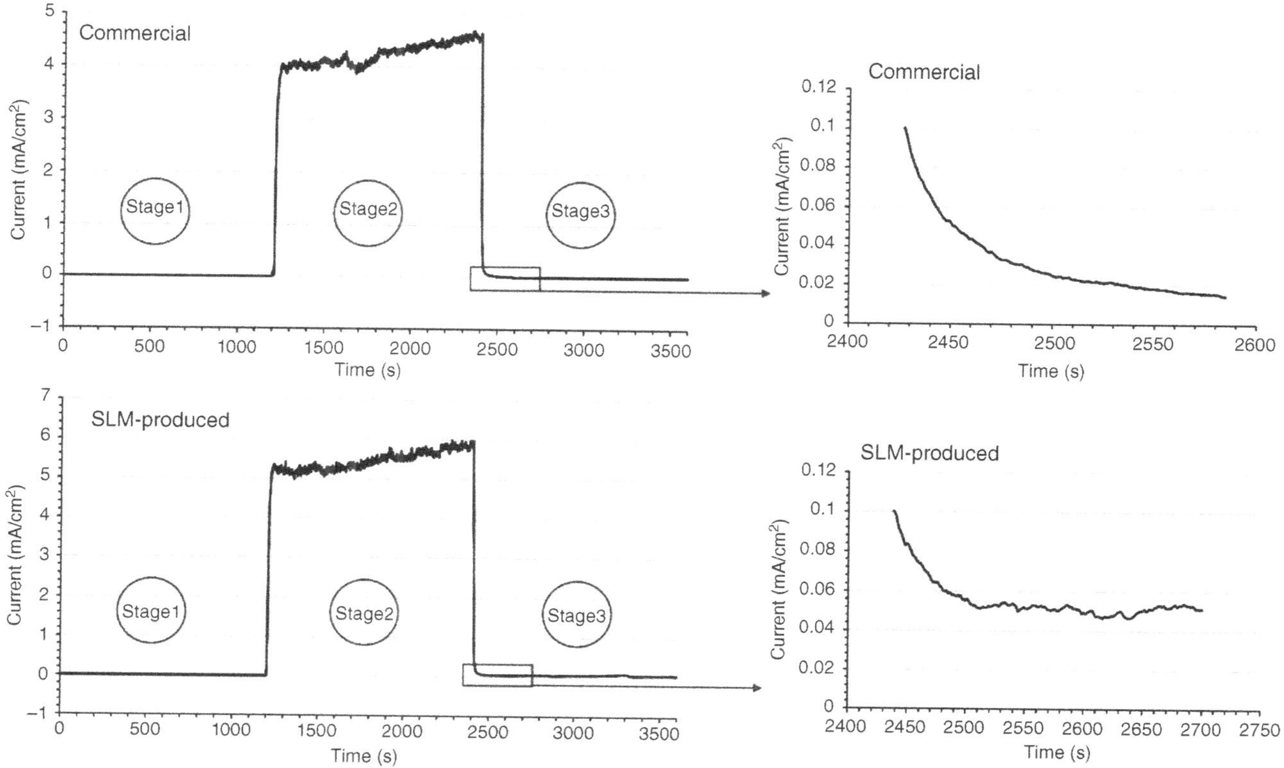

Figure 6.8 Three-stage potentiostatic polarization measurement for evaluating the re-passivation ability under erosion–corrosion conditions for commercial and SLM-produced 316L SS. During stage 1, the pump is switched off for 20 minutes. At stage 2, the de-passivation process starts by switching the pump on for 20 minutes. During this stage the slurry 0.6 M NaCl solution containing 2.5 wt% sand particles continually impinged the surface of the specimens at a constant flow speed of 14 m/s. At stage 3, the pump is switched off and the re-passivation process occurs. *Source:* Laleh et al. [49] / With permission of Elsevier.

remained high, indicating a weaker repassivation ability of the SLM-produced 316L SS. This suggests that when the depassivation occurs on the SLM-produced 316L SS due to impingement, the repassivation is less easy compared to the commercial specimen and therefore erosion corrosion continues to occur. A reason that the SLM-produced 316L SS has weaker repassivation ability could be related to the presence of the SLM process-induced pores within the specimen. When depassivation and corrosion occur inside these pores, it would lead to chemistry changes inside these pores and they would be more difficult to repassivate, similar to the situation of corrosion occurring in growth pits and crevices [49]. Another possible contributing factor to the weaker repassivation ability of the SLM-produced 316L SS could be that higher residual stresses existed in the SLM-produced specimen. Because of the high thermal gradients associated with SLM processing, the SLM-produced 316L SS is expected to exhibit a high level of residual stress [49]. Nevertheless these results clearly show very different effects of the flowing fluid on SLM-produced and conventional 316L SS specimens. Despite the higher pitting corrosion resistance and microhardness exhibited by SLM-produced 316L SS, it showed an unusually weaker erosion-corrosion resistance under the jet impingement condition [49]. These results suggest the need for assessment and monitoring of materials performance when they are used in complex environmental conditions.

6.2.2 Monitoring Corrosion Inhibitor Film Damage by Flowing Liquid

A method of corrosion control in oil and gas flowlines is the application of corrosion inhibitors, either through continuous injection or through batch treatment. An issue is that flowing fluids can remove the inhibitor films from the pipeline surface, therefore the monitoring of the persistency of corrosion inhibitors is required. Using the electrochemical cell and a RCE shown in Figure 6.3, corrosion inhibitor film damage could be monitored using electrochemical methods in particular EIS [21, 42, 50]. EIS data allow the determination of electrochemical parameters and their changes, providing information on variations in electrode interfacial structure, electrode capacitance and charge transfer kinetics under the effect of fluid flow [50]. Figure 6.9 shows typical impedance changes when an inhibitor film was damaged by flow [21], after an electrode with inhibitor film were transferred from a solution containing 50 ppm imidazoline into inhibitor-free brine where the inhibitor film may deteriorate. The continuous decrease in the diameter of Nyquist plots and the gradual disappearance of the high-frequency phase angle shift, as shown in Figure 6.9a and b, suggest that the inhibitor film was gradually removed by flow-induced surface shear stress. In these experiments, EIS has been shown to be a valuable technique for understanding the mechanism of inhibitor film formation and destruction and for evaluating the film persistency of CO_2 corrosion inhibitors [21]. However EIS has limitations in

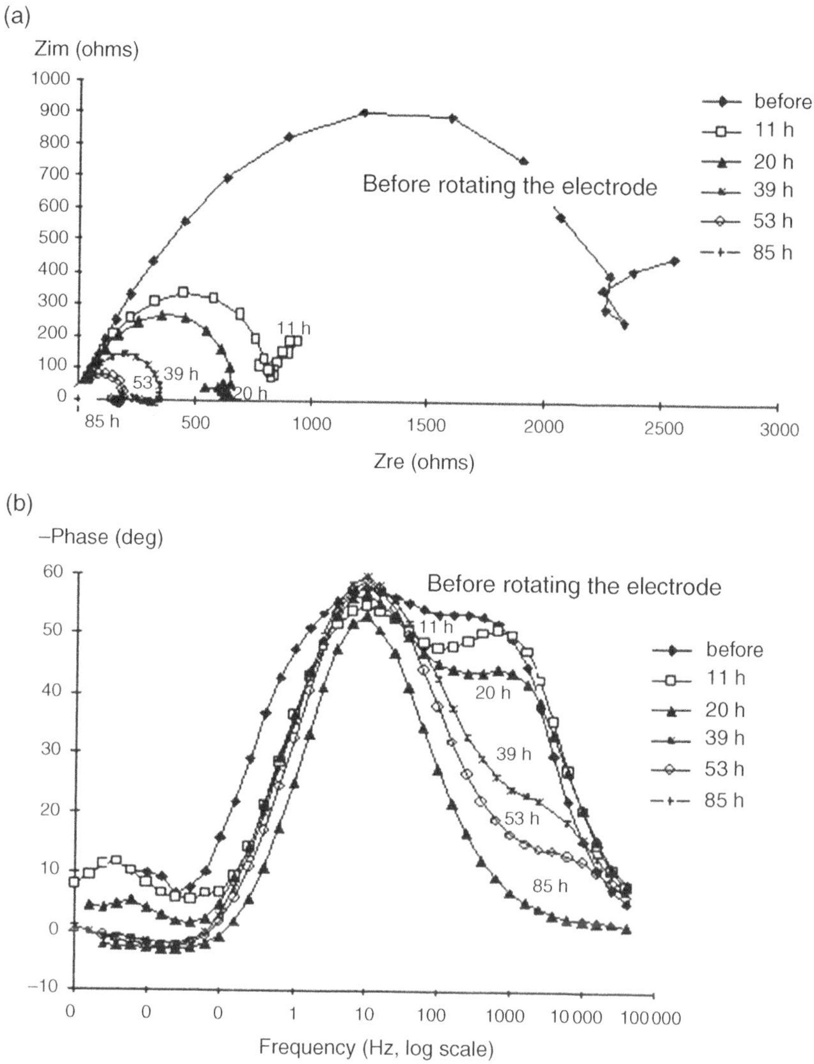

Figure 6.9 EIS plots before and after rotating the inhibitor filmed electrode (1000 rpm) in an inhibitor-free 3% NaCl brine. (a) Nyquist plots; (b) Bode plots [21]. *Source:* Tan et al. [21] / With permission of Elsevier.

monitoring rapid electrode interfacial changes on a continuous basis because EIS measurement takes quite a long time (often more than one hour). EIS is a steady-state technique, in principle it cannot be used in corrosion systems where parameters such as potential are rapidly changing. For these reasons, EIS has difficulties in studying rapid corrosion electrochemical process, e.g. corrosion inhibition processes immediately after inhibitor addition [21].

Electrochemical noise analysis (ENA) is a method that has been used for monitoring dynamic changes in electrode interfacial processes by measuring the electrochemical noise resistance (R_n) on a continuous basis. R_n is the ratio of the standard deviation of voltage noise and the standard deviation of current noise and is equivalent to the polarization resistance (R_p) [22, 23, 51–55]. In a typical experiment illustrated as shown in Figure 6.4 [22, 23], a dual mild steel electrode with an inhibitor imidazoline film was quickly transferred into an inhibitor-free 3% NaCl solution with 1000 rpm stirring. Electrode potential and current noise were recorded on a continuous basis, in conjunction with linear polarization measurements in order to monitor rapid changes in electrode processes due to inhibitor film destruction due to flow induced by flow. After the electrode transfer, as shown in Figure 6.10, the current noise amplitude apparently increased and this can be observed in the raw noise record (Figure 6.10a), and is more clearly shown in Figure 6.10b, where DC trend has been removed from the raw noise data. R_n decreased rapidly after electrode transfer, as shown in Figure 6.10c, indicating rapid degradation of the inhibitor film. Although R_n values in Figure 6.10 shows fluctuations, the trend line follows the breakdown process of the inhibitor film. This result suggests that electrochemical resistance could be a convenient method for monitoring inhibitor film breakdown and so to evaluate inhibitor film persistency [22, 23]. In another typical experiment, this continuous electrochemical noise resistance measurement method was used successfully for studying the formation process of chromate conversion coatings on aluminum [56]. These experiments demonstrate the usefulness of RCE and electrochemical methods in monitoring flow induced corrosion inhibitor film damages; however it should be noted that a limitation of the EIS and ENA data such as those in Figures 6.9 and 6.10 is that they do not reveal localized damage of inhibitor films [22].

6.2.3 Probing Localized Damage of Inhibitor Film and Corrosion

A common limitation of the potentiostatic polarization, EIS and ENA measurements in monitoring flow induced corrosion and inhibitor film damages is that they donot provide information regarding the location, size and depth of any localized damage. This issue can be overcome if the electrode array-based WBE probe device is employed in these experiments. Using the setup shown in Figure 6.5, the WBE method was used in conjunction with electrochemical noise resistance (R_n) measurements to detect and monitor flow-induced localized damage of inhibitor films by mapping local electrochemical parameters including corrosion potential, galvanic current, and electrochemical noise resistance from corroding surfaces exposed to highly resistive and inhomogeneous multiphase hydrocarbon/produced water mixtures [43, 45]. Figure 6.11 shows typical corrosion potential, galvanic current, noise resistance, and corrosion rate distribution

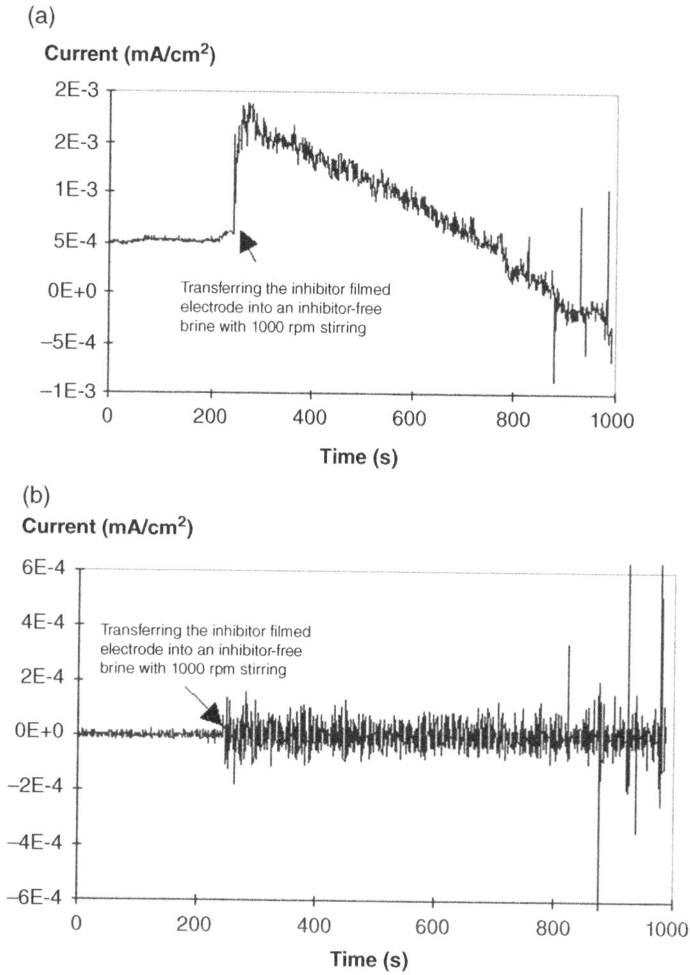

Figure 6.10 Monitoring dynamic changes in inhibitor film by measuring the electrochemical noise resistance (R_n) on a continuous basis. (a) Original current noise records after transferring the inhibitor filmed electrode into inhibitor-free brine with 1000 rpm stirring; (b) current noise (after moving average removal treatment); (c) change of electrochemical noise resistance. *Source:* Tan et al. [22] / With permission of Elsevier.

maps measured from a WBE exposed to a multiphase water–kerosene–carbon dioxide mixture for 720 hours. It is clear that over the WBE surface there is a nonuniform distribution of corrosion potential, galvanic current, noise resistance, and thus a nonuniform corrosion rate distribution. Localized corrosion concentrated mainly over an area (wire 10) where there was a turbulent flow that locally damaged the inhibitor film. The noise resistance values over the areas near

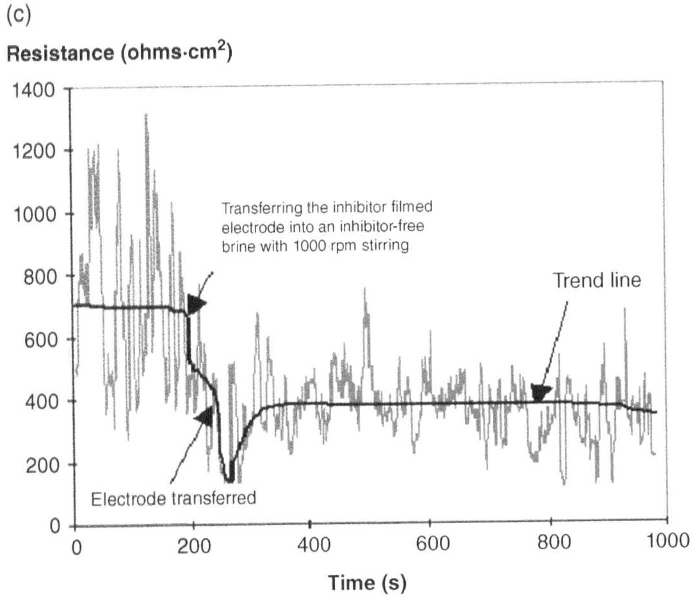

Figure 6.10 (Continued)

wire 10 (~ 10^3 ohms) are much smaller than the majority of the electrode surface (~ 10^7 ohms), indicating local inhibitor film damage arisen from flow impact on wire 10 [43]. The damaged areas clearly exhibited low noise resistance, more negative potential, and correspondingly positive galvanic current, which indicate electrons move out from this area to cathodic areas. These electrochemical parameters were used to calculate the rates of localized corrosion and their distribution over a WBE surface. As shown in Figure 6.11, electrochemically "calculated" corrosion maps correlated clearly with microscopically "observed" corrosion depth maps. This experiment suggests that the combined WBE-R_n technique is capable of measuring CO_2 corrosion under multiphase conditions [43].

Figure 6.12 shows more localized inhibitor film damage arisen from more intense turbulent flow, which locally damaged the inhibitor film on a WBE surface [45]. In the noise resistance distribution map of Figure 6.12, noise resistance values in the anodic areas were much lower than in other areas, resulting in a similar pattern to other maps in Figure 6.12. The clear correlation between the noise resistance map and other maps in Figure 6.12 suggests that in the anodic areas around wire 91 not only was galvanic corrosion occurring, but uniform corrosion was also occurring with a significantly higher corrosion rate than in other areas. This suggests that both uniform and galvanic corrosion components contribute to the localization of corrosion. It also confirms that the

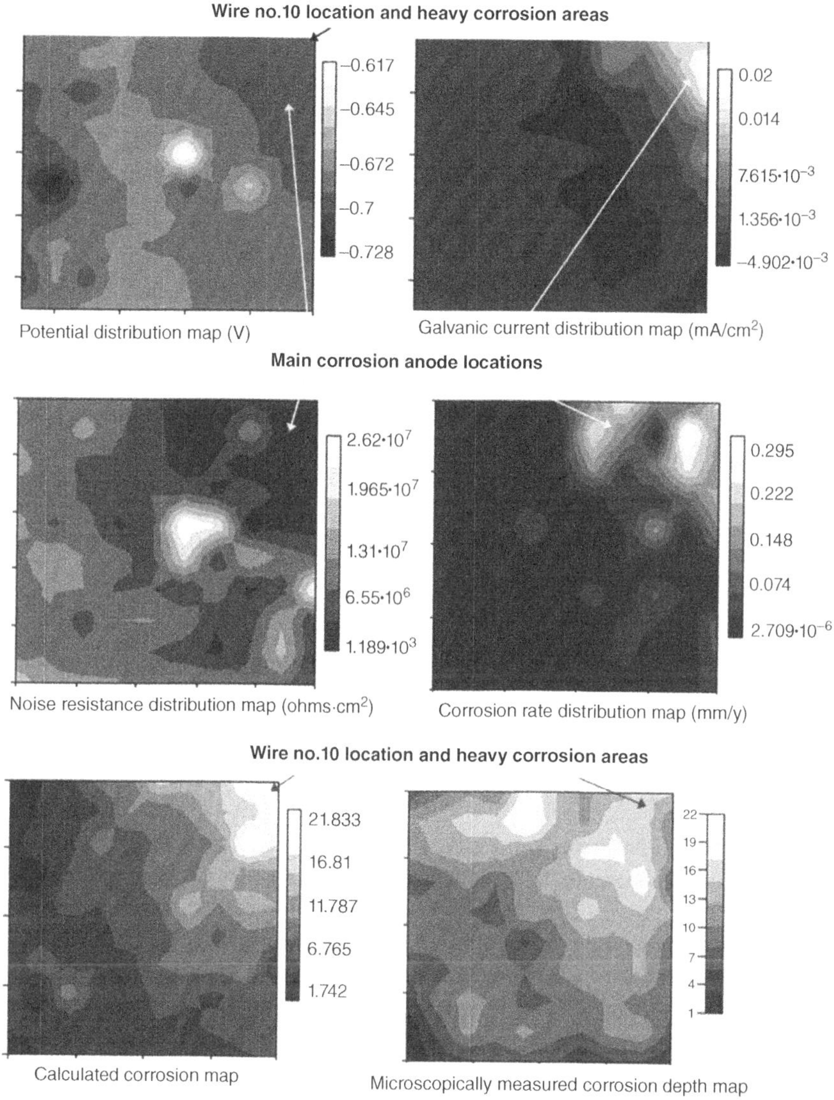

Figure 6.11 Corrosion potential, galvanic current, noise resistance, and corrosion rate maps measured from a WBE exposed to the water–kerosene–carbon dioxide mixture [43, 45], and "Observed" and "calculated" corrosion depth maps and values (in μm) over a WBE surface after exposure to a water–kerosene–carbon dioxide mixture for 917 hours [45]. *Source:* Tan et al. [45] / With permission of Elsevier.

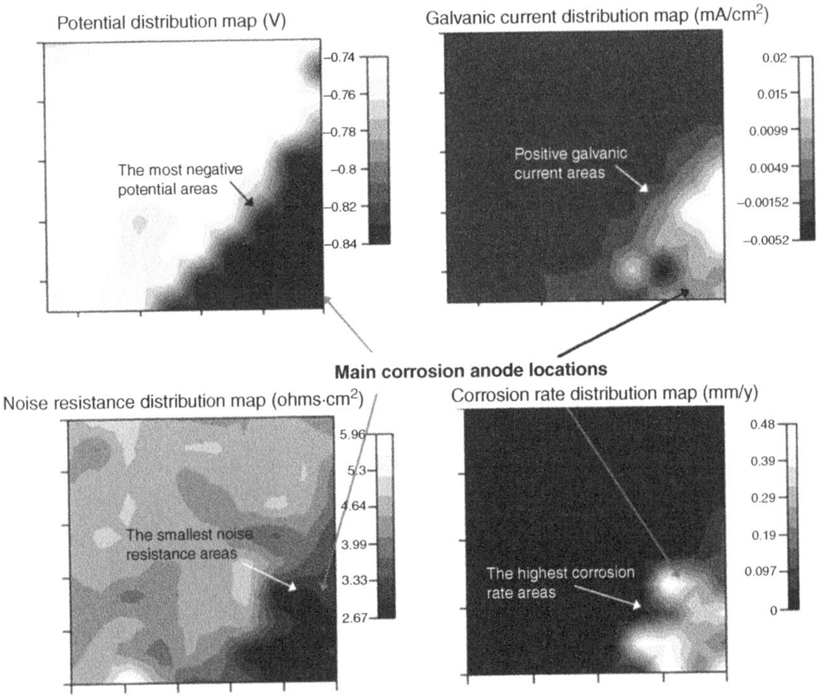

Figure 6.12 Corrosion potential, galvanic current, noise resistance, and corrosion rate distribution maps measured from a WBE after exposure to a simulated CO_2 corrosion cell. *Source:* Tan et al. [43] / With permission of IOP.

inhibitor film on the areas around wire 91 was locally damaged by turbulent flow and became less protective, which is confirmed by the corrosion rate distribution map of Figure 6.12 showing corrosion rates on the anodic areas (wire 91) significantly higher than that in other areas. To confirm the accuracy of the corrosion rate distribution maps, total accumulative corrosion depths and the distribution of these depths were calculated from the corrosion rate maps. If the corrosion rate distribution maps such as those in Figure 6.12 are accurate, this total corrosion depth distribution map should correlate quantitatively with the real corrosion depth and its distribution, which is measurable with a microscope at the end of the total exposure period. As shown in Figure 6.13, the calculated corrosion depth map gives similar, although not identical, corrosion depth values to those measured microscopically [43].

Figure 6.14 shows less localized inhibitor film damage arisen from less intense turbulent flow, which only generally damaged the inhibitor film on a WBE surface [43, 45]. After 408 hours exposure, as shown in Figure 6.14, corrosion of the

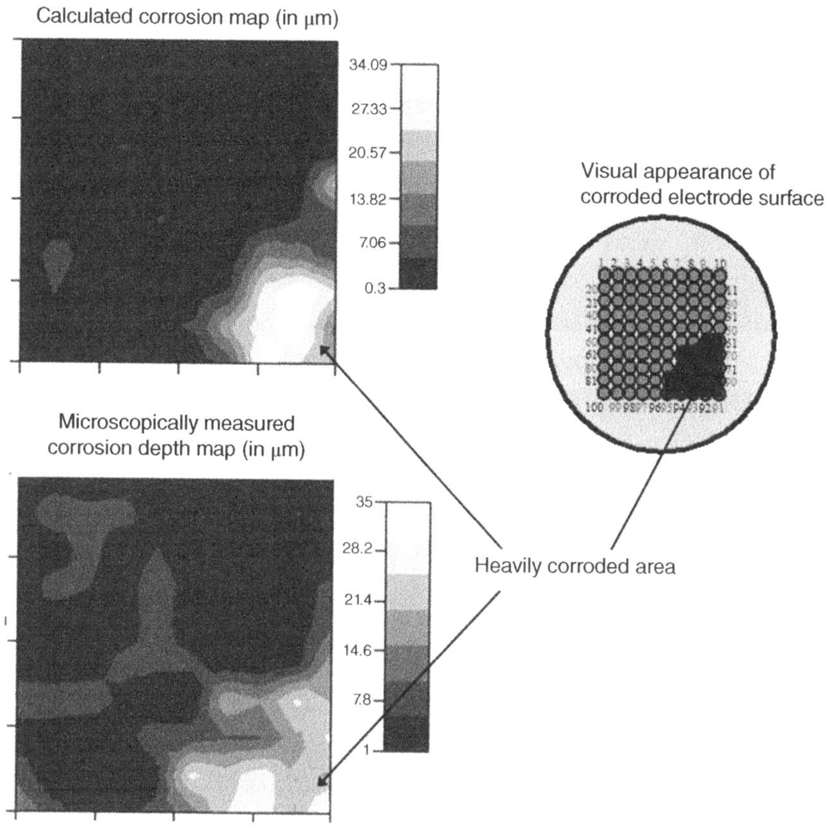

Figure 6.13 Observed and calculated corrosion depth maps and values (in μm) over a WBE surface, after exposure to a simulated CO_2 corrosion environment for 241 hours [43, 45]. *Source:* Tan et al. [43] / With permission of IOP.

electrode surface was less localized and had the appearance often referred to as general corrosion [43, 45]. Compared to the maps shown in Figure 6.12, maps obtained from this general corrosion system show different characteristics. In this system electrochemical parameters recorded from different locations of the electrode surface were similar, for instance, the maximum potential difference between anodic and cathodic zones were only about 30 mV. Furthermore, across the entire duration of the test, the majority of the electrode surface was under corrosion typically as shown in Figure 6.14. Corrosion rate values recorded from different locations of the electrode surface were relatively close. The range of corrosion rates observed across the entire probe surface was also narrower. These observations are consistent with the fact that the electrode surface exhibited a

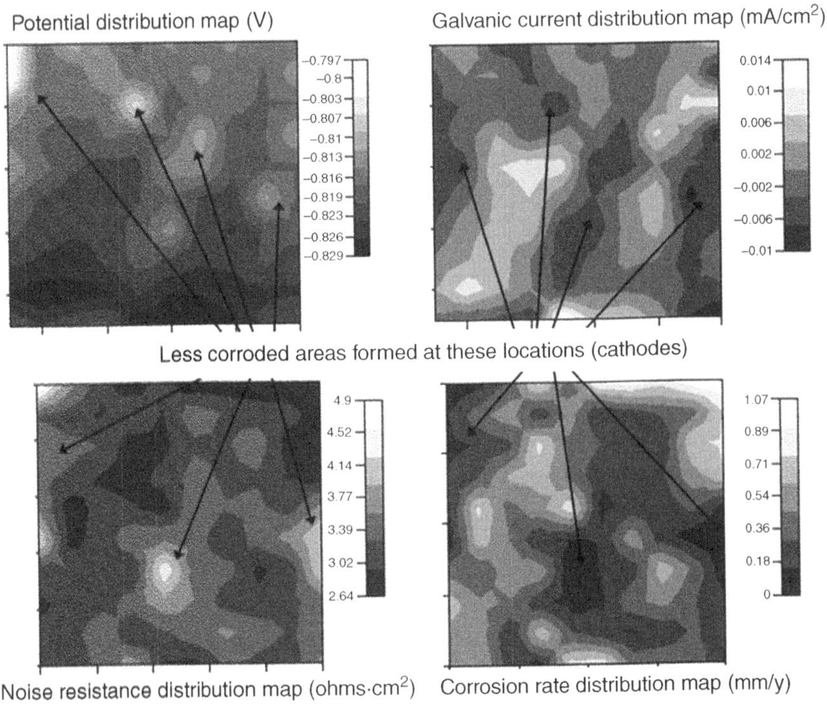

Figure 6.14 Corrosion potential, galvanic current, noise resistance, and corrosion rate distribution maps measured from a WBE after exposure to a CO_2 corrosion cell. *Source:* Tan et al. [43] / With permission of IOP.

lesser degree of localization in its corrosion pattern. In general, the corrosion potential, galvanic current, noise resistance, and corrosion rate distribution maps in Figure 6.14 correlate well with each other [43, 45]. Again to confirm the accuracy of the corrosion rate distribution maps such as those shown in Figure 6.14, accumulative corrosion depths and their distribution were calculated from the corrosion rate maps obtained at various stages of exposure. The corroded electrode surface was observed and measured under a microscope at the end of the experiment to obtain an observed corrosion depth map. Figure 6.15 shows observed and calculated corrosion depth maps that have generally similar features and corrosion depth values. This result demonstrates that the electrochemical method also works well even if the corrosion is not particularly localized. These data confirm the usefulness of the electrode array probes as a unique approach to probing corrosion inhibitor behavior under the complex and changing effects of flow conditions [43, 45].

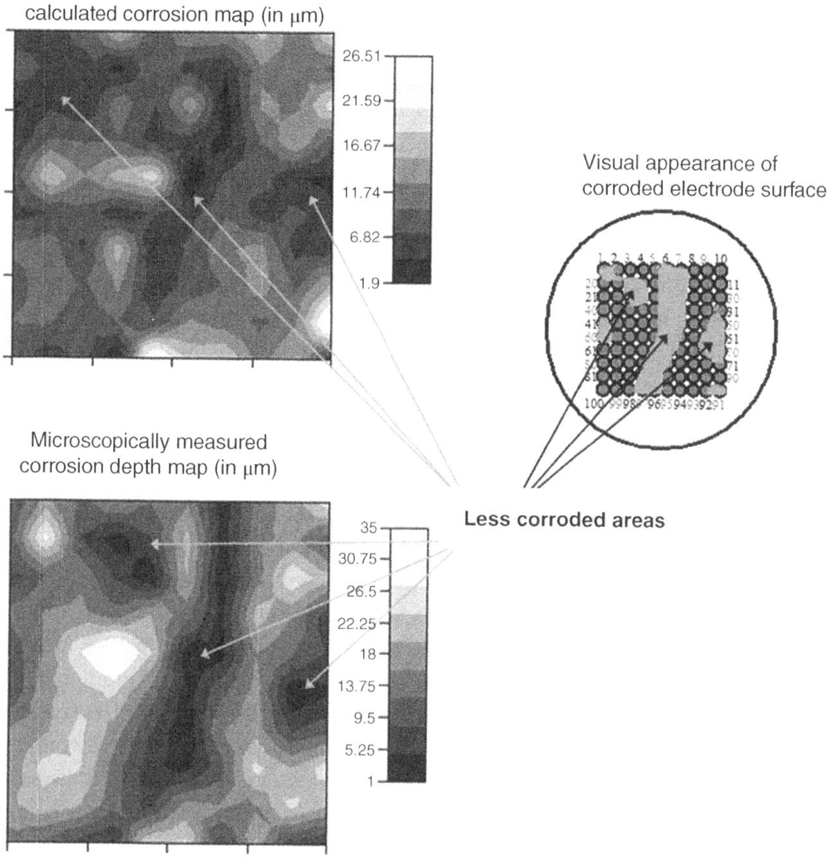

Figure 6.15 Observed and calculated corrosion depth maps and values (in μm) over a WBE surface, after exposure to a simulated CO_2 corrosion environment for 408 hours [43]. *Source:* Tan et al. [43] / With permission of IOP.

6.3 Probing Localized Corrosion Mechanisms Impacted by Flow and Erosion

6.3.1 Assessing Factors Affecting FAC and Erosion Corrosion

FAC and erosion corrosion are two challenging forms of localized corrosion that are affected by a complex combination of electrochemical, mass-transfer, fluid dynamic and mechanical factors [10, 15, 57–60], often acting synergistically and resulting in metal loss rates significantly higher than either erosion or corrosion acting alone. Erosion corrosion is an electrochemical–mechanical process involving the impingement of particles, bubbles, or cavitation, which cause the mechanical wear on the metal surface, often leading to loss of passivation and localized corrosion. FAC is considered to be distinctly different from erosion corrosion and is primarily a corrosion process aided by chemical dissolution and mass transfer [61], and often shows a distinguishable corrosion pattern with a typical "flow mark" [58, 59], while erosion corrosion often has a crater-like appearance [15, 57] due to mechanical impacts such as impingement of particles. Although FAC and erosion corrosion have been studied extensively over the past decades, the *in situ* probing of FAC and erosion corrosion under natural conditions remained a technological challenge. For instance FAC and erosion corrosion are believed to be significantly affected by competition between fresh metal surface generation through depassivation and repassivation [62] due to combined local electrochemical, interfacial chemistry, and confined mass transfer conditions, probing these localized, dynamic, and often complex processes and their mechanisms are difficult because of the lack of *in situ* monitoring tools capable of probing these localized processes. Traditionally FAC and erosion corrosion are assessed in industry through periodic inspection using nondestructive test methods, electrical resistance probes [63], and iron concentration measurements [64]. In the laboratory they are typically evaluated electrochemically employing methods such as current transient and potentiodynamic polarization measurements, and devices such as rotating disc electrode [65, 66] and slurry jet impingement loops [62, 67, 68]. Conventionally electrochemical study of erosion corrosion involves measuring currents under polarisation. For instance, Luo and co-workers have carried out extensive research on the erosion-enhanced corrosion of metals and alloys by detecting current transients caused by particle impact while holding the electrode in the passive potential range [10, 15, 62, 69]. This method is very useful for understanding passive film breakdown and repassivation under anodic polarization; however it is unable to do so under natural conditions because no current transient is detectable under the natural open-circuit potential (OCP) condition. In an effort to probe practical passive behavior under OCP conditions, scratching electrodes were used to simulate the effect of sand impingement on corrosion process through the measurement of electrochemical noise generated during the scratching process [10]. The scratching technique also has a limitation since it only measures overall electrochemistry over the whole electrode surface rather than probing the localized electrode processes at the scratch. This illustrates limitations in traditional electrochemical methods for probing localized and dynamic interfacial processes, and the need for new technologies able to address these limitations [46].

Figure 6.6a shows a schematic diagram of a typical experimental setup designed by Xu et al. in an electrochemical cell to generate flow dynamic conditions required for simulating and probing FAC and erosion–corrosion processes and mechanisms [46–48]. A specially designed WBE probe and a steel coupon (electrode) are mounted in the test cell with the propeller positioned in the middle of their surfaces. In a typical experiment, the WBE consists of 100 closely packed but isolated square-shaped electrodes of X65 pipeline steel (2.24 mm × 2.24 mm each) embedded in epoxy resin. The gaps between neighboring electrodes were kept very small (0.2 ± 0.05 mm). Steel coupon was also made of the same steel with a surface dimension of 10 mm × 10 mm. Computational fluid dynamic simulation has also been used in combination with multi-electrode array in a similar way as those used to study FAC of a simulated pipeline elbow [70–72]. Sodium chloride solutions (NaCl, 0.5 M) without and with 10% sand particles (by weight) were used as the test fluids that were stirred at 1000 and 2000 rpm to create completely turbulence flow conditions [46]. The WBE current maps were obtained by regularly mapping local currents over the WBE, in the same way as that described previously in references [73, 74]. Figure 6.16 shows photographs of the corroded and eroded surfaces of the WBEs and steel coupons after 20 hours exposure to the electrochemical cell under different flow conditions. As shown in Figure 6.16a and b, a distinguishable corrosion pattern with a typical FAC "flow mark" [58, 59] is observable on both WBE and coupon surfaces, while a distinctively different crater-like erosion–corrosion pattern [15, 57], as shown in Figure 6.16c, is observable on both WBE and coupon surfaces. Differences in characteristic corrosion patterns in Figure 6.16b and c are more obvious if examining the local 3D surface profilometry images. These results demonstrate that the WBEs were able to simulate the steel coupon in FAC and erosion–corrosion behavior under flow conditions, suggesting that the WBE surface allowed sufficient electrochemical integration and chemical interactions between electrodes. This experiment provides direct evidence that the WBE surface is able to simulate a conventional one-piece steel in FAC and erosion–corrosion behavior [46].

The WBE probe is able to perform near real-time FAC and erosion–corrosion monitoring if current mapping is carried out frequently, and display as a video. The WBE probe current density distribution maps of the type shown in Figure 6.17 provide more details on FAC and erosion–corrosion initiation and propagation processes with high spatial (mm) and temporal (10 minutes) resolutions. Different characteristics can be observed from WBE maps in Figure 6.17. As shown in Figure 6.17a, at stirring speed of 1000 rpm corrosion initiated from a few isolated anodic sites, indicated by the randomly distributed positive galvanic currents, among a large number of cathodic sites. Some of the anodic sites merged forming larger localized corrosion zones. These are similar to localized corrosion processes reported previously from corrosion in static NaCl solutions [75]. When the solution stirring speed was increased to 2000 rpm, as shown in Figure 6.17b, very different corrosion initiation and propagation characteristics are clearly visible. A longer incubation time of four hours was needed for the formation of stable local anodes. This suggests that under this higher flow rate condition, the distinctive separation of anodic and cathodic areas takes longer time to occur, probably due

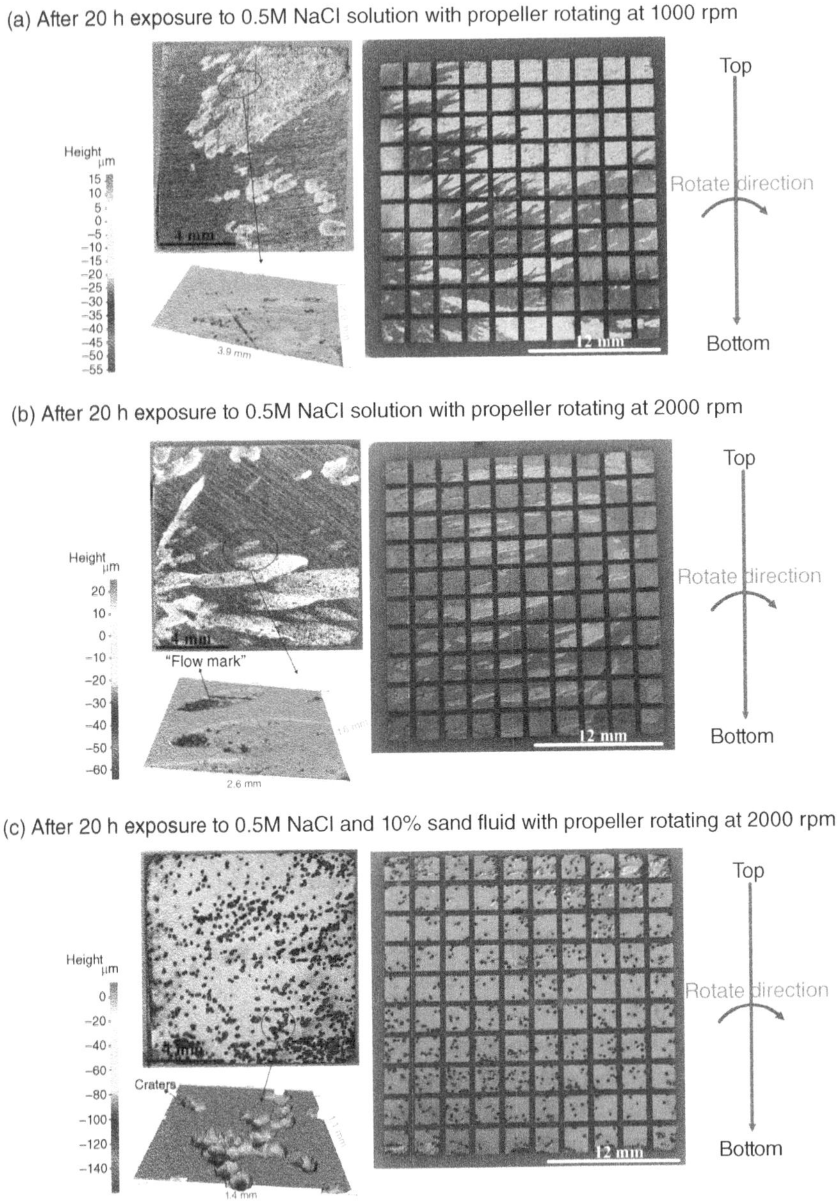

Figure 6.16 Photos of the WBEs and coupons after 20 hours exposure to different FAC (a, b) and EC (c) conditions, and local 3D surface profilometry images of typical coupon areas [46]. *Source:* Xu and Tan [46], ELSEVIER.

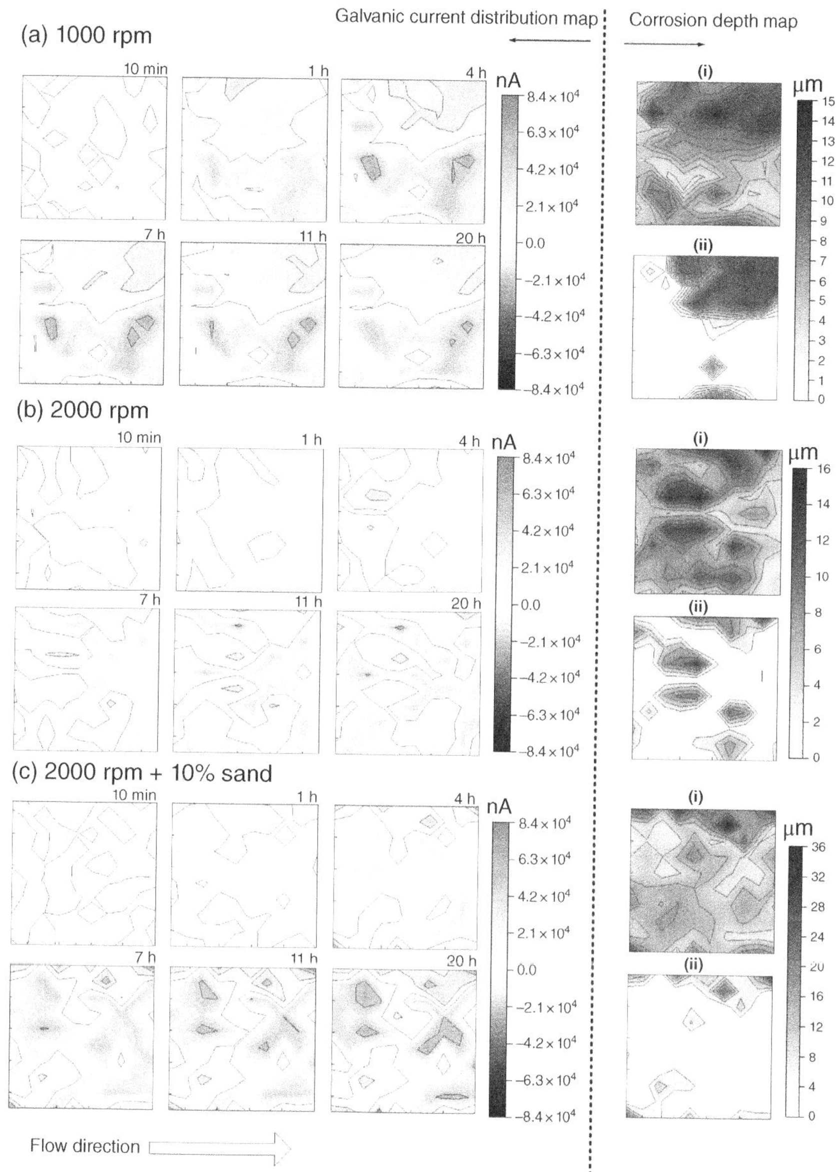

Figure 6.17 The WBE current distribution maps and the accumulated corrosion depth maps obtained from (i) surface profilometry measurement and (ii) Faraday's law calculation. Maps from electrolytes stirring at (a) 1000 rpm; (b) 2000 rpm; and (c) 2000 rpm + 10% sand [46]. *Source:* Xu and Tan [46]/ With permission of Elsevier.

to enhanced mass-transfer on electrode surfaces. Another interesting observation is that anodes propagated along the electrolyte flow direction, forming narrow and long anodic areas. This is in agreement with the typical narrow and long FAC "flow mark" shown in Figure 6.16b. These results suggest that FAC propagation was controlled by the interfacial solution chemistry (e.g. local pH) that was influenced by electrolyte flow. The corrosion controlling factor was probably flow-induced mass transfer in which the interfacial chemistry such as the protective oxide or the local higher pH was changed, causing corrosion anodes to propagate in the flow direction. When 10% of sand was added to the electrolyte under 2000 rpm stirring, as shown in Figure 6.17c, the anodes appear to distribute randomly, probably at sand collision sites, and propagate along the flow direction to form mesa type corrosion. This is in agreement with the crater-like erosion–corrosion appearance shown in Figure 6.16c. This suggests that the controlling factor for erosion corrosion was both the mechanical impact and the mass transfer induced local surface condition changes. These results indicate that the WBE is able to acquire near real-time FAC and erosion–corrosion rates and mechanism information if measurements are carried out frequently [46].

The accumulated corrosion depth maps shown in Figure 6.17 were calculated using the Faraday's law and measured using a surface profilometry. These maps especially the high corrosion sites are well matched, suggesting that galvanic currents are the main contributors of localized corrosion, although corrosion depth losses from profilometry measurements are slightly larger. However it can also be seen that when sand particles were added to the fluid, as shown in Figure 6.17c, the corrosion depths obtained from surface profilometry measurements were considerably larger than that calculated using Faraday's law, suggesting more metal losses due to erosion under the impact of sand. Nevertheless the WBE has been able to effectively simulate and visualize the FAC and erosion–corrosion processes and provide a means of estimating FAC and erosion–corrosion patterns and kinetics [46].

Computational fluid dynamic simulation has been used in combination with multi-electrode array to study FAC of a simulated pipeline elbow, which enabled Zhang et al. to detect different corrosion behavior due to varying hydrodynamics of fluid flow at the elbow [70–72]. Zeng et al. [6, 72] used array electrodes arranged in pipe elbow shapes, in conjunction with CFD simulation, to study erosion–corrosion behavior of an X65 pipeline elbow. Erosion–corrosion rates at different locations of a pipeline elbow were quantified, including pure corrosion rate, pure erosion rate, erosion-enhanced corrosion rate, and corrosion-enhanced erosion rate. The relative percentage contributions of these erosion–corrosion rates were determined to identify the dominant influence factor at different locations of the elbow. The distribution of erosion–corrosion rates were found to be in good agreement with the hydrodynamics and sand particles distribution at the elbow [6]. This work was extended to the probing of the erosion–corrosion behavior of stainless steel pipe elbows. It was demonstrated that more negative corrosion potential and pitting potential, higher passive current density, and higher erosion–corrosion rate would occur at the outer wall of elbow outlet. The electrodes at the outer wall of elbow inlet and at the inner wall act as cathodes while the electrodes at the outer wall of elbow outlet as anodes after coupling. The passivation behavior and galvanic effect of stainless steel elbow were found to be associated with the hydrodynamics and sand particles distribution at the elbow [72]. Zhong et al. used array electrodes and CFD simulation to correlate local flow accelerated corrosion rates and local hydrodynamics at a gradual contraction of steel pipe. Results showed that the distribution of corrosion rate measured by the array electrode at the gradual pipe contraction was in good accordance with the distribution of hydrodynamics parameters including flow velocity and wall shear stress. The presence of an imidazoline derivative inhibitor greatly mitigated corrosion at the gradual pipe contraction through hindering the mass transfer process and charge transfer process of corrosion [76]. Li et al. used array electrodes made with steel wires within a steel/copper alloy couple immersed in static and flowing seawater. Results showed that the potential and current density distributions of steel wires in both static and flowing seawater were inhomogeneous, while the electrochemical inhomogeneities of steel wires immersed in flowing seawater were significantly reduced [77]. Using the WBE, Deng et al. investigated the corrosion of X70 pipeline steel in marine thermocline where dramatic changes of temperatures, pH values, dissolved oxygen contents, and nutrients occur, affecting corrosion [78]. Results indicated that the steel's galvanic currents increased with experimental time, and the galvanic corrosion got worse vertically from the top to the bottom of the marine thermocline simulator [78]. It should be noted, however, that the electrode array used by some researchers had a large inter-electrode spacing of approximately 6 mm between electrodes. Unfortunately, such a large inter-electrode spacing would lead to insufficient chemical interactions between electrodes located at different areas and therefore, as explained in a review of the electrode array method [73, 74], insufficient electrochemical integration.

6.3.2 Exploring FAC and Erosion–Corrosion Mechanisms

Erosion corrosion occurs most frequently in flowlines involving the impingement of sand particles, bubbles, or cavitation, generating crater-like and mesa-like damages on pipeline surfaces [11, 15, 57, 58, 60]. FAC, which is distinct from erosion corrosion by a distinguishable "flow mark" corrosion feature [6, 58], is believed to be primarily a corrosion process aided by chemical dissolution and mass transfer [61]. Erosion corrosion is believed to be induced by the mechanical wear of the passive film by the impingements of the sand particles [57, 62], causing the fresh metal matrix to be directly exposed to the corrosive medium and therefore the acceleration of corrosion [35]. FAC and erosion corrosion have been studied extensively in the past, both erosion corrosion and FAC are believed to be affected by a complex combination of electrochemical, mass-transfer, fluid dynamic and mechanical factors that often act synergistically, resulting in metal loss rates significantly higher than either erosion or corrosion acting alone [60, 61]. For passive metals, erosion-enhanced corrosion is believed to be the main contributor to the synergy of erosion and corrosion [12, 15]. For active metals, however, erosion-accelerated corrosion is normally thought to be due to the removal of corrosion products from metal surface by erosion and the enhancement of transportation of corrosives such as dissolved oxygen or carbon dioxide to the metal surface by flow [6, 79, 80].

Over the past several decades, various techniques have been employed to understand the mechanisms and synergism of erosion and corrosion [9, 10, 14, 31, 61, 81–83]. The gravimetric and the electrochemical measurements are the most commonly used methods in erosion–corrosion studies [14, 19, 72]. The total metal weight loss caused by both erosion and corrosion can be obtained from the gravimetric measurement, while the weight loss caused by electrochemical corrosion can be estimated using electrochemical measurements such as linear polarization resistance (LPR) and electrochemical impedance spectroscopy (EIS) techniques [14, 15]. For instance, using gravimetric and LPR measurements, Islam and co-workers [81, 82] studied the synergism of the erosion and corrosion of steels occurring during a cyclic erosion–corrosion process. Experiments were carried out by applying high–speed erosive particles with the velocity higher than 36 m/s on the steel surface, and then the eroded steel was immersed into the static corrosive solution for corrosion tests. After several circles, both erosion-enhanced corrosion and corrosion–enhanced erosion was observed. They attributed the corrosion enhanced erosion to the removal of the work hardened layer and the embedded sand particles [81] and the erosion-enhanced corrosion to the removal of the corrosion product film, the breakdown of the cementite network, the increasing of the surface roughness, and the formation of the sub-surface cracks [81]. Using electrochemical polarization techniques, Luo and co-workers [9, 10, 67] studied the erosion–corrosion behavior of the X65 pipeline steel in corrosive slurries. It was found that erosion began to happen only when the anodic current was applied on the steel surface and when the anodic current value was over a threshold and that the surface hardness degradation occurred when the steel surface was under a constant anodic current polarization [84]. This finding is significant since it suggests a major interaction between electrochemical corrosion processes on the mechanical erosion processes, suggesting that the decrease of the surface hardness caused by the corrosion can significantly reduce the critical velocity of the sand particles to induce erosion [10]. This finding indicates that active corrosion can result in the degradation of the surface hardness and make the erosion much easier to happen [9, 10, 84], suggesting that there could be dynamic change of the metal's surface microstructure interacting with hydrodynamic and electrochemical effects. However, these interesting experiments were carried out under anodic polarization, the actual effects of this phenomenon on the initiation and propagation of erosion corrosion under natural open-circuit corrosion conditions were not unclear.

It should be noted, however, that although these gravimetric and electrochemical measurements can be used to evaluate the synergy of erosion and corrosion, they can only provide general or averaged corrosion and erosion rates related to the overall weight losses of the whole metal surface. Electrochemical measurements using conventional electrodes provides only averaged electrochemical data over the whole electrode surface and have difficulties in probing local electrochemical processes associated with FAC and erosion corrosion, especially under natural open circuit corrosion conditions [46, 73]. There is a need for more direct evidences in order to gain more detailed understanding of the localized, dynamic and often complex electrochemical, interfacial chemistry and confined mass transfer processes associated with FAC and erosion–corrosion initiation and propagation. Although considerable studies on the behavior of localized corrosion initiation and propagation have been reported on various metal surfaces exposed to numerous environmental conditions in the historical literature [85–88], relatively less study has been carried out on the behavior of the initiation and propagation of erosion corrosion and FAC that could be more complex due to the combined effects of mechanical, electrochemical, and fluid dynamic factors. FAC and erosion–corrosion initiation and propagation processes are often not well investigated most likely due to limitations of conventional techniques in probing electrode processes occurring under natural open-circuit corrosion conditions. There are knowledge gaps in FAC and erosion corrosion. For instance the crater-like appearance of erosion corrosion is generally considered to be due to the mechanical impacts such as impingement of sand particles on metal surface; however, an unanswered question is that why erosion corrosion does not have a more general corrosion appearance given the random nature of sand particle impact on steel surface. This crater-like appearance of erosion corrosion suggests the existence of more complex and localized interactions between erosion and corrosion processes that were not understood. Technological capability to probe such localized interactions is required in order to understand this crater-like erosion corrosion pattern.

A relatively new approach to probing FAC and erosion corrosion is the WBE method that has been used in conjunction with gravimetric and electrochemical measurements to achieve further understanding of processes associated with FAC and erosion–corrosion initiation and propagation. A typical test setup shown in

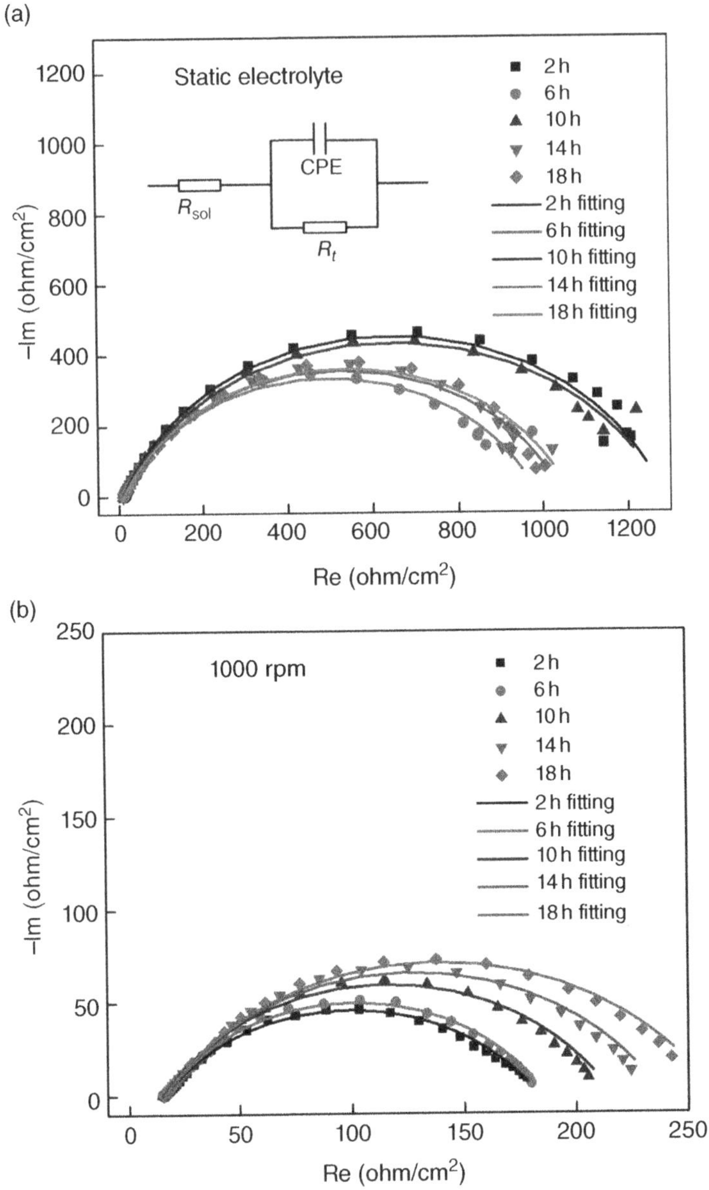

Figure 6.18 EIS measurement and fitting results in different test conditions: (a) static electrolyte, (b) electrolyte under 1000 rpm stirring, (c) electrolyte under 2000 rpm stirring, and (d) electrolyte under 1000 rpm stirring with 10% sand addition [47]. *Source:* Xu and Tan [47]/ With permission of Elsevier.

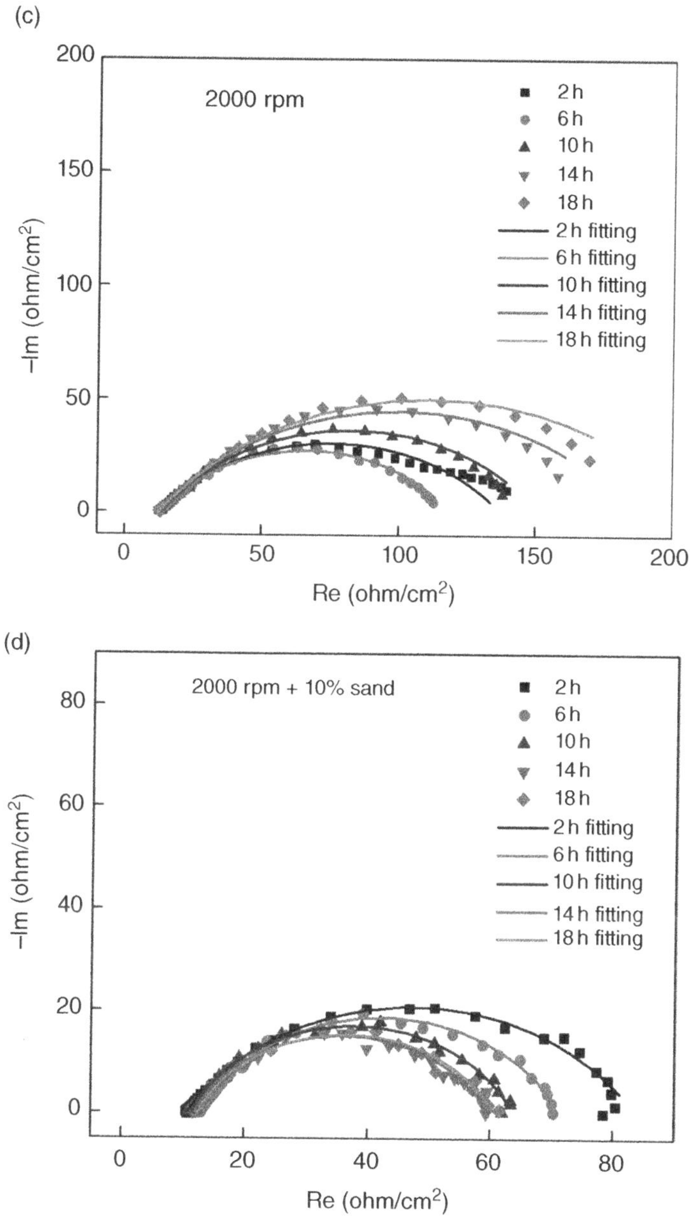

Figure 6.18 (Continued)

Figure 6.6a has been used for FAC and erosion–corrosion experiments [46, 47]. Corrosion rates and total metal losses of steel coupon electrodes were measured using EIS methods and weight-loss measurements under simulated FAC and erosion–corrosion conditions. It can be seen from Nyquist plots from EIS measurements in Figure 6.18 that when the electrolyte was rotated from static state to 1000 rpm and when sand was added to the electrolyte under 2000 rpm rotation, a dramatic acceleration of corrosion occurred [47], indicating the solution flow sand particles significantly enhanced corrosion. Major differences in the surface morphologies of the steel coupon electrodes after 20 hours of tests are clearly visible in Figure 6.19 that shows three typical corrosion patterns, which could be attributed to different corrosion–erosion mechanisms. The corrosion pattern in Figure 6.19a is featured by well-known general corrosion behavior of steel in static NaCl solutions [89]. When a steel coupon electrode was exposed to an electrolyte under stirring with the rotation speed 1000 rpm, as shown in Figure 6.19b, a clear "flow mark" brown rust pattern can be seen on the coupon and can be

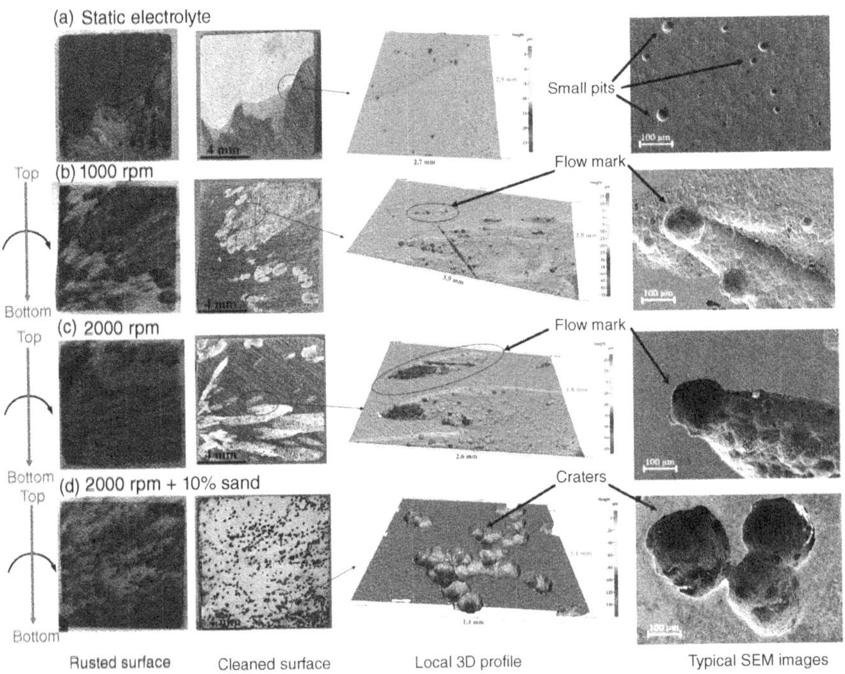

Figure 6.19 Photos, 3D profile, and SEM images of the coupon electrodes in different flowing conditions: (a) static electrolyte, (b) electrolyte under 1000 rpm stirring, (c) electrolyte under 2000 rpm stirring, and (d) electrolyte under 1000 rpm stirring with 10% sand addition [47]. *Source:* Xu and Tan [47], ELSEVIER.

observed more clearly from both the 3D surface profile and SEM image. An interesting observation is that a small pit can be observed at the tip of the "flow mark," that could be the initial anodic site. With the rotation speed of the electrolyte increased to 2000 rpm, it can be seen from Figure 6.19c that the "flow mark" corrosion morphology along the flow direction become a more dominate corrosion pattern. These are in agreement with corrosion morphologies such as the long and narrow "flow mark" observed in some industry pipelines [58, 59, 90]. When the sand particles were introduced into the solution, the corrosion pattern changed from "flow mark" to pitting and mesa type corrosion as shown in Figure 6.19d. Many small and deep craters are distributed on the steel coupon. It can also be seen from the 3D surface profile in Figure 6.19d that the depth of the pits reached nearly 150 μm after 20 hours of test, indicating serious localized corrosion on the steel surface. An interesting observation from Figure 6.19d is that craters were closely connected to each other, which can be clearly seen from both the 3D surface profile and the SEM images. These observations provide useful information on different corrosion–erosion mechanisms occurring on steel coupon electrodes exposed to different flow and erosion conditions [47].

In order to further understand the initiation and propagation processes and mechanisms of FAC and erosion corrosion on steel electrodes exposed to different flow conditions, electrode array–based corrosion probes have also been employed as shown in Figure 6.6a [47, 48]. Results indicate that both FAC and erosion corrosion initiated from small anodic sites randomly distributed over electrode surfaces. The propagation of FAC is due to flow-generated anolyte transportation along the flow direction in the boundary layer from the initial anodic sites to the downstream, causing the "flow mark" corrosion appearance. The propagation of erosion corrosion appears to be due to the impingement of the initial anodic sites by sand particles, inducing the "crater-like" corrosion feature [47, 48]. As shown in Figure 6.20, the corrosion patterns shown on the WBEs are almost identical to those shown on coupons in Figure 6.19, with "flow mark" corrosion pattern

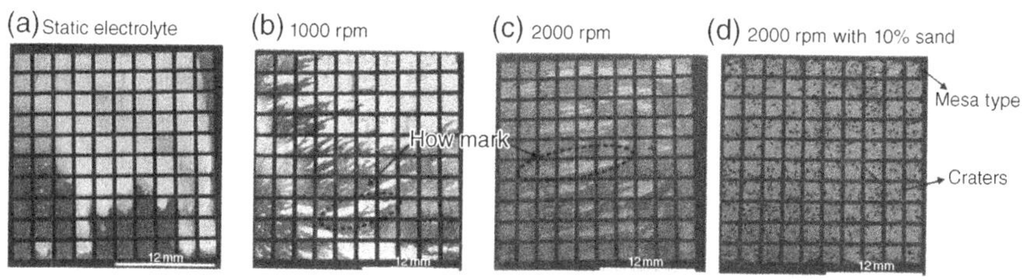

Figure 6.20 The photo of the WBEs after 20 hours test in different flowing conditions: (a) static electrolyte, (b) electrolyte under 1000 stirring, (c) electrolyte under 2000 rpm stirring, and (d) electrolyte under 1000 rpm stirring with 10% sand addition. *Source:* Xu and Tan [47], ELSEVIER.

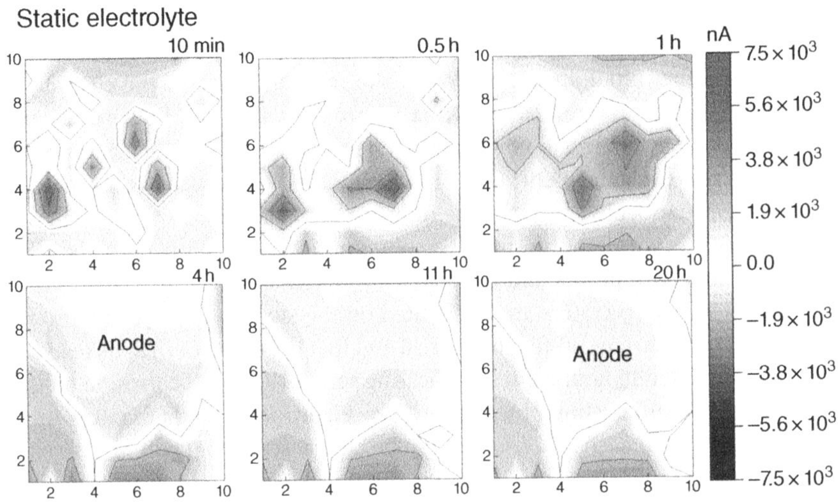

Figure 6.21 The current distribution maps measured from the WBE exposed to the static electrolyte at different test durations. *Source:* Xu and Tan [47]/ With permission of Elsevier.

appeared on the WBE in both 1000 and 2000 rpm conditions and craters and mesa type erosion corrosion occurred on the WBE in the 2000 rpm with sand condition, similar to previous observations [46].

Figures 6.21 and 6.22 present current maps showing the "baseline" corrosion initiation and propagation behaviors on the WBEs in the static electrolyte and exposed to different flowing conditions [47, 48]. It is shown that after 10 minutes of immersion of the WBE in the electrolyte major anodic sites immediately appeared on the WBE surface, suggesting the initiation of corrosion was a random process with the probable initiation sites being "active" spots on the steel surface such as impurities, defects, and grain boundaries. The propagation of the anodic areas after approximately four hours could be due to local chemistry changes and the diffusion of oxygen, and kept unchanged until the end of the 20 hours test. Similar random initiation behaviors, as shown in Figure 6.22, are observable on WBEs exposed to different flowing conditions. As shown in Figure 6.22a and b, both maximum anodic current and total anodic current had dramatic rising when the electrolyte was stirred at 1000 and 2000 rpm compared with that in the static electrolyte condition. The main anodic sites formed after 0.5 hour of immersion and gradually propagated along the flow direction to form large anodic areas. The maximum anodic current and the total anodic current on the WBE surface are presented in Figure 6.23. It can be seen from Figure 6.23a, it took one hour for the maximum anodic current to climb to the peak value under 1000 rpm condition,

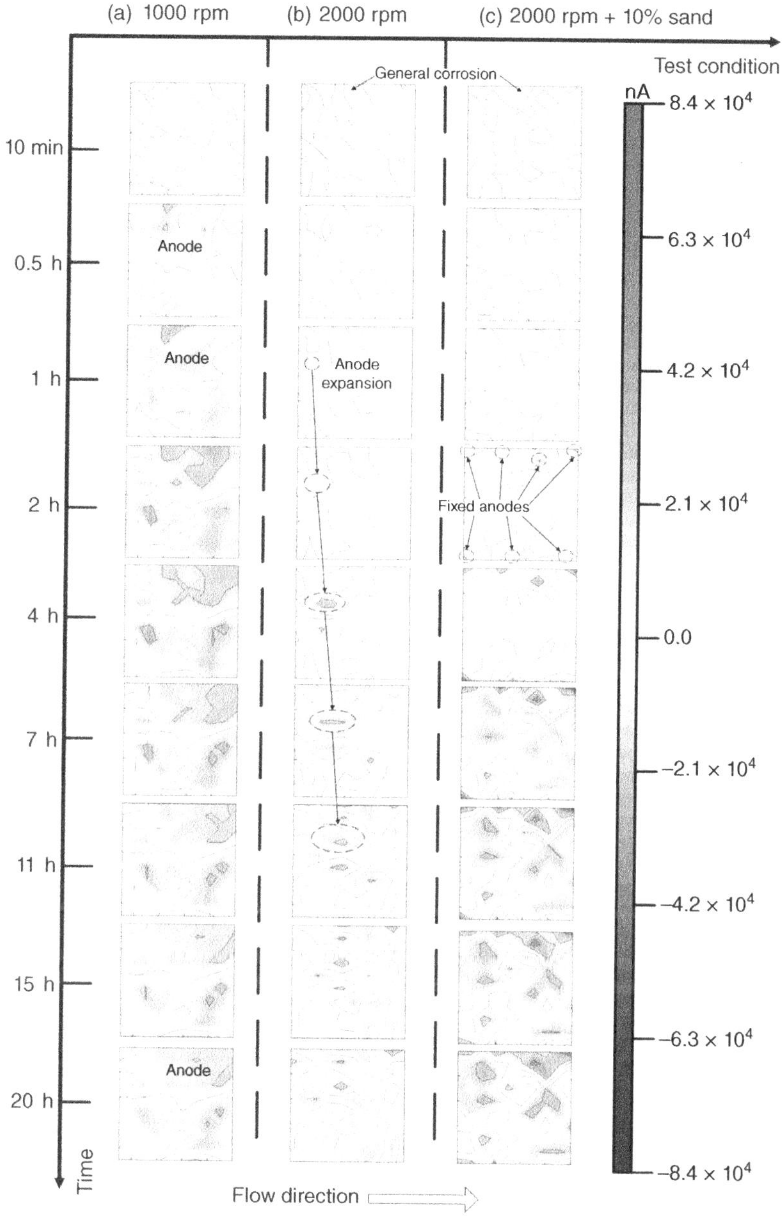

Figure 6.22 The current distributions maps measured from the WBE exposed to electrolytes under different flowing conditions: (a) electrolyte under 1000 rpm stirring, (b) electrolyte under 2000 rpm stirring, and (c) electrolyte under 2000 rpm stirring with 10% sand addition. *Source:* Xu and Tan [47]/ With permission of Elsevier.

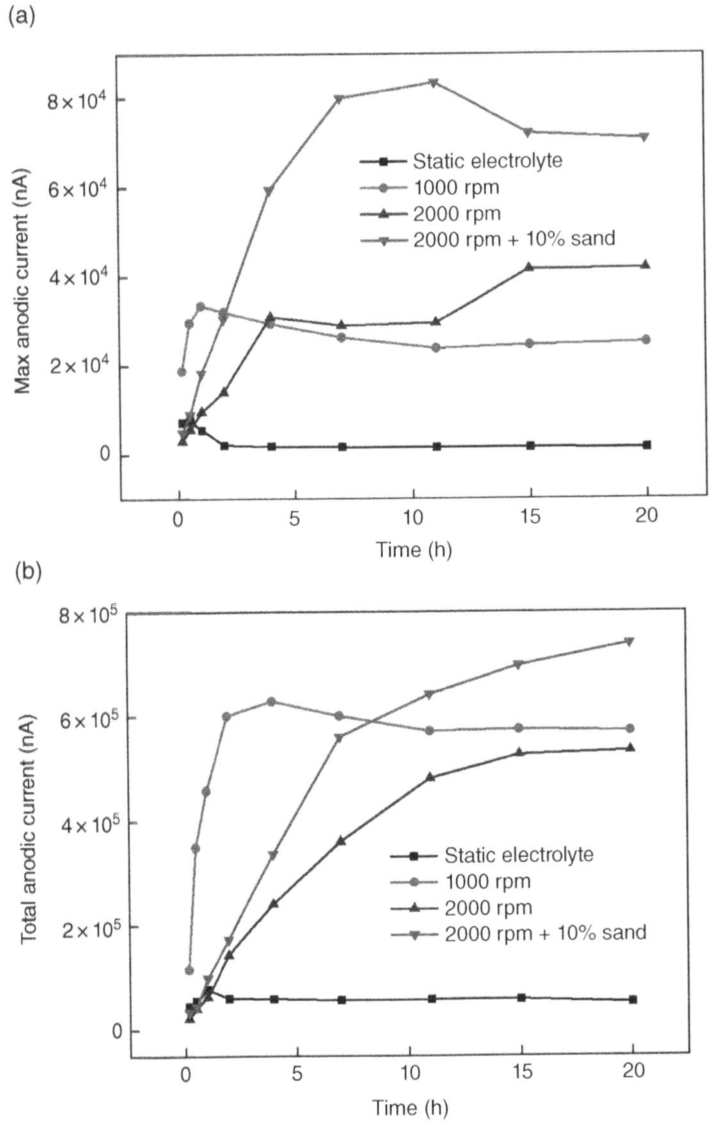

Figure 6.23 (a) The max anodic current and (b) the total anodic current obtained from the WBE measurement results shown in Figure 6.22. *Source:* Xu and Tan [47]/ With permission of Elsevier.

which is longer than that in the static electrolyte. As shown in Figure 6.23b, the total anodic current reached the highest value after four hours of immersion and gradually decreased to a stable status. Under 2000 rpm without sand condition, the total anodic current also kept increasing in the first 15 hours suggesting a longer incubation time for the steel surface to form the significant anodic area in a higher flow rate condition. The fact that stable anodes took longer time to form is in agreement with the enhanced transportation of corrosion products, slowing down the localization of chemistry on steel surface, most likely in a boundary layer between the flowing liquid and the solid steel surface. It is well-known that there is a boundary layer between the flowing liquid and the solid steel surface due to the viscous force, and that the thickness of the boundary layer would decrease dramatically with the increasing of the flowing speed under a turbulence flowing condition, leading to significantly higher corrosion rates in FAC and EC systems [57, 91].

It can be seen in Figure 6.22b and c that the galvanic current was tiny at the beginning of the FAC and erosion–corrosion experiments, indicating that the initiation of FAC and erosion corrosion are nearly general corrosion processes with initial anodes randomly distributed. The propagation of corrosion is indicated by the gradual appearance of major anodic sites after one hour of test in 2000 rpm electrolyte, and more importantly, the major anodic areas began to expand along the flow direction in the following periods. The expansion of the anodic areas can be clearly identified from the variation of the current distribution maps, which shows the major anodic sites gradually expanded from the left side to the right side of the WBE. The initiation and propagation of the major anodic sites clearly shows the formation process of the "flow mark" in the 2000 rpm condition. Previously similar corrosion pattern transformation was reported by Wharton that in the flow dynamic conditions the shape of the pit on the steel surface would change from a typical semi-circle to a "comet tail" shape [57]. Wharton attributed the formation of this form of corrosion to the transportation of the anolyte from the tip of the pits to the downstream along the flow direction. Although the transportation of the analyte can explain the formation of the "flow mark" corrosion pattern, there is not enough evidence to support this opinion only based on traditional electrochemical measurements. These maps shown in Figure 6.22 are probably the first direct evidence of local anode movement over steel surface due to flow. Most likely local chemistry changed at the electrolyte/steel interfacial layer, leading to movement of lower pH anodic regions [47, 48].

With the adding of sand particles in the electrolyte, the corrosion pattern further transformed to pitting and mesa type of corrosion, as shown in Figures 6.22c and 6.23a, it took 11 hours for the maximum anodic current to climb to the peak value in the 2000 rpm with 10% sand added condition. The maximum anodic current had a sharp increase in the first seven hours and the total anodic current kept increasing in the 20 hours of test, suggesting an even longer incubation period for the formation of significant galvanic corrosion in the sand containing condition. Interestingly, in Figure 6.22c, almost no obvious anodic expansion could be found along the flow direction, which suggests the local chemistry movement in the boundary layer was not the main mechanism of erosion–corrosion propagation. It seems the mechanical impingements of the sand particles are the main factor significantly influencing the corrosion propagation. It is known that the initiation of erosion corrosion on passive metals is induced by the mechanical wear of the passive film by the impingements of the sand particles [62, 57], causing the fresh metal matrix to be directly exposed to the corrosive medium and therefore the acceleration of corrosion [35]. While for active metals erosion-accelerated corrosion is normally thought to be due to the removal of corrosion products from metal surface by erosion and the enhancement of transportation of corrosives such as dissolved oxygen or carbon dioxide to the metal surface by flow [6, 79, 80]. However, these appear not explaining the crater-like and mesa-like erosion–corrosion damages on steel surface [11, 15, 57, 58, 60], as well as the distinguishable "flow mark" FAC corrosion feature [58, 59] because a more general corrosion appearance would be expected given the random nature of flow and sand particle impact on "active" steel surface. There could be more complex localized interactions between erosion and corrosion processes [47].

It should be noted that the currents shown in the WBE current maps are "galvanic" currents flowing between each "wire" and the whole integrated WBE, and therefore corrosion rates calculated from these currents do not necessarily predict the total metal loss, especially when general corrosion and erosion corrosion occurs simultaneously with localized corrosion. Figure 6.24 shows corrosion depths measured from both surface profilometry method (Figure 6.24 map 1) and calculated from the WBE anodic currents (Figure 6.24 map 2). As the corrosion depth measured from the 3D profile is calculated from the total volume loss, this measurement result represented the total metal loss of each wire. The corrosion depth differences between maps 1 and 2 are shown as map 3 in Figure 6.24. In electrolyte under 1000 (Figure 6.24a) and 2000 without sand addition (Figure 6.24b), nearly no erosion happened in both conditions suggesting the difference between the two maps is mainly induced by the internal current. It can be seen from map 3 in electrolyte under 1000 and 2000 rpm without sand that the locations having the highest anodic current charge are well matched to the most serious metal loss area indicating the galvanic current is the main contribution to the localized corrosion. As shown in Figure 6.24c, when sand particles were added into the solution, the difference between map 1 and map 2 became more significant. The increasing of the difference can be caused by both the internal current and erosion. To further understand the detailed influence of the cathodic and anodic electrochemical corrosion reactions on erosion–corrosion processes, statistical analysis on data from the WBE has been conducted [47]. A general trend that was found is that the erosion metal loss is affected by the galvanic currents flowing into or out of a wire in a WBE. For cathodic wires in the WBE-recorded larger cathodic currents, smaller steel thickness losses due to erosion are

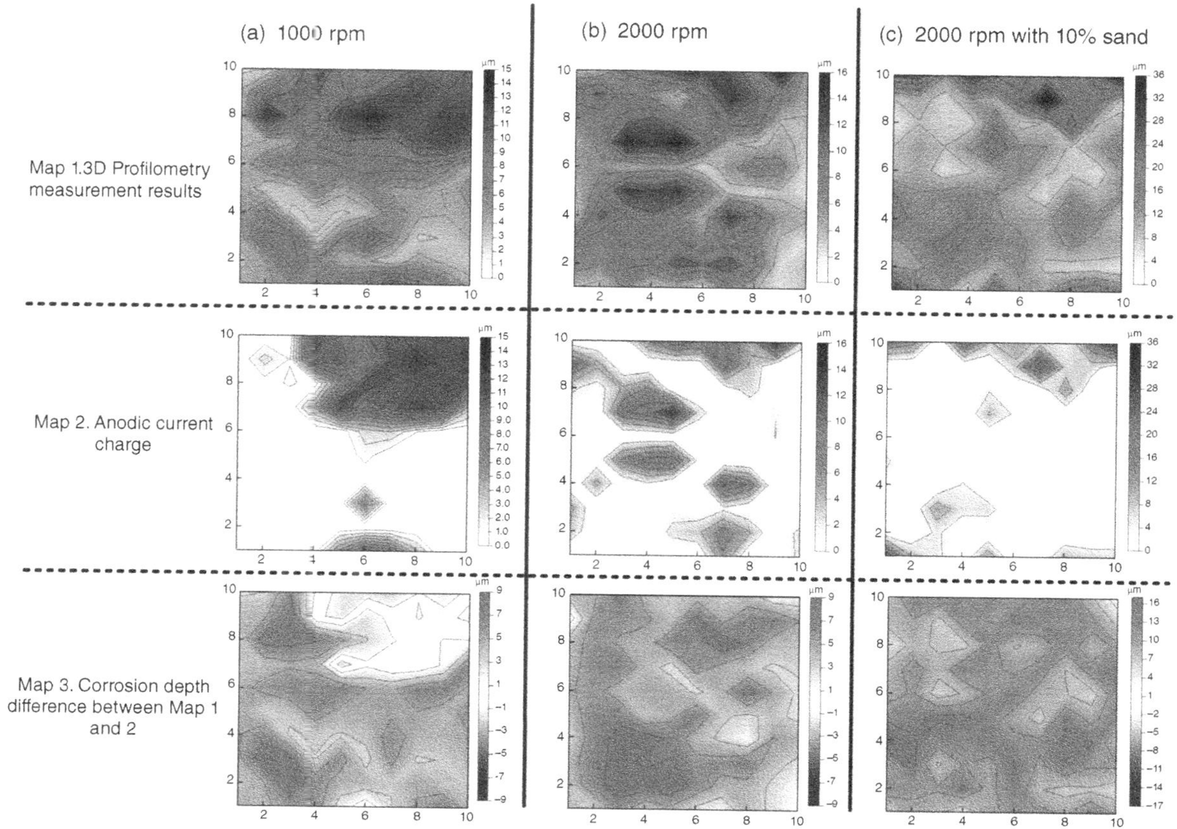

Figure 6.24 The corrosion depth maps measured by 3D profilometry (Map 1), calculated from WBE anodic currents (Map 2), and the depth measurement differences (Map 3) between Map 1 and Map 2 in different test conditions: (a) electrolyte under 1000 rpm stirring, (b) electrolyte under 2000 rpm stirring, and (c) electrolyte under 1000 rpm stirring with 10% sand addition [47]. *Source:* Xu and Tan [47].

registered, while for wires registered zero current and anodic currents larger steel thickness losses due to erosion are registered although the data are somewhat scattered [47]. These trends show interactions between electrochemical anodes and mechanical erosion.

The effect of cathodic (negative) currents on erosion could be explained by the fact that cathodic reaction could change the chemistry of the boundary layer at cathodic areas, probably leading to the formation of more protective layer on these cathodic areas, leading to less erosion on these areas. This could explain that fact that erosion often has the appearance of localized attack, rather than general metal losses. On the other hand, at wires where zero and anodic currents were registered, high-erosion-caused steel loss are generally observed, suggesting that erosion prefers to occur on anodic areas of the WBE surface. These are probably first evidence showing effects of electrode corrosion cathode and anodes on erosion under open-circuit potential natural corrosion conditions. Previously Luo and co-workers [9, 10, 67] studied the erosion–corrosion behavior of the X65 pipeline steel in corrosive slurries and found that erosion rate had a significant increase with the applied anodic current increasing on the steel surface. They explained this effect as the surface hardness degradation occurred when the steel surface was under a constant anodic current polarization [84]. However, these findings were obtained from steel surface under rather large anodic polarization, not under natural open-circuit potential natural corrosion conditions [47]. Based on these findings and analysis, as shown in Figure 6.25, models are proposed to describe localized corrosion propagation from the initial pits in the flow direction with a relative deep pit appeared at the tip of the "flow mark" [47].

Overall, the WBE method, with its capability of probing and visualizing FAC and erosion–corrosion processes with high spatial and temporal resolutions, has been successfully used in conjunction with traditional electrochemical and gravimetric methods to probe the initiation and propagation processes and mechanisms of FAC and erosion corrosion on steel electrodes exposed to different flow conditions. It is shown that both FAC and erosion corrosion initiate from small anodic sites randomly distributed over steel electrode surfaces. Experimental evidences indicate that the propagation of FAC is due to flow-generated anolyte transportation along the flow direction in the boundary layer from the initial anodic sites to the downstream, causing the "flow mark" corrosion appearance, while the propagation of erosion corrosion appears to be due to the impingement of the initial anodic sites by sand particles, inducing the "crater-like" corrosion feature. Mechanical erosion was found to concentrate on electrochemical corrosion anodes, which less erosion occurred on corrosion cathodes, suggesting interactions between erosion and corrosion.

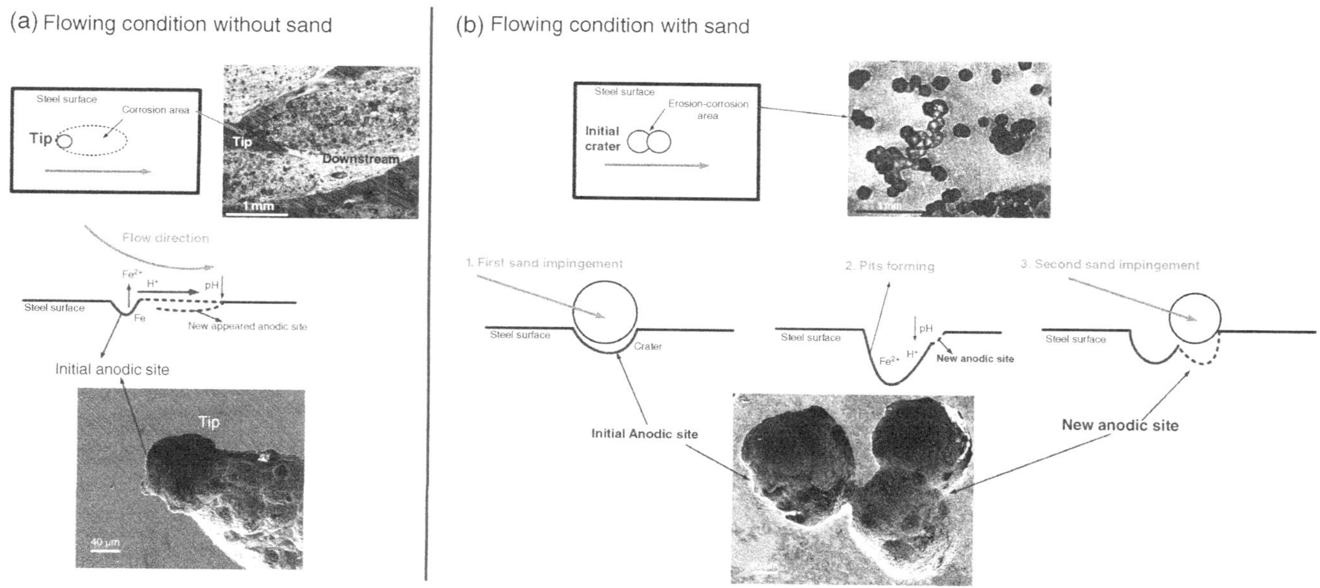

Figure 6.25 Schematic diagram of the erosion–corrosion propagation and the local SEM image in the flowing conditions (a) without sand particles and (b) with sand particles. *Source:* Xu and Tan [47], ELSEVIER.

References

1 Xu, Y., Liu, L., Zhou, Q. et al. (2020). An overview of major experimental methods and apparatus for measuring and investigating erosion-corrosion of ferrous-based steels. *Metals 10*: 180. https://doi.org/10.3390/met10020180.

2 Matsumura, M. (2012). *Erosion-Corrosion: An Introduction to Flow Induced Macro-Cell Corrosion*. Hiroshima, Japan: Bentham Science Publishers, Hiroshima University.

3 Zhu, M., Sun, L., Ou, G. et al. (2016). Erosion corrosion failure analysis of the elbow in sour water stripper overhead condensing reflux system. *Engineering Failure Analysis 26*: 93–102.

4 Muthanna, B.G.N., Amara, M., Meliani, M.H. et al. (2019). Inspection of internal erosion-corrosion of elbow pipe in the desalination station. *Engineering Failure Analysis 102*: 293–302.

5 Zhang, M., Liu, Y., Wang, W. et al. (2016). The fatigue of impellers and blades. *Engineering Failure Analysis 62*: 208–231.

6 Zeng, L., Zhang, G.A., and Guo, X.P. (2014). Erosion–corrosion at different locations of X65 carbon steel elbow. *Corrosion Science 85*: 318–330.

7 Tian, B.R. and Cheng, Y.F. (2008). Electrochemical corrosion behavior of X-65 steel in the simulated oil sand slurry I: effects of hydrodynamic condition. *Corrosion Science 50*: 773–779.

8 Dong, C.F., Xiao, K., Li, X.G., and Cheng, Y.F. (2010). Erosion accelerated corrosion of a carbon steel–stainless steel galvanic couple in a chloride solution. *Wear 270*: 39–45.

9 Guo, H.X., Lu, B.T., and Luo, J.L. (2006). Non-Faraday material loss in flowing corrosive solution. *Electrochimica Acta* 51: 5341–5348.

10 Guo, H.X., Lu, B.T., and Luo, J.L. (2005). Interaction of mechanical and electrochemical factors in erosion–corrosion of carbon steel. *Electrochimica Acta 51*: 315–323.

11 Tang, X., Xu, L.Y., and Cheng, Y.F. (2008). Electrochemical corrosion behavior of X-65 steel in the simulated oil–sand slurry II: synergism of erosion and corrosion. *Corrosion Science 50*: 1469–1474.

12 Li, Y., Burstein, G.T., and Hutchings, I.M. (1995). The influence of corrosion on the erosion of aluminium by aqueous silica slurries. *Wear 186*: 515–522.

13 Zhang, G.A., Xu, L.Y., and Cheng, Y.F. (2009). Investigation of erosion–corrosion of 3003 aluminum alloy in ethylene glycol–water solution by impingement jet system. *Corrosion Science 51*: 283–290.

14 Malka, R., Nešić, S., and Gulino, D.A. (2007). Erosion–corrosion and synergistic effects in disturbed liquid-particle flow. *Wear 262*: 791–799.

15 Lu, B.T., Luo, J.L., Mohammadi, F. et al. (2008). Correlation between repassivation kinetics and corrosion rate over a passive surface in flowing slurry. *Electrochimica Acta* 53: 7022–7031.

16 Hu, X. and Neville, A. (2009). CO_2 erosion–corrosion of pipeline steel (API X65) in oil and gas conditions – a systematic approach. *Wear 267*: 2027–2032.

17 Owen, J., Ramsey, C., Barker, R., and A. (2018). Neville erosion-corrosion interactions of X65 carbon steel in aqueous CO_2 environments. *Wear 414–415*: 376–389.

18 Barik, R.C., Wharton, J.A., Wood, R.J.K. et al. (2005). Erosion and erosion–corrosion performance of cast and thermally sprayed nickel–aluminium bronze. *Wear 259*: 230–242.

19 Tan, K.S., Wharton, J.A., and Wood, R.J.K. (2005). Solid particle erosion–corrosion behaviour of a novel HVOF nickel aluminium bronze coating for marine applications – correlation between mass loss and electrochemical measurements. *Wear 258*: 629–640.

20 Stack, M.M. and Stott, F.H. (1993). An approach to modeling erosion-corrosion of alloys using erosion-corrosion maps. *Corrosion Science 35*: 1027–1034.

21 Tan, Y.J., Bailey, S., and Kinsella, B. (1996). Investigations on the formation and destruction processes of corrosion inhibitor films using electrochemical impedance spectroscopy. *Corrosion Science* 38: 1545–1561.

22 Tan, Y., Bailey, S., and Kinsella, B. (1996). The monitoring of the formation and destruction of corrosion inhibitor films using electrochemical noise analysis (ENA). *Corrosion Science* 38 (10): 1681–1695.

23 Tan, Y., Kinsella, B., and Bailey, S. (1997). Monitoring batch treatment inhibitor performance continuously using electrochemical noise analysis. *British Corrosion Journal* 32 (3): 212–216.

24 Bailey, S., Tan, Y., and Kinsella, B. (1997). Use of electrochemical impedance spectroscopy for study of CO_2 corrosion prevention by batch treatment inhibitor films. *British Corrosion Journal* 32 (1): 49–55.

25 Kinsella, B., Tan, Y., and Bailey, S. (1998). Studies of CO_2 corrosion product scales using electrochemical impedance spectroscopy and surface characterisation techniques. *Corrosion: Journal of Science and Engineering* 54 (10): 835–842.

26 Toor, I.U., Irshad, H.M., Badr, H.M., and Samad, M.A. (2018). The effect of impingement velocity and angle variation on the erosion corrosion performance of API 5L-X65 carbon steel in a flow loop. *Metals 8*: 402.

27 Zhao, W., Wang, C., Zhang, T. et al. (2016). Effects of laser surface melting on erosion–corrosion of X65 steel in liquid–solid jet impingement conditions. *Wear 362–363*: 39–52.

28 Zhang, G.A. and Cheng, Y.F. (2010). Electrochemical characterization and computational fluid dynamics simulation of flow-accelerated corrosion of X65 steel in a CO_2-saturated oilfield formation water. *Corrosion Science 52*: 2716–2724.

29 Zhao, Y., Zhou, F., Yao, J. et al. (2015). Erosion–corrosion behavior and corrosion resistance of AISI 316 stainless steel in flow jet impingement. *Wear 328–329*: 464–474.

30 Aribo, S., Barker, R., Hu, X., and Neville, A. (2013). Erosion–corrosion behaviour of lean duplex stainless steels in 3.5% NaCl solution. *Wear 302*: 1602–1608.

31 Ukpai, J.I., Barker, R., Hu, X., and Neville, A. (2013). Exploring the erosive wear of X65 carbon steel by acoustic emission method. *Wear 301*: 370–382.

32 Karimi, S., Shirazi, S.A., and McLaury, B.S. (2017). Predicting fine particle erosion utilizing computational fluid dynamics. *Wear 376–377*: 1130–1137.

33 Pasha, A., Ghasemi, H.M., and Neshati, J. (2016). Study of the pitting corrosion of superduplex stainless steel and X-65 carbon steel during erosion-corrosion by cyclic polarization technique. *Corrosion Engineering, Science and Technology 51*: 463–471.

34 Azarian, N.S., Ghasemi, H.M., and Monshi, M.R. (2015). Synergistic erosion and corrosion behavior of AA5052 aluminum alloy in 3.5 wt% nacl solution under various impingement angles. *Journal of Bio- and Tribo-Corrosion 1*: 1–10.

35 Mohammadi, F., Luo, J., Lu, B., and Afacan, A. (2010). Single particle impingement current transients for prediction of erosion-enhanced corrosion on 304 stainless steel. *Corrosion Science 52*: 2331–2340.

36 Liu, L., Xu, Y., Xu, C. et al. (2019). Detecting and monitoring erosion-corrosion using ring pair electrical resistance sensor in conjunction with electrochemical measurements. *Wear 428–429*: 328–339.

37 Xu, Y., Li, K., Liu, L. et al. (2016). Experimental study on rebar corrosion using the galvanic sensor combined with the electronic resistance technique. *Sensors 16*: 1451.

38 Vieira, R.E., Parsi, M., Zahedi, P. et al. (2017). Electrical resistance probe measurements of solid particle erosion in multiphase annular flow. *Wear 382–383*: 15–28.

39 Fan, C.L. and Mclaury, B.S. (2011). Evaluation of electrical resistance probes to detect pipeline erosion and sand production in low liquid loading flow conditions. *In Proceedings of the SPE International Conference*, OH, USA (17–19 August 2011).

40 Gan, F., Tian, G., Wan, Z. et al. (2016). Investigation of pitting corrosion monitoring using field signature method. *Measurement 82*: 46–54.

41 Huang, Y., Xu, Y., Li, B. et al. (2016). Novel electrical resistance method to measure underdeposit corrosion and its inhibition in pipeline steels. *Corrosion Engineering, Science and Technology 51*: 211–222.

42 Kinsella, B., Tan, Y., Edward, J., and Bailey, S. (1997). The dual cylinder electrode, an important tool for electrochemical corrosion measurements and for the evaluation of inhibitor films Australia-New Zealand International Business Academy. In: *Proceedings of the 10th Asia Pacific Corrosion Control Conference: Proceedings* (ed. B. Widyanto), Bali, Indonesia (27–31 October 1997). Indonesian Corrosion Association, Bandung, Indonesia (Conference paper).

43 Tan, Y.J., Bailey, S., Kinsella, B., and Lowe, A. (2000). Mapping corrosion kinetics using the wire beam electrode in conjunction with electrochemical noise resistance measurements. *Journal of the Electrochemical Society* 147: 530–540.

44 Pine Research Rotating Ring Disk Electrode Fundamentals – Pine Research Instrumentation Store, https://pineresearch.com/shop/kb/theory/hydrodynamic-electrochemistry/rotating-electrode-theory/, Last Updated: 5/23/22 by Neil Spinner (Accessed 3 September 2022).

45 Tan, Y.J., Bailey, S., and Kinsella, B. (2001). Mapping non-uniform corrosion using the wire beam electrode method I. Multi-phase carbon dioxide corrosion. *Corrosion Science* 43: 1905–1918.

46 Xu, Y. and Tan, M.Y. (2018). Visualising the dynamic processes of flow accelerated corrosion and erosion corrosion using an electrochemically integrated electrode array. *Corrosion Science* 139: 438–443.

47 Xu, Y. and Tan, M.Y. (2019). Probing the initiation and propagation processes of flow accelerated corrosion and erosion corrosion under simulated turbulent flow conditions. *Corrosion Science* 151: 163–174.

48 Xu, Y., Liu, L., Xu, C. et al. (2020). Electrochemical characteristics of the dynamic progression of erosion-corrosion under different flow conditions and their effects on corrosion rate calculation. *Journal of Solid State Electrochemistry* 26: 1–14. https://doi.org/10.1007/s10008-020-04795-9.

49 Laleh, M., Hughes, A.E., Xu, W. et al. (2019). Unexpected erosion-corrosion behaviour of 316L stainless steel produced by selective laser melting. *Corrosion Science* 155: 67–74.

50 Mansfeld, F. and Lorenz, W.J. (1991). Electrochemical impedance spectroscopy: application in corrosion science and technology. In: *Techniques for Characterisation of Electrodes and Electrochemical Processes* (ed. R. Varma and J.R. Selman). John Wiley & Sons.

51 Eden, D.A., Hladky, K., John, D.G., and Dawson, J.L. (1986). Electrochemical noise resistance. Corrosion 86, Paper 274, NACE, Houston.

52 Cottis, R.A., Turgoose, S., and Newman, R. (1999). Corrosion Testing Made Easy: Electrochemical Impedance and Noise. NACE International.

53 Mansfeld, F. and Xiao, H. (1993). Electrochemical noise analysis of iron exposed to NaCl solutions of different corrosivity. *Journal of the Electrochemical Society* 140: 2205.

54 Bertocci, U., Gobrielli, C., Huet, F., and Keddam, M. (1997). Noise resistance applied to corrosion measurements: I. Theoretical analysis. *Journal of the Electrochemical Society* 144: 31–37.

55 Cottis, R.A. (2001). The interpretation of electrochemical noise data. *Corrosion* 27: 265.

56 Tan, Y.J., Bailey, S., and Kinsella, B. (2002). Studying the formation process of chromate conversion coatings on aluminium using continuous electrochemical noise resistance measurements. *Corrosion Science* 44: 1277–1286.

57 Wharton, J.A. and Wood, R.J.K. (2004). Influence of flow conditions on the corrosion of AISI 304L stainless steel. *Wear* 256: 525.

58 Wang, Q.-Y., Bai, S.-L., and Liu, Z.-D. (2013). Surface characterization and erosion–corrosion behaviour of Q235 steel in dynamic flow. *Tribology Letters* 53: 271.

59 Ilman, M.N. and Kusmono, K. (2014). Analysis of internal corrosion in subsea oil pipeline. *Case Studies in Engineering Failure Analysis* 2: 1.

60 Lu, B.T., Lu, J.F., and Luo, J.L. (2011). Erosion–corrosion of carbon steel in simulated tailing slurries. *Corrosion Science* 53: 1000.

61 Kain, V. (2014). Flow accelerated corrosion: forms, mechanisms and case studies. *Procedia Engineering* 86: 576.

62 Lu, B.T., Luo, J.L., Guo, H.X., and Mao, L.C. (2011). Erosion-enhanced corrosion of carbon steel at passive state. *Corrosion Science* 53: 432.

63 Kishawy, H.A. and Gabbar, H.A. (2010). Review of pipeline integrity management practices. *International Journal of Pressure Vessels and Piping* 87: 373.

64 NACE International (2012). Monitoring corrosion in oil and gas production with iron counts. *Standard: NACE SP0192.*

65 Walsh, F.C., Kear, G., Nahlé, A.H. et al. (2017). The rotating cylinder electrode for studies of corrosion engineering and protection of metals – an illustrated review. *Corrosion Science* 123: 1.

66 Dong, C.F., Xiao, K., Li, X.G., and Cheng, Y.F. (2010). Erosion accelerated corrosion of a carbon steel–stainless steel galvanic couple in a chloride solution. *Wear* 270: 39.

67 Lu, B.T. and Luo, J.L. (2006). Synergism of electrochemical and mechanical factors in erosion-corrosion. *The Journal of Physical Chemistry. B* 110: 4217.

68 Zheng, Z.B., Zheng, Y.G., Zhou, X. et al. (2014). Determination of the critical flow velocities for erosion–corrosion of passive materials under impingement by NaCl solution containing sand. *Corrosion Science* 88: 187.

69 Lu, B.T., Luo, J.L., and Ma, H.Y. (2007). A theoretical model on electrochemical response of passivated metals to solid particle impingement. *Journal of The Elctrochemical Society* 154: C159.

70 Zhang, G.A., Zeng, L., Huang, H.L., and Guo, X.P. (2013). A study of flow accelerated corrosion at elbow of carbon steel pipeline by array electrode and computational fluid dynamics simulation. *Corrosion Science* 77: 334.

71 Zeng, L., Zhang, G.A., Guo, X.P., and Chai, C.W. (2015). Inhibition effect of thioureidoimidazoline inhibitor for the flow accelerated corrosion of an elbow. *Corrosion Science* 90: 202.

72 Zeng, L., Shuang, S., Guo, X.P., and Zhang, G.A. (2016). Erosion-corrosion of stainless steel at different locations of a 90°elbow. *Corrosion Science* 111: 72.

73 Tan, Y.J. (2009). Sensing electrode inhomogeneity and electrochemical heterogeneity using an electrochemically integrated multielectrode array. *Journal of the Electrochemical Society* 156: C195.

74 Tan, Y. (2013). *Heterogeneous Electrode Processes and Localized Corrosion*, 1–246. Hoboken, NJ: Wiley.

75 Tan, Y. (1998). Monitoring localised corrosion processes and estimating localised corrosion rates using a wire-beam electrode. *Corrosion* 54: 403.

76 Zhong, X., Shang, T., Zhang, C. et al. (2020). In situ study of flow accelerated corrosion and its mitigation at different locations of a gradual contraction of N80 steel. *Journal of Alloys and Compounds* 824: 153947.

77 Li, Z., Liu, J., Zhang, L. et al. (2019). Electrochemical inhomogeneities of steel in steel/copper alloy couple during galvanic corrosion in static and flowing seawater. *Materials and Corrosion* 70: 726–737.

78 Deng, P., Li, Z., Li, X. et al. (2020). Vertical galvanic corrosion of pipeline steel in simulated marine thermocline. *Ocean Engineering* 217: 107584.

79 Satoh, T., Shao, Y., Cook, W.G. et al. (2007). Flow-assisted corrosion of carbon steel under neutral water conditions. *Corrosion* 63: 770–780.

80 Song, G.D., Jeon, S.H., Son, Y.H. et al. (2018). Galvanic effect of magnetite on the corrosion behavior of carbon steel in deaerated alkaline solutions under flowing conditions. *Corrosion Science* 131: 71–80.

81 Islam, M.A. and Farhat, Z. (2017). Erosion-corrosion mechanism and comparison of erosion-corrosion performance of API steels. *Wear* 376–377: 533–541.

82 Islam, M.A., Farhat, Z.N., Ahmed, E.M., and Alfantazi, A.M. (2013). Erosion enhanced corrosion and corrosion enhanced erosion of API X-70 pipeline steel. *Wear* 302: 1592–1601.

83 Tan, K.S., Wood, R.J.K., and Stokes, K.R. (2003). The slurry erosion behaviour of high velocity oxy-fuel (HVOF) sprayed aluminium bronze coatings. *Wear* 255: 195–205.

84 Guo, H.X., Lu, B.T., and Luo, J.L. (2006). Response of surface mechanical properties to electrochemical dissolution determined by in situ nanoindentation technique. *Electrochemistry Communications* 8: 1092–1098.

85 Frankel, G. (1998). Pitting corrosion of metals a review of the critical factors. *Journal of the Electrochemical Society* 145: 2186–2198.

86 Williams, D., Newman, R., Song, Q., and Kelly, R. (1991). Passivity breakdown and pitting corrosion of binary alloys. *Nature* 350: 216.

87 Punckt, C., Bölscher, M., Rotermund, H.H. et al. (2004). Sudden onset of pitting corrosion on stainless steel as a critical phenomenon. *Science* 305: 1133–1136.

88 Parvizi, R., Hughes, A.E., Tan, M.Y. et al. (2017). Probing corrosion initiation at interfacial nanostructures of AA2024-T3. *Corrosion Science* 116: 98–109.

89 Tan, Y. (1998). Monitoring localized corrosion processes and estimating localized corrosion rates using a wire-beam electrode. *Corrosion* 54: 403–413.

90 Qiao, Q., Cheng, G., Li, Y. et al. (2017). Corrosion failure analyses of an elbow and an elbow-to-pipe weld in a natural gas gathering pipeline. *Engineering Failure Analysis* 82: 599–616.

91 Efird, K.D., Wright, E.J., Boros, J.A., and Hailey, T.G. (1993). Correlation of steel corrosion in pipe-flow with jet impingement and rotating cylinder tests. *Corrosion* 49: 992–1003.

7

Localized Corrosion Induced by Metallurgical Heterogeneities

Many engineering structures are composed of metallurgically nonuniform structural components such as multi-metal segments, heterogeneous alloy parts, and weldments. When these structures are exposed to a corrosive environment, localized corrosion could occur, initiating from nano-, micro-, or mm-scale metallurgical heterogeneity and propagating to macroscale localized corrosion. The detection and prediction of such multiscale localized corrosion require techniques that are able to probe and model corrosion processes over a wide range of length and time scales. This chapter provides an overview of various electrochemical and surface analytical techniques that could be used for acquiring corrosion data over multiple time and length scales including corrosion probes comprised of multi-length scale and metallurgical heterogeneous electrode arrays. Localized corrosion of weldments and additively manufactured stainless steels is selected as a typical example of multiscale corrosion affected by heterogeneous metallurgy.

7.1 Multiscale Corrosion Induced by Metallurgical Heterogeneity

7.1.1 Localized Corrosion Over Multi-Time and Length Scales

Corrosion can proceed through a series of electrochemical processes occurring over various metallurgical structures on an electrode surface, permitting the localization of surface chemical conditions and the propagation of corrosion from nano- or micro-scale to macro-spatial scales [1, 2]. Although some types of localized corrosion such as galvanic corrosion of weldments can occur in a pre-existing macro-sized galvanic cell, in most cases localized corrosion is initiated from pre-existing atomic, nano-, micro- or millimeter-sized electrode inhomogeneities and

Localized Corrosion in Complex Environments, First Edition. Mike Yongjun Tan.

heterogeneities such as compositional or metallurgical nonuniformities existing on the surface of a metal or alloy. Take the corrosion of a high-strength aluminum aircraft alloy AA2024 as an example, the addition of strength-enhancing alloying elements such as copper leads to a complex heterogeneous alloy microstructure with hardening precipitates, intermetallic particles (IMPs), crystallographic defects, and deformation-induced defects such as dislocations. With such a broad range of compositionally and structurally complex microstructure, a wide spectrum of complicated localized corrosion mechanisms can take place on AA2024 over wide ranges of length and time scales such as trenching [3–7], intergranular corrosion [3, 6–13], and pitting corrosion [10–22]. Figure 7.1 shows sample images of a corroded AA2024 surface with the scale bar length changing from 200 to 10 and 2 μm in order to visualize corrosion at different scales [3].

Multiscale localized corrosion also occurs on steels and stainless steels in engineering structures such as seawater desalination pipe. For instance, although stainless steels like 316 stainless steel (SS316) and 2205 duplex stainless steel (SS2205) are widely used in seawater desalination systems because of their excellent general corrosion resistance in seawater and brine, multiscale corrosion can occur on weldments of stainless pipes exposed to the highly corrosive seawater reverse osmosis desalination environments due to adverse effects by welding process that could result in modifications of local metallurgical composition and microstructures. In the seawater feed pre-treatment zone, for instance, feed seawater salinity can vary from 32 000 to 44 000 mg/L. In the pre-treatment zone, seawater is performed to eliminate the undesirable constituents by adding various chemicals such as sodium hypochlorite (NaClO) for the prevention of microorganism growth, ferric chloride ($FeCl_3$) as a flocculant, and sulfuric acid (H_2SO_4) for the adjustment of pH. Corrosion on the stainless steel weldments exhibits different corrosion behavior in each zone that is affected by the surface morphology, microstructure, and surface composition [23–25]. Similar multiscale corrosion also occurs on carbon steel weldments [26, 27] such as on pipelines where local damage to protective surface scale occur. Figure 7.2 shows photos of some typical pits on a piping in a water injection line where serious pitting corrosion occurred. Similar localized attack also occurred on brass alloys water distributor pipes due to dezincification [28]. Although the processes and mechanisms of aluminum alloys shown in Figure 7.1 and of steel pipe in Figure 7.2 are different, they have something in common: localized corrosion initiates due to the formation of macroscopic local anodes and cathodes over a metal surface where local chemistry at an metal–solution interface is developed. The most common locations for local anodes to form are metal surface metallurgical defects, flaws, scratches, impurities, mill scales, precipitated phases, grain boundaries, dislocation arrays, localized stresses, selective dissolution, and chemical (such as chloride)-induced damages in passive films.

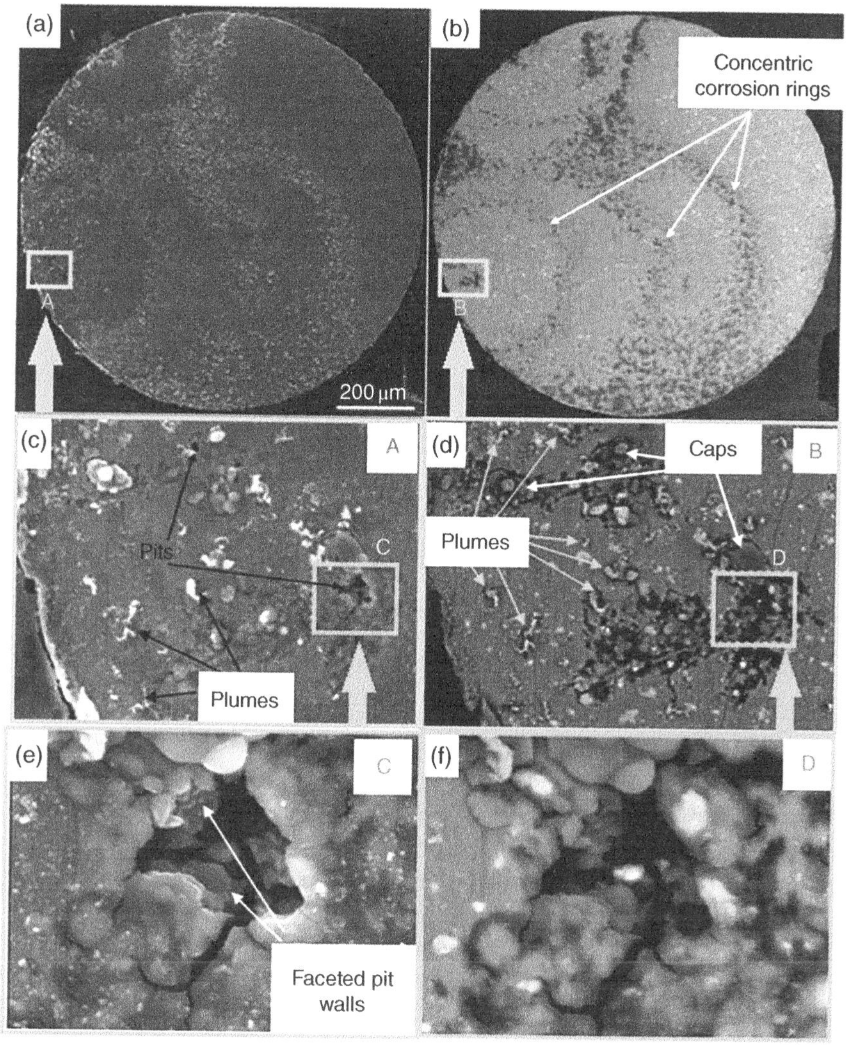

Figure 7.1 Some sample images of AA2024 corrosion over different scales after 120 minutes of immersion in open-to-air 0.05 M NaCl solution. (a) SE-SEM of an AA2024 wire and (b) BSE-SEM images showing the whole wire surface. (c) SE-SEM and (d) BSE-SEM images of two adjacent anodes surrounded by concentric corrosion rings, labeled "A" and "B" in (a) and (b), respectively. (e) SE-SEM and (f) BSE-SEM images of pits labeled "C" and "D" in (c) and (d), respectively. Note that the scale bar length in (c) and (d) is 10 μm and in (e) and (f) is 2 μm [3]. *Source:* Parvizi et al. [3], ELSEVIER.

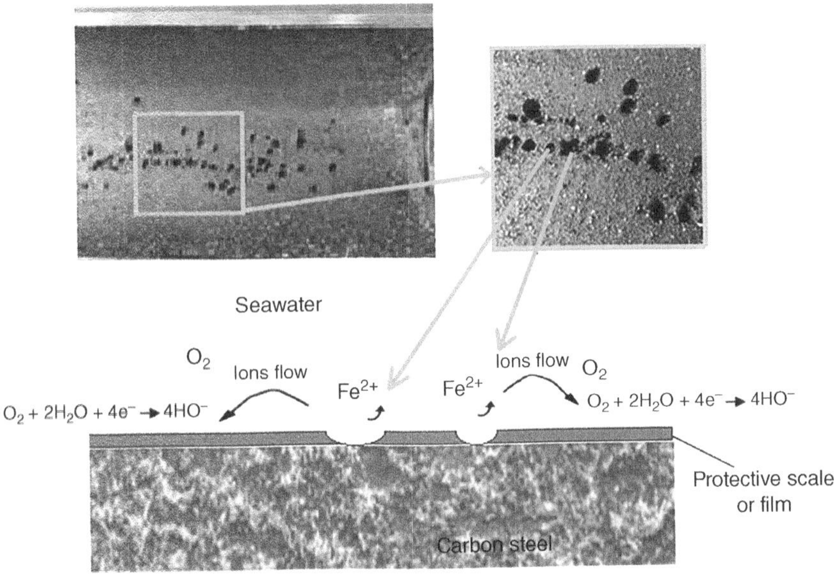

Figure 7.2 Photos of some typical pits on the steel pipe in a water injection line, and a model illustrating localized corrosion due to the formation of macroscopic local anodes and cathodes over a single piece of metal with damaged surface protective film.

The localized corrosion behavior of complex alloys could be significantly affected by interactions between a wide range of microstructures including defects and IMPs varying from nano- to micro- or even macro-scales in size. Localized corrosion processes can be dynamic, initiating from micro-sized event and propagating into macro-sized corrosion problems over different periods of time. This attribute suggests that in order to probe and fully understand localized corrosion of complex alloys, there is a need of characterization techniques that have appropriate temporal and spatial resolutions for probing and visualizing the propagating electrochemical heterogeneity at various scales. Unfortunately conventional methods generally have limitations in measuring electrochemical corrosion at multiple scales, and therefore probing localized corrosion processes occurring on metal surfaces at different spatial and temporal scales remains a technological challenge.

Corrosion models are important corrosion prediction tools; the accuracy of model prediction relies upon available corrosion data especially those related to localized corrosion. Various multiscale models and computational methods have been proposed over past decades to fill the gaps in the experimental characterization of the corrosion system particularly at the lower end of the scales [3, 29]. Unfortunately multiscale corrosion models reported in the literature are usually based on limited number of 'retrospective' field inspection and exposure testing data that generally do not have sufficient temporal resolution required to model the dynamic early stages of localized corrosion and provide little spatial information about corrosion patterns and the underlying localized corrosion mechanisms. In an extensive review on multiscale modeling of localized corrosion, it is pointed out that existing models lack the resolution required for effective corrosion prediction and control and that there are gaps in the experimental characterization of the corrosion system [29].

7.1.2 Techniques for Probing Localized Corrosion Over Different Scales

Traditionally visual, optical, and scanning electron microscopic (SEM) observations of corrosion coupons are the most common approach to determining localized corrosion over different length scales. Figure 7.1 shows a sample of such approach to visualizing corrosion at different scales with the scale bar length changing from 200 (Figure 7.1a and b), to 10 (Figure 7.1c and d) and 2 μm (Figure 7.1e and f) [3]. Visual inspection of metal surface can determine the sizes and shapes of major macro-sized inhomogeneities such as surface scratches, while a more detailed examination of a corroded surface can be carried out through an optical microscope and SEM as shown in Figure 7.1. More advanced surface analytical techniques such as the scanning tunneling electron microscope (STM) and atom probe tomography (APT) [9] are able to examine inhomogeneities in the nano- and atomic dimensions. Unfortunately localized corrosion testing using corrosion coupons is slow and does not have sufficient temporal resolutions for probing multiscale localized corrosion. It is cumbersome and requires periodic removal of test specimen from the corroding environment, which may alter the progress of heterogeneous electrochemical processes on the coupon surface. Some levels of temporal resolutions can be added to certain advanced surface analytical techniques by capturing surface changes with time. For instance, X-ray radiography techniques such as X-ray photoelectron spectroscopy (XPS), Auger electron spectroscopy (AES), X-ray absorption and fluorescence measurements could be employed for *in situ* study of chemical and physical changes during electrochemical and corrosion processes [30]. Ryan et al. used focused ion beam/secondary ion mass spectroscopy to find a depletion of chromium during the processing of a stainless steel that triggered a high-rate electrochemical dissolution at the edge of the sulfide inclusion [31], although these results were challenged by Schmuki et al. by using Auger microscopy and STM [32]. Unfortunately most of the advanced surface analytical instruments used in characterizing the microstructure and composition of metal surfaces operate in high vacuum and therefore they cannot produce real *in situ* data from an undisturbed corroding surface in aggressive media under practical environmental conditions.

Localized corrosion data with higher spatial and temporal resolutions could be acquired through *in situ* measurement of heterogeneous electrochemical parameters related to local corrosion processes such as local electrode potential and local electrochemical kinetics. However, conventional electrochemical techniques have major limitations in measuring local electrochemical parameters, in detecting local chemistry changes and in determining local electrochemical kinetics. For instance electrochemical techniques such as linear polarization and AC impedance spectroscopy do not measure the rates of localized corrosion at a specific location over an electrode surface. In order to add spatial resolutions, scanning probes such as the scanning reference electrode technique (SRET), the scanning Kelvin probe (SKP), the scanning Kelvin probe force microscopy (SKPFM), scanning vibrating electrode technique (SVET), local electrochemical impedance spectroscopy (LEIS), and scanning electrochemical microscopy (SECM) have been developed and used for scanning localized corrosion. The SRET, as illustrated in Figure 7.3, is probably the simplest scanning probe technique that has been widely applied in studying various forms of localized corrosion such as pitting initiation, intergranular corrosion, surface film breakdown, stress corrosion cracking, and galvanic corrosion. SKPFM is probably the most widely used scanning probe that combines the SKP with atomic force microscopy (AFM) and uses small probes and operates at small distances from the corroding metal surface, and thus SKPFM has an improved lateral resolution of better than 0.1 μm compared to the classical SKP of 100 μm. In a typical application, Schmutz and Frankel [33] studied the surface inhomogeneity and localized corrosion of aluminum alloy AA2024-T3 using SKPFM by detecting large differences in the Volta potential of intermetallic particles and the matrix phase that resulted in a potential map showing the locations of the particles. A linear relationship was found between the Volta potential measured in air and the corrosion potential measured in aqueous solution, indicating that Volta potential could be used to determine the practical nobility of the surface [33]. This relationship forms a basis for the application of the Volta potential in understanding corrosion processes and has been used in various studies [34–42].

A purpose of probing corroding surface with high spatial and temporal resolution is to identify localized corrosion initiation and its relationship with surface inhomogeneity in order to inform corrosion mitigation and management. Although electrode inhomogeneity could origin from the eruption of a passive film, mostly they should be a natural existence that presents on all practical electrode surfaces, regardless of its shape and sizes. It is logical to expect that electrode inhomogeneities are the weakest locations over an electrode surface, where local electrochemical activities such as localized corrosion could initiate. Indeed many experimental evidences indicate that electrode inhomogeneities such as intermetallic particles, second phases, or impurities behave as the vulnerable or susceptible sites for the initiation of electrochemical heterogeneity, causing the formation

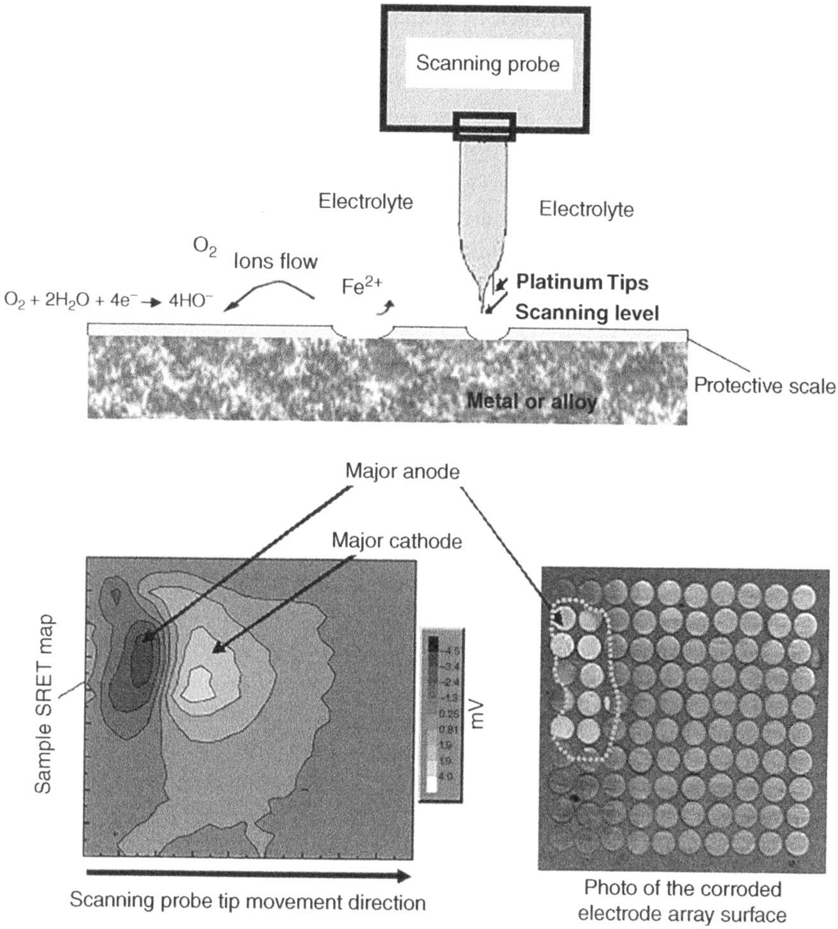

Figure 7.3 Illustrating localized corrosion mapping using the scanning reference electrode technique (SRET) and an electrode array. *Source:* Tan [1, 2], IOP Publishing.

of pitting precursors [1–3]. For instance in stainless steels, the composition, density, and size of the sulfide inclusions have been reported to have a critical effect on the probability of initiation of pitting corrosion [43]. Intermetallics in aluminum alloys were found to weaken the passive film and form key sites for pit nucleation [44]. The β-phase structures were found to be responsible for the poor corrosion resistance of aluminum–magnesium alloys [45]. Pitting potential decreases as the surface roughness increases [46], aiding the nucleation of metastable pitting [47, 48]. It is therefore necessary to possess the capability of probing electrode inhomogeneity and its effect on localized corrosion.

The SRET, as illustrated in Figure 7.3, is probably the simplest scanning probe technique that has the capability of probing electrode inhomogeneity and its effect on localized corrosion such as pitting initiation, intergranular corrosion, surface film breakdown, stress corrosion cracking, and galvanic corrosion [49–53]. The sensitivity of SRET is highly dependent on corrosion current values and distribution over metal surface, electrolyte conductivity, and the probe proximity over corrosion sites. Since SRET only scans the currents in solution phase not exactly at the metal–solution interface, they have limited resolution to localized corrosion processes that have anodic and cathodic sites not well separated. The movement of scanning tips could interfere with electrochemical processes occurring over an electrode surface, especially when a corrosion product forms over an electrode surface. The SVET technique is developed to improve the sensitivity of detecting local electrochemical and corrosion events by operating with a nonintrusive scanning, vibrating probe that measures the electric field generated by localized electrode processes over an electrode–solution interface [54, 55]. The LEIS [56] is another scanning probe technique that can measure the ac impedance distribution maps over an electrode surface that could detect electrochemical heterogeneity and localized corrosion activities over the electrode surface, while traditional ac impedance of this electrode gave little indication of its presence [55–57]. The scanning electrochemical microscope (SECM) [58, 59] is a tool that enables us to perform difficult tasks of detecting localized chemistry changes by means of variously designed scanning probes that detect amperometrically surface-generated electroactive ions or molecules in the solution phase as a function of spatial location with an electrochemically sensitive or ion-selective ultramicroelectrode tip. It has been applied for topography mapping, surface modification, and redox reactivity imaging [58, 59], as well as making high-resolution chemical concentration maps of corroding metal surfaces [48, 60].

Scanning probes could also be used in conjunction with traditional optical microscope such as SEM and EDS to provide chemical and topographical information that are often critical for explaining electrochemical heterogeneity and localized corrosion activities. For instance, using scanning micro-reference electrode technique combined with SEM and EDS, Shao et al. [21] found that the micro-pitting on AA2024–T3 initiated immediately after its exposure to NaCl solution at open-circuit potential and that most of the micro-pits were found to be associated with the second-phase particles, specially the S phase particles. Some corrosion active nuclei were found to develop with time while others disappeared. Rough surface finishes [47] and non-metallic inclusions in stainless steels [16, 28] have also been reported to nucleate metastable pitting. However due to their limited temporal resolutions, scanning probes may be unable to observe dynamic initiating and terminating pits. A method of detecting dynamic metastable pitting events is the recording and analysis of current transients, often referred to as electrochemical noise, that are supposed to be due to evolution of metastable corrosion pits. Electrochemical noise analysis (ENA) is a technique proposed and utilized for detecting localized corrosion and for measuring corrosion rates [61–64]. A more detailed overview of these techniques can be found in reference [2].

It should be noted that scanning probe techniques including SRET, SVET, LEIS and SECM can detect ionic currents, carried by ions in the electrolyte phase, flowing over a corroding metal surface, however they are unable to measure the currents flowing exactly at the metal-solution interface. For this reason, they may not be able to accurately detect all ionic currents, especially those flowing very close to the metal–solution interface. Scanning probe techniques commonly operate in a relatively specific and localized area, and thus, in many circumstances, the scan image does not necessarily represent the full details of an electrode process that involves different reactions occurring simultaneously over distinctively separated electrode areas. For instance in practical corrosion studies, it is difficult to ensure that a scanning tip is correctly positioned over a pit precursor unless the precursor is generated by the probe tip itself. This implies that successful imaging of a natural pit initiation process by scanning probes could depend upon 'luck' in experiments. Figure 7.4 shows an overview diagram of different types of techniques plotted against their respective temporal and spatial resolution regimes, showing a comparison of the spatial vs. temporal resolutions of these methods in detecting

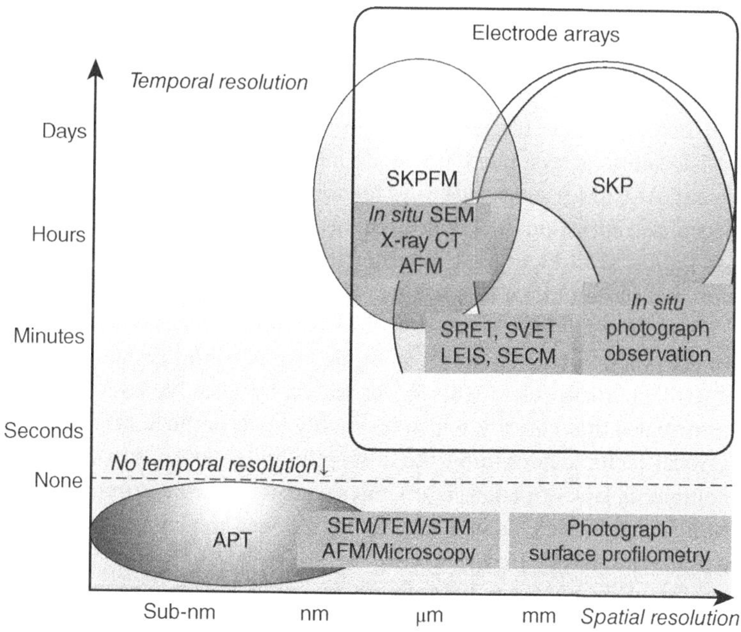

Figure 7.4 Overview diagram of different types of corrosion and hydrogen mapping techniques plotted against their respective temporal and spatial resolution regimes.

corrosion and other chemical or electrochemical processes such as hydrogen distribution in the microstructures of high-strength steels, providing different levels of information of varying spatial and temporal resolutions on corrosion and hydrogen embrittlement behavior of steels.

7.1.3 Three-Dimensional Techniques for Characterizing Multiscale Localized Corrosion

Over the past decade, techniques with three-dimensional (3D) spatial resolutions are more frequently used in corrosion research. 3D analysis of localized corrosion is advantageous in revealing the active corrosion occurring below the visible surface and in providing more insights into the development of localized corrosion, which is not achievable by traditional two-dimensional (2D) techniques. One such technique is atom probe tomography (APT) [9]. In a recent work, the unique capabilities of APT supported by other high-resolution techniques have been used observe entangled oxide networks co-located with dislocation structures piled up at corroded intermetallic particles of an AA2024-T3 alloy. It was revealed that dislocation arrays act as pathways for corrosive species and promote structural degradation at interfacial regions, providing a new insight into corrosion initiation at the nano-scale [9]. TEM results confirm the existence of residual dislocation segments with corrosion products in the vicinity of different IMPs. APT results have shown the accumulation of elements together with metal hydrides, oxides/hydroxides within the cores of piled up dislocations. These results suggest that localized corrosion was triggered by high-density nano-scale dislocation structures. Dislocation arrays piled up at de-alloyed nano-scale precipitates ($Al_{20}Mn_3Cu_2$ and Al_2CuMg) act as pathways for corrosive species, promoting the initiation of localized corrosion [9]. Unfortunately this technique has no temporal resolution, as indicated in Figure 7.4. Another three-dimensional characterization technique that has temporal resolution is X-ray computed tomography (CT) analysis. In a typical study [65], a number of techniques described above have been used in conjunction with X-ray CT analysis to reveal multiscale localized corrosion of 316L stainless steel (316L SS) produced by selective laser melting (SLM). It was revealed that porosity, which inevitably exists in the SLM-produced parts, is a critical factor determining the susceptibility to localized corrosion. Specimens containing lack-of-fusion (LOF) pores were found to be extremely susceptible to localized corrosion, as indicated by their lower breakdown potentials measured in polarization tests [65].

316L SS produced by SLM has been found by some researchers to possess improved pitting corrosion resistance [66–69] that was attributed to the absence of harmful manganese sulfide (MnS) inclusions in SLM-produced 316L SS [66] since MnS inclusions are considered to be the preferential sites for pit initiation [70–74].

However, there are also reports showing a deterioration in pitting corrosion resistance of SLM-produced 316L SS that was generally attributed process-induced defects especially porosities [75–77], which has been identified as the most critical issue affecting the localized corrosion behavior of SLM-produced materials [78–80]. X-ray CT has been used to probe the effects of such defects on the corrosion behavior [65], in particular to overcome technological difficulties in obtaining direct experimental evidence to validate the role of SLM process-induced defects in pit initiation/propagation [81, 82]. In a typical study [65], electrochemical polarization measurements were first used to probe the overall localized corrosion susceptibility. Figure 7.5a shows the potentiodynamic polarization curves for the SLM-produced and commercial 316L SS specimens recorded in 0.6 M NaCl solution at room temperature. Breakdown potentials (E_b) acquired from the polarization curves are presented in Figure 7.5b. The breakdown potential is an electrochemical parameter that has been commonly used to evaluate the susceptibility of a metallic material to the formation of a stable pit in aggressive environments A nobler breakdown potential is indicative of a material less susceptible to localized corrosion in the given environment [83]. As shown in Figure 7.5, the SLM-produced specimens show rather different behaviors with some show a nobler E_b and some show less noble E_b compared to their commercial counterpart, indicating large variabilities in localized corrosion resistance. The difference in localized corrosion behavior appeared to be linked to microstructural differences, especially LOF pores in SLM-produced 316L SS specimens [65].

In order to gain experimental evidences and insight on the role of LOF pores in the localized corrosion susceptibility of SLM-produced specimens, 2D surface analytical techniques were used to relate localized corrosion and pores on the corroded surfaces. As shown in Figure 7.6a and b, pits in stainless steels tend to grow under the metal surface, leaving a porous metallic surface, which is known as lacy cover [84, 85], while no visible pits are present in high-density specimens (P175-S600 and P150-S400 with densities around 99.5%), as shown in Figure 7.6c and d. This suggests that pits tend to form at the sites of LOF pores. To further confirm the role of the LOF pores in the localized corrosion performance, local electrochemical measurements were carried out by performing local polarization measurement over selected surface areas while other areas are masked by a coating layer. Figure 7.7 display potentiodynamic polarization curves for regions with LOF pores and regions with very few pores, together with corroded surface images. The polarization curves clearly show a remarkably different localized corrosion behavior of the different types of pores. The regions containing mostly spherical pores (Figure 7.7a) show very noble breakdown potentials. In contrast, the regions containing LOF pores (Figure 7.7b) show much lower E_b relative to those of Figure 7.7a, indicating their higher susceptibility to localized corrosion. These results suggest that pores and especially LOF pores contribute significantly

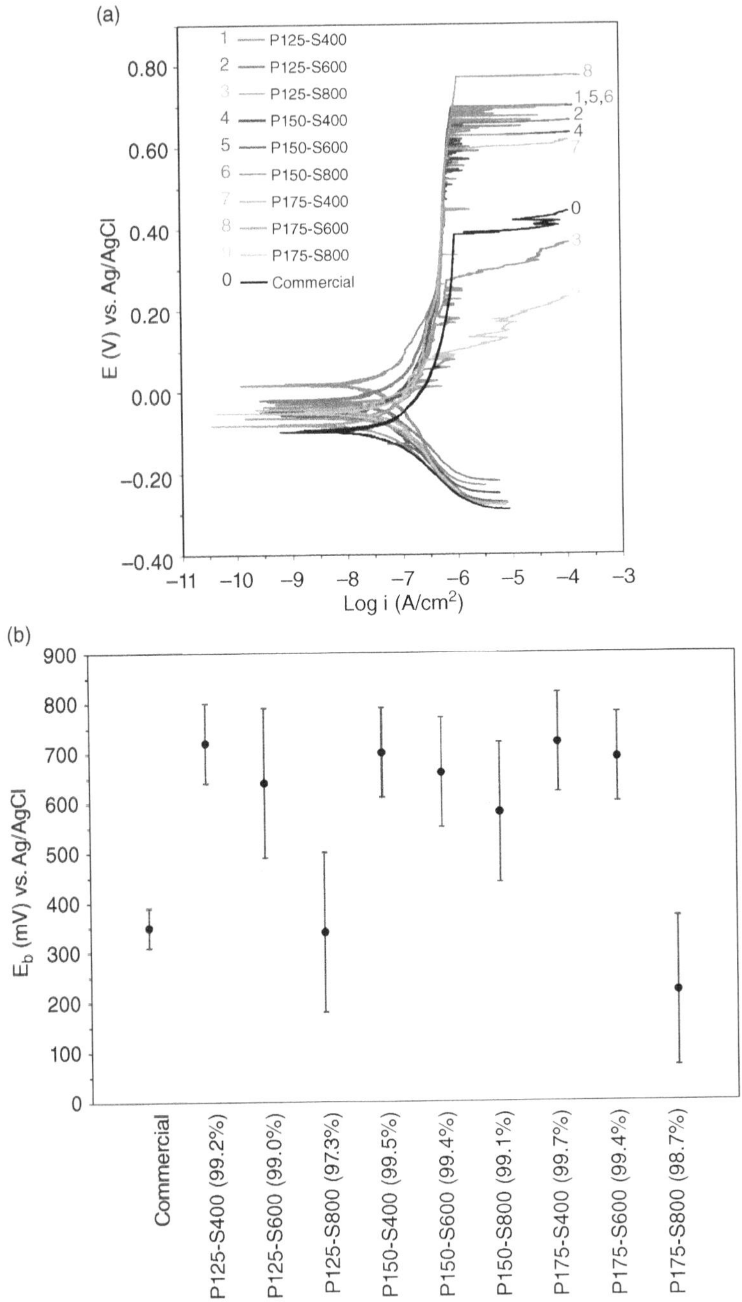

Figure 7.5 (a) Potentiodynamic polarization curves of the SLM-produced and commercial 316L SS specimens recorded in 0.6 M NaCl solution at room temperature and (b) values of breakdown potentials for the commercial and SLM-produced 316L SS specimens acquired from the polarization curves. *Source:* Laleh et al. [65] / With permission of Elsevier.

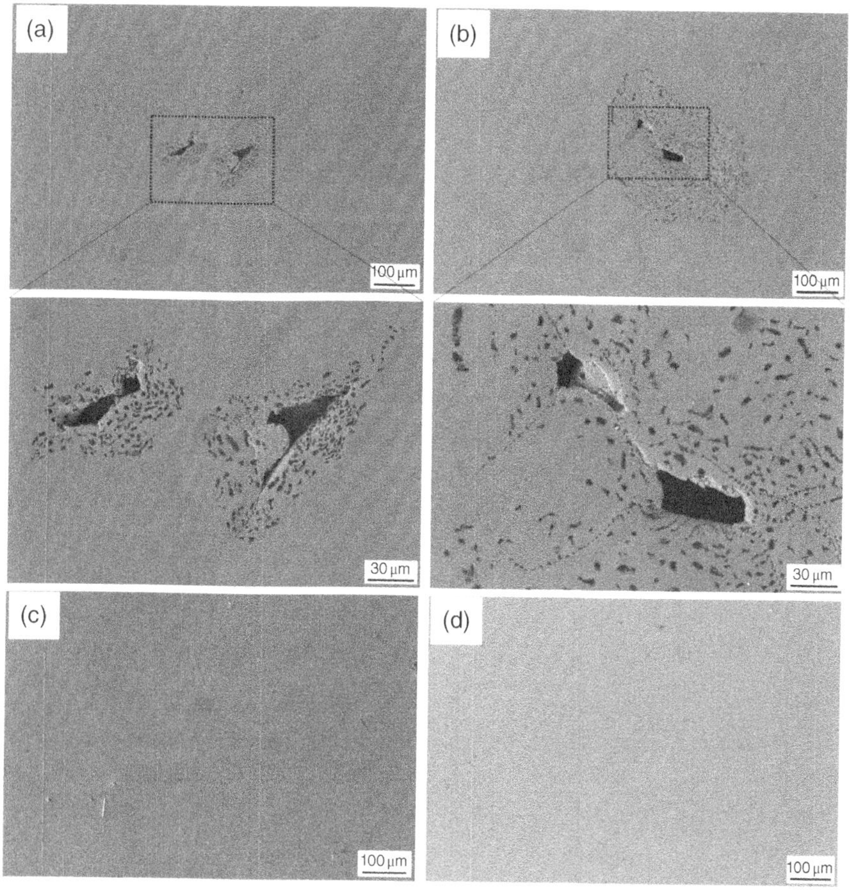

Figure 7.6 SEM images from the corroded surface of the (a) P175-S800, (b) P125-S800, (c) P175-S600, and (d) P150-S400 specimens after potentiodynamic polarization tests, indicating the LOF pores acted as the pit formation sites. Details about the test specimens are in reference. *Source:* Laleh et al. [65], ELSEVIER.

to the localized corrosion susceptibility of the SLM-produced 316L SS specimens [65]. However, it should be noted that there are still gaps between this electrochemical and 2D surface analytical analysis because direct visualization of the actual effect of LOF pores on localized corrosion within the body of the specimens is not available in 2D surface analytical and electrochemical techniques [65].

In order to gain more direct experimental evidences, 3D analysis would be advantageous in revealing active corrosion processes occurring at certain microstructures or defects, especially those below the visible surface [65]. 3D X-ray CT experiments were therefore employed and combined with Data Constrained Modelling (DCM [86]) to visualize the localized corrosion propagation within the

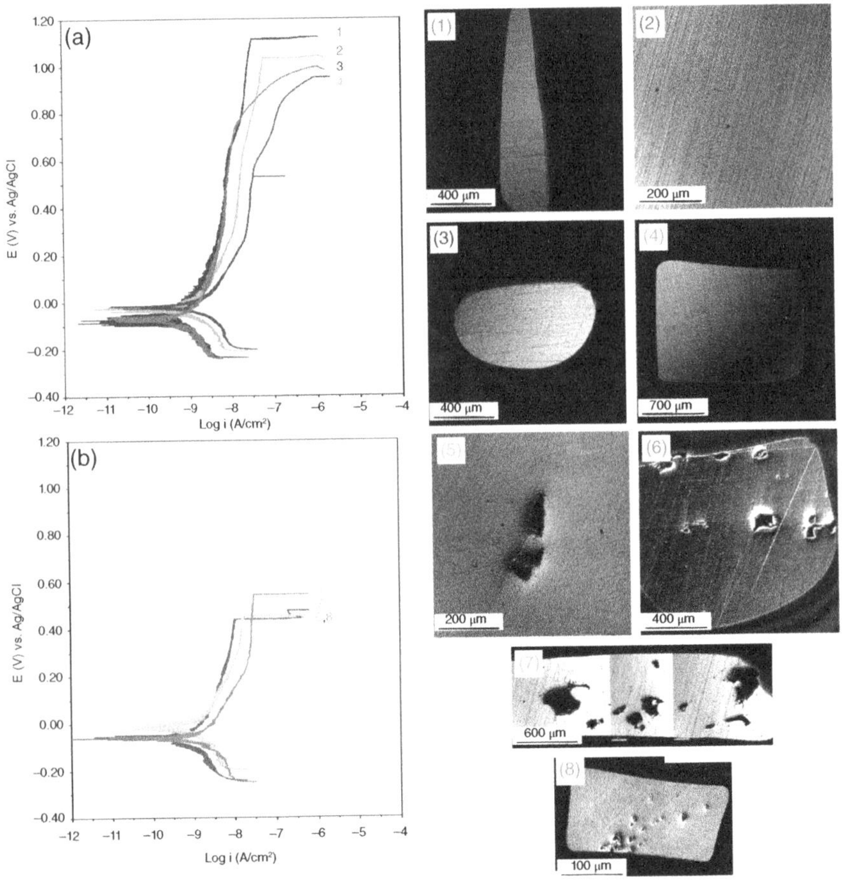

Figure 7.7 Potentiodynamic polarization curves recorded in 0.6 M NaCl solution from different regions of SLM-produced 316L SS specimens: (a) regions containing small pores and (b) regions containing big LOF pores, together with images of the regions that the measurements were done. *Source:* Laleh et al. [65] / With permission of Elsevier.

SLM-produced specimens. Figure 7.8 shows the 3D volumetric reconstruction of different components for the lowest-density specimen (P125-S800) prior to and after immersion into ferric chloride solution. Figure 7.8a and Figure 7.8b show the shape of the tip of the 316L SS sample before and after corrosion, respectively. Figure 7.8c and d show some of the internal structure. In Figure 7.8c the 316L SS (in light gray) has been made transparent to show internal distribution of void and oxide structures within this high-porosity specimen. Many irregular-shaped LOF structures are clearly visible inside the material and form vertical structures of the 316L SS specimens. These LOF structures are presented without the 316L SS

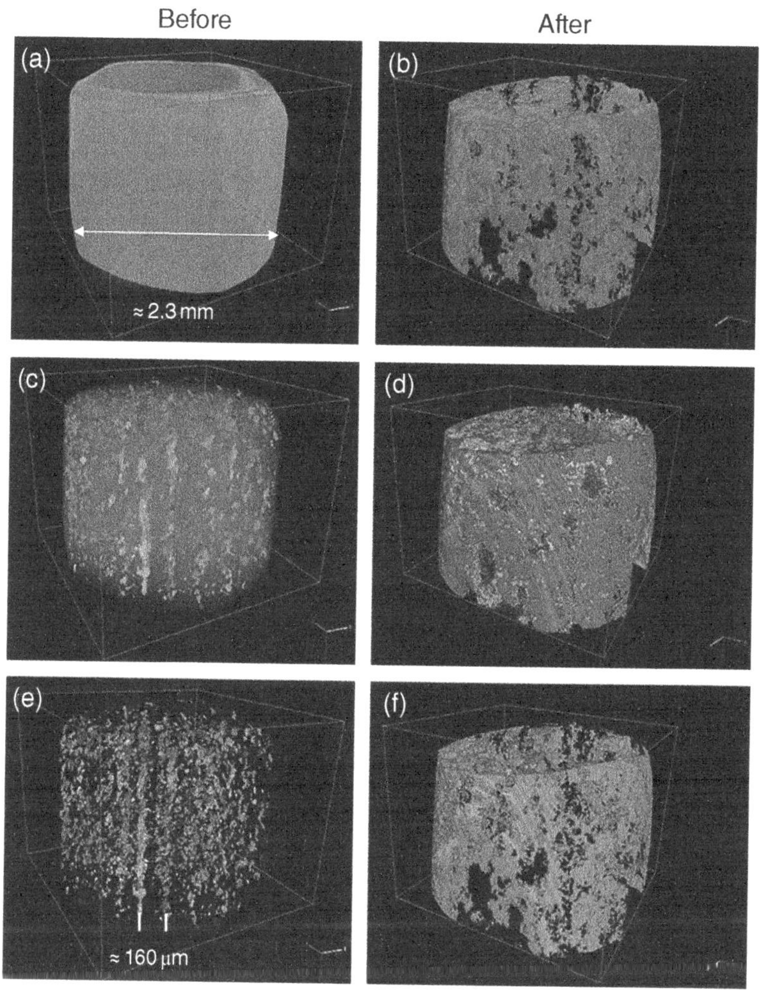

Figure 7.8 DCM X-ray CT volumetric reconstructed images of the lowest-density SLM-produced specimen (P125-S800): (a,c,e) prior to and (b,d,f) after one-week immersion into corrosive ferric chloride solution. (a) and (b) represent the whole sample prior to and after exposure. (c) and (d) show the LOF structures prior to and after corrosion, respectively. (e) shows the LOF clusters in different colors, which are clearly aligned vertically within the sample while (f) shows the remaining 316L SS after the corrosion experiment, which is essentially a shell. *Source:* Laleh et al. [65], ELSEVIER.

component in Figure 7.8e. In this way, it can be seen that corrosion of most of these LOF structures do not propagate throughout the whole sample. After corrosion it can be seen in Figure 7.8d and f that most of the interior of the specimen has been etched away after immersion into the ferric chloride solution, indicating

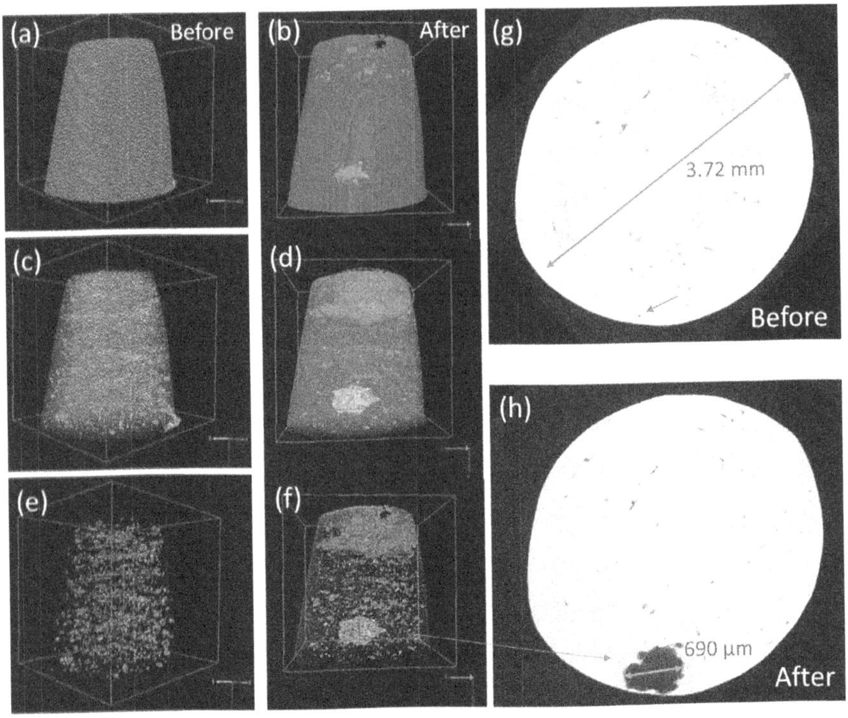

Figure 7.9 DCM CT volumetric reconstructions of P125-S600 specimen (a,c,e) before and (b,d,f) after one-week immersion into the ferric chloride solution, showing the development of two pits. (a) shows the shape of the sample prior to corrosion, whereas (b) shows the same perspective after corrosion. The salmon color indicates where voids from pits intersect the surface of the sample. In (c) and (d) selective transparency has been applied to the 316L SS revealing the internal LOF structures within the specimen before and after corrosion. After corrosion two large pits are evident in the sample (light blue and wheat volumes) as indicated by the red ellipses. In (e) and (f) the 316L SS has been completely removed to reveal the LOF structures (refer to supplementary material 2 to see the 3D movie LOF structure in this specimen). Reconstructed slices in the plane of the bottom pit are shown in (g) and (h). In the slice presented in (h) the pit has a diameter of approx. 690 microns. The red arrow in (g) indicates an LOF structure that may have been involved in pit initiation as discussed in more detail in Figure 7.10. *Source:* Laleh et al. [65], ELSEVIER.

a very weak resistance of this specimen to corrosion in this solution. These results suggest that LOF structures on the external surface played a critical role for the penetration of corrosive electrolytes into the body of the specimen and internal propagation of corrosion within the interior of the specimen [65]. To further demonstrate the role of the LOF structures in the propagation of corrosion, a high-density specimen was investigated using 3D X-ray CT technique. Figure 7.9 shows the DCM images of this specimen before and after immersion in the ferric

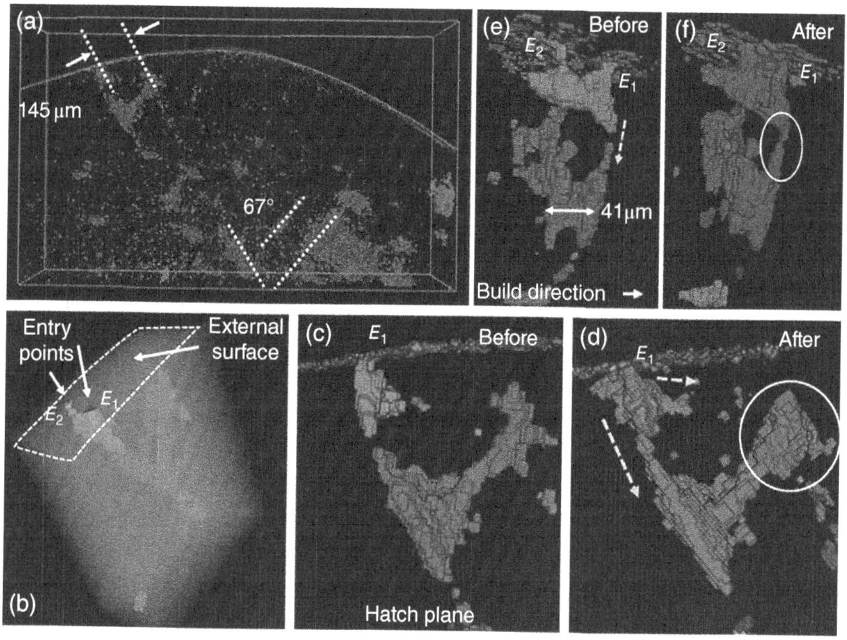

Figure 7.10 DCM CT volumetric reconstructions of P125-S600 specimen (a,b,c,e) before and (d,f) after one-week immersion into the ferric chloride solution, showing the development of corrosion within a LOF structure. (a) General overview of the region containing the target LOF structure along with other LOF structures. At the top of (a) the hatch spacing is 145 μm and propagates well into the sample. The hatch layer above can be seen to be at angle 67° with respect to the first layer. The LOF structure of interest is next to the label "145 μm." (b) The external surface is indicated by the dashed white line. The entry points into the LOF structure are also indicated by the labels E_1 and E_2. (c) and (d) are for the LOF structure prior to and after corrosion, respectively; only entry point E_1 can be seen from this perspective. (c) shows two major LOF structures in off-white and cyan. (d) shows that after corrosion these structures are connected via the LOF structure itself indicating propagation of corrosion within the structure, it also shows enlargement of the structures indicating etching not associated with the LOF structure. (e) and (f) shows the LOF structure within the powder bed plane before and after corrosion. The spacing between two linear features is around 40 μm, which is slightly larger than the powder bed thickness of 30 μm. Entry points E_1 and E_2 are also shown. (f) after corrosion all these independent features have the same color indicating that they are joined. The white ellipse indicates where the corrosion has propagated from the off-white LOF structure to the cyan LOF structure [65].

chloride solution. Figure 7.10 shows DCM volumetric reconstructions for an LOF structure in the specimen both before and after one-week immersion into the ferric chloride solution. The LOF structure of interest in Figure 7.10a is displayed at higher magnification in Figure 7.10b–f. These results indicate that propagation of corrosion occurs along LOF structural features. However, corrosion is not merely confined to propagation along the LOF structures as the data also shows enlargement of the structures (e.g. indicated by white circle in Figure 7.10d) [65].

It is clear by comparing X-ray DCM CT images in Figures 7.8–7.10 that the extent of corrosion is substantially greater for the lower-density specimen relative to higher-density one. For the lower-density specimen, there are numerous big LOF structures inside the specimen, as shown in Figure 7.8c and e. Thus when pitting propagates from the surface, it has a higher chance of encountering an adjacent LOF structure inside the material and thus rapidly forms a much larger pit by incorporating the void of the LOF structure. However, for the pit developed in the higher density specimen, there are far fewer LOF structures adjacent to the pit initiation site, thus pitting must proceed merely via a normal process without the assistance of internal voids, which facilitate internal attack. Thus in both cases, propagation is probably a combination of grain etch out and grain boundary attack, but in the case of lower density specimens, there is also the contribution of the LOF structures to its total volume. Therefore, the internal microstructure of the SLM-produced specimens will have an influence on localized corrosion development; the presence/absence of internal LOF structures adjacent to the propagation site significantly influences the proceeding of corrosion. These results suggest that the presence of internal porosity associated with LOF structures facilitates corrosion, justifying the hypothesis that the differences in the localized corrosion resistance of the SLM-produced 316L SS specimens is based on the attack within the LOF structures. For the occluded volumes in the LOF structure, corrosion may begin in a similar fashion to the open pores, but because the diffusion path to the external electrolyte is long, then the generation of metal cations results in a build-up of aggressive anions (Cl^-), greater acidity and the corrosion accelerates [87]. Therefore, the presence of occluded regions inside the irregular-shaped pores can have a strong influence on the localized corrosion susceptibility of the SLM-produced 316L SS not experienced by conventionally manufactured 316 SS. A similar explanation has been proposed by Otero et al. [88] for stainless steels prepared by powder metallurgy [65].

Overall this typical case indicates that 3D X-ray CT technique is able to show that the localized corrosion resistance of the SLM-produced 316L SS is significantly affected by the pore characteristics. It was revealed that the type of the pores in the SLM-produced specimens is a key factor controlling the susceptibility to localized corrosion. The irregular-shaped LOF structures have a detrimental effect on both corrosion initiation and propagation as determined by polarization measurements with the support of the 2D and 3D analysis. 3D mapping of microstructure and porosity using X-ray CT provided more insights into localized corrosion development in SLM-produced 316L SS. CT analysis showed that the LOF structures intersected the outer surface of the material and provided potential initiation sites for corrosion. Moreover, the irregular-shaped pores of LOF structures facilitated the propagation of corrosion playing a significant role in the localized corrosion susceptibility of the SLM-produced specimens.

7.2 Various Probes Designed for Probing Multiscale Corrosion

7.2.1 2D and 3D Electrode Arrays for Probing Multiscale Corrosion

Another approach to probing multiscale localized corrosion is through the use of variously designed 2D and 3D electrochemically integrated multi-electrode array probes. Although the electrode arrays such as those shown in Figures 2.4–2.6 have enabled measurement of localized activities at reasonably high temporal resolutions, conventionally designed electrode arrays have a major drawback in spatial resolutions due to the use of relatively large size wires. Usually 1000 μm (1 mm) diameter wires are used in electrode arrays to visualize mm-scale localized corrosion processes over an approximately $10 \times 10\,mm^2$ or $4 \times 15\,mm^2$ electrode areas by mapping currents crossing closely packed wires. Such integrated electrode array would allow effective chemical and electrochemical interaction between wires located at different areas visualized; however, electrochemical activities within each individual wire are not visible. This suggests that conventionally designed electrode arrays would not be able to probe early corrosion initiation processes that are usually controlled by microstructural features of an alloy. For instance, heterogeneous alloys like AA2024 have many local galvanic cells of various sizes over an individual wire surface because of the metallurgical heterogeneity. Obviously, an electrode array made of mm-sized wires has a limitation in sensing 'internal' galvanic currents flowing between μm-sized microstructural features with a metal wire. In this case, the electrode array only be able to detect a portion of the total galvanic current flowing over the whole electrode surface, i.e. galvanic currents flowing crossing wires. In order to address this issue, the size of wires used for an electrode array has been considered in order to achieve desirable accuracy and resolution of the galvanic current mapping [3]. For instance, for studying the initiation stages of AA2024-T3 localized corrosion where the corrosion cell size is in micron range, micron-sized wires need to be used [3]. Figure 7.11 shows a schematic illustration of a typical experimental setup used for *in situ* monitoring of localized corrosion/inhibition behaviors of the AA2024 alloy at mm- and μm-scales using the WBE probes made from 1 mm and 100 μm AA2024 wires [3]. A conventionally designed electrode array made of 1 mm diameter aluminum alloy AA2024-T3 wires was used to probe co-operative

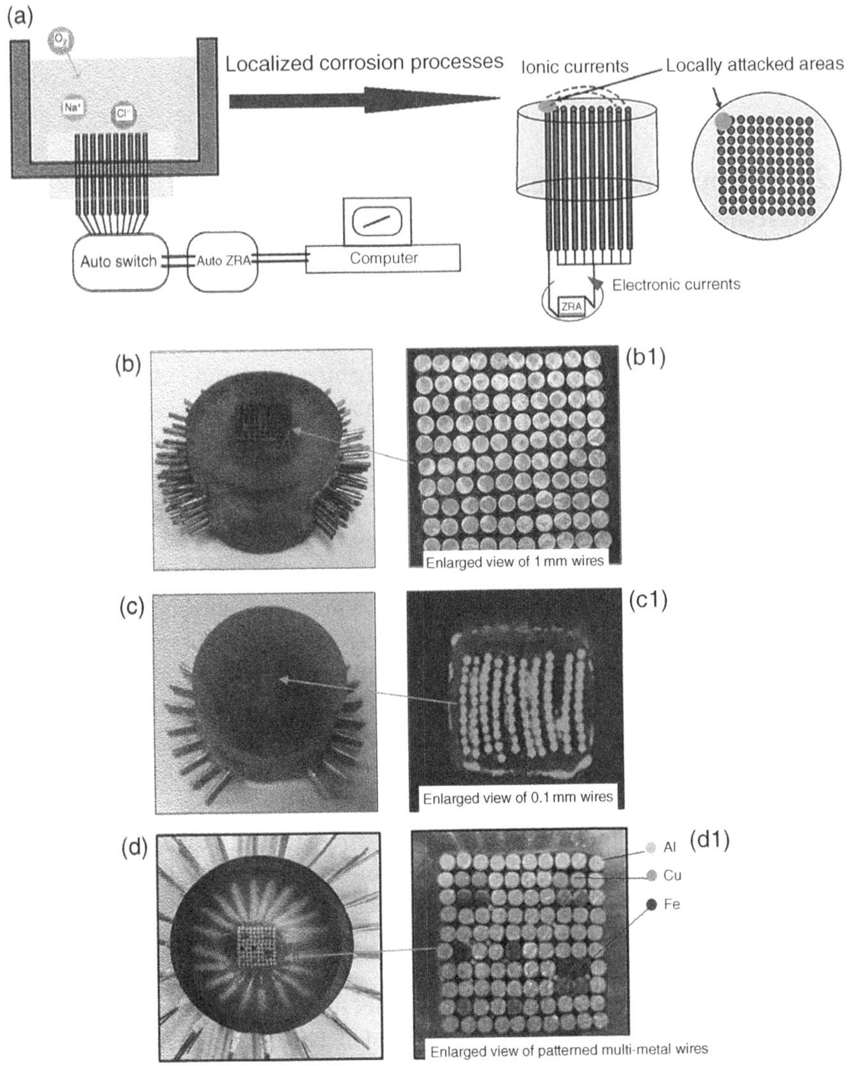

Figure 7.11 (a) Schematic of the test setup for *in situ* monitoring of corrosion processes using arrayed electrode, (a1) real setup designed for video capturing of the corrosion advancement on WBE surface, (b)-(b1) electrode arrays made from 1 mm AA2024-T3 wires. (c)-(c1) electrode arrays made from 0.1 mm AA2024-T3 wires. (d)-(d1) electrode arrays made from pure Al, Cu, and Fe wires, named "simulated arrayed multi-electrode." Note that each 1 mm pin shown in (b), (c), and (d) are connected to its defined socket of data cables shown in (a). *Source:* Parvizi et al. [3] / With permission of ELSEVIER.

corrosion in the millimeter scale. An electrode array made of 100 μm diameter AA2024-T3 wires was used to reveal early stage corrosion initiation activities over individual wire surfaces in the micro-meter scale. A patterned electrode array was made from combination of 1 mm pure Al, Cu, and Fe wires to imitate complex microstructures in order to simulate and probe the processes and mechanisms of co-operative corrosion occurring across different metallurgical structural features [3].

A combination of variously designed and patterned electrode arrays and surface analytical techniques are shown to be able to provide unique insight on different stages of localized corrosion at different spatial and temporal scales. Electrode arrays made from 1mm and 100 μm diameter aluminum alloy AA2024-T3 wires have been used to visualize *in situ* micrometer to millimeter scale electrochemical corrosion processes occurring on a typical structurally complex alloy, AA2024, while surface analytical techniques including SEM-EDS have been used to image micrometer to micrometer scale *ex situ* post corrosion structural features [3]. Figure 7.12 shows electrochemical corrosion processes occurring over a 100 μm wire array electrodes that were able to reveal early stage corrosion initiation activities over individual wire surfaces in the micro-meter scale, while the conventionally designed 1 mm wire electrode array was able to probe co-operative corrosion in the millimeter scale. A patterned electrode array was made to probe the processes and mechanisms of co-operative corrosion occurring across different metallurgical structural features. A combination of variously designed and simulated or patterned electrode arrays and surface analytical techniques are shown to be able to provide unique insight on different stages of localized corrosion at different spatial and temporal scales [3]. These results suggests that the combined use of variously sized electrode arrays and advanced surface analytical techniques is a practical approach to facilitating the visualization and probing of heterogeneous metallurgical, electrochemical, and chemical processes occurring on metal and alloy surfaces at different spatial, from nanometer to centimeter, and temporal scales [3]. This electrode array can also be used in combination with scanning electrochemical techniques. As shown in Figure 7.13, localized corrosion and inhibition of aluminum processes were mapped over mm- to cm-scale using the electrode array in combination with scanning reference electrode technique (SRET) [89].

Although the multi-electrode array approach is able to measure corrosion processes over different length scales individually, anyone of them is only able to show corrosion processes over a narrow window of the length scales. Electrode arrays developed and used by many researchers over the past two decades are unable to measure these dynamic early stage corrosion processes occurring over multiscale, because, generally, previous electrode arrays were designed to measure localized corrosion only over cm length scales and minutes in temporal scales.

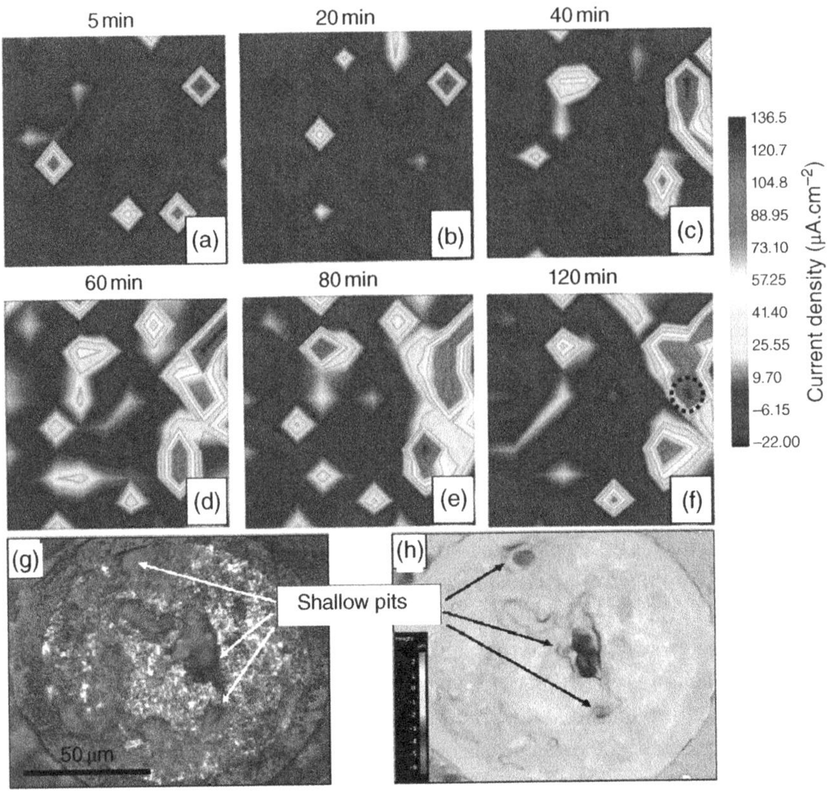

Figure 7.12 (a–f) Galvanic current density distribution maps of the electrode array (made from 100 μm AA2024-T3 wires) represented in Figure 7.11(c-c1) during 120 minutes of exposure to open-to-air 0.05 M NaCl solution. (g) Optical image and (h) colored contour plot of the most anodic wire at the test termination. The color bar in (h) indicates the depth profile of the corroded surface (depth of micro-pits) across the heavily corroded wire. *Source:* Parvizi et al. [3], ELSEVIER.

The design of electrode array probes for use in complex heterogeneous material-environmental conditions where multiscale and dynamic localized corrosion occurs remains major challenges and obstacles [90]. Fundamental technological obstacles and knowledge gaps associated with current electrode arrays can be illustrated in Figure 7.14 that shows microstructures of a heterogeneous alloy consisting of a wide range of grain sizes, defects, and intermetallic particles (IMPs) varying from nano- to micro- or even macro-scales in size. The initiation of corrosion is usually due to galvanic activities between the micron-sized intermetallic particles and the matrix. If conventional 1 mm wires/electrodes are cut from this plate to make a cm-sized multi-electrode array with mm-sized wires, as shown on the top right corner of Figure 7.14, the current measurable by an

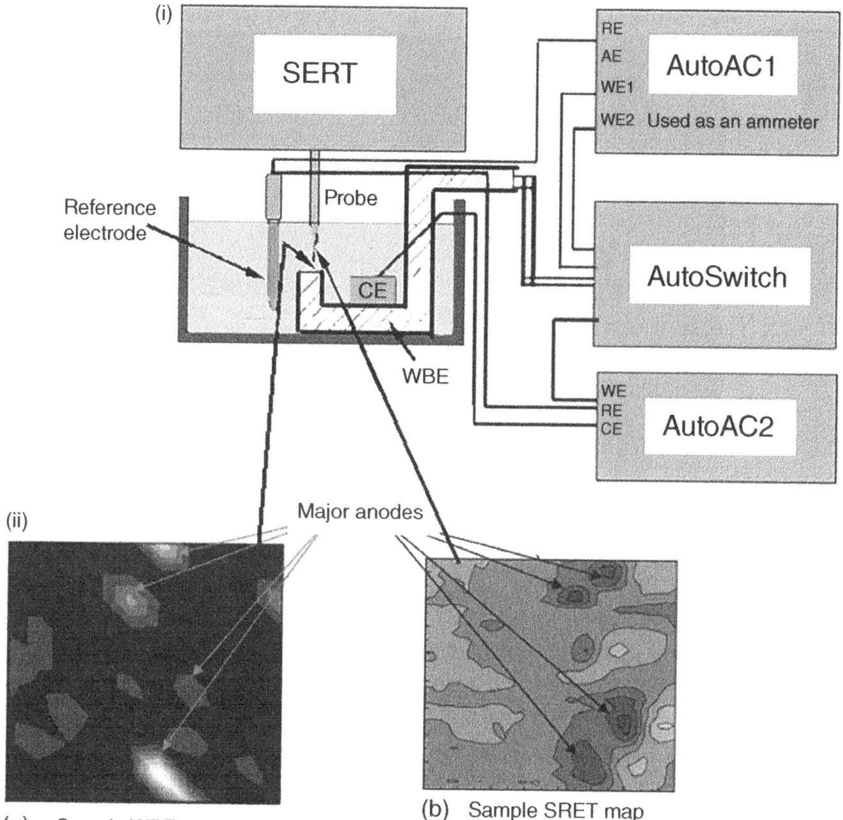

Figure 7.13 (i) The schematic setup of experiment using the electrode array (i.e. the WBE) in combination with SRET, and photos showing the WBE and its working surface [89]. (ii) (a) The WBE and (b) SRET map obtained after two hours, immersion of an AA2024-T3 WBE in 0.005MNaCl+10ppm $CeCl_3$ solution. *Source:* [89] / With permission of Elsevier.

ammeter (Ammeter 1) would be only that driven by potential differences between these two mm-sized electrodes (mm electrodes 1 and 2). For the alloy plate, the overall distribution of micro-scaled intermetallic particles on the plate over a larger scale (e.g. mm-size) could be considered random. In the extreme case that these two mm-sized electrodes could have exactly the same amount of intermetallic particles over their surfaces, there will be no measurable current flowing through the Ammeter 1. Since "internal" currents flowing between micron-sized intermetallic particles and alloy matrix within each mm-sized electrode are not measurable using an "external" ammeter. Therefore measurements using the conventional 1mm sized electrode array would overlook information related to

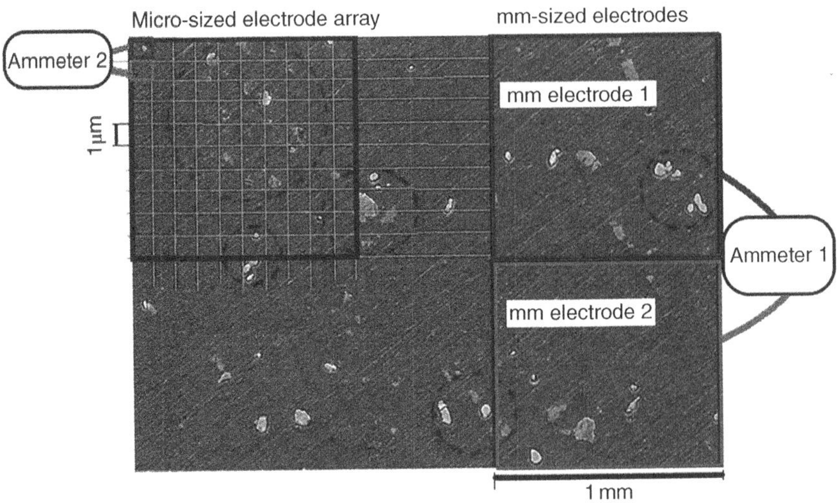

Figure 7.14 Illustrating galvanic current measurements over a heterogeneous alloy surface containing many intermetallic particles.

corrosion initiation processes occurring within each individual wire surface. This suggests a major limitation of the conventional mm-sized electrode array in probing this stage of localized corrosion initiation processes occurring between intermetallic particles and matrix. Similarly, this conventional 1 mm electrode array would not be able to detect corrosion initiation from micro-scale sized scratch/damage in the passive film due to the impingement of high-velocity multi-phase corrosive fluids/particles. This problem becomes even more complex if the actual corrosion surface has a complex shape and is exposed to dynamic flow.

This limitation could be overcome if micro-sized electrodes are used to make an electrode array, as shown on the top left corner of Figure 7.14, then micron-sized intermetallic particles would be distributed over different electrode surfaces (μm-electrode 1&2). If the IMP size on μm-electrode 1 is 10 μm and there is no IMP on μm-electrode 2, then galvanic currents flowing between the intermetallic particle and matrix would be partially measurable by the Ammeter 2. In this manner, the measurement capability of the electrode arrays is extended from mm to μm length scale. This measurement limit could be further extended if smaller electrodes and electrodes with selected metallurgy are used to fabricate electrode arrays. Given the fact that different stages of localized corrosion on heterogeneous alloy surfaces are driven by different surface features of various sizes, the combined use of variously designed and patterned electrode arrays and surface analytical techniques (to identify microstructural features) would provide unique insights on different stages of localized corrosion [3]. However this approach still has several unsolved issues: (i) A small-scale electrode array (e.g. μm-electrodes) could make small-scale corrosion initiation process visible; however, because of its small size, it would not be able to continue visualizing large-scale events that are over the electrode size limit. A possibility to overcome this problem is to use a significant large number of electrodes in an electrode array, however this makes electrode array fabrication too difficult and measurement not practical. (ii) The flat surface of conventionally designed electrode arrays would not be able to simulate the corrosion environment such as the erosion corrosion occurring on a complex surface such as the ebb of a pipeline. (iii) The fabrication of small-sized electrode arrays is a manufacturing challenge. It has been extremely difficult to fabricate a small-scale electrode array using μm-electrodes [3].

A solution to this issue is to develop an electrochemically integrated multiscale and three-dimensional electrode array (M3D array) method in order to enable real time probing and monitoring of complex multiscale and dynamic localized corrosion processes with the unprecedented capabilities of the M3D array for probing localized corrosion initiation from micro-scale and propagation to macro-scale under complex erosion–corrosion–inhibition conditions, and in acquiring corrosion data for validating selected major models for multiscale corrosion prediction. Figure 7.15 illustrates a M3D array designed for probing erosion corrosion processes in small-scale flow simulation devices and jet impingement cells. In Figure 7.15, the M3D is designed specifically for probing erosion corrosion of stainless steel 316L under jet impingement conditions. It simulates a critical corrosion condition that commonly occurs in industrial systems such as oil and gas, oil sand, water, mineral processing, paper milling and power generation where high-velocity multi-phase corrosive fluids, slurry, and steam can damage passive films on pipelines, pumps, valves, and other infrastructural components. Under the impact of high-velocity fluids and particles, passive metals such as stainless steel 316L are subjected to the simultaneous effects of mechanical impacts on the surface and electrochemical corrosion, causing localized corrosion. It is a situation that is highly complex and dynamic with multiple time and length scale for localized corrosion processes. As shown in the cross-section view, the surface of the M3D array is three-dimensional to simulate jet impingement erosion-corrosion conditions. In order to measure multi-scale corrosion processes, the M3D consists multiscale electrodes. As shown in the top-view, the electrode array consists three 'sub-arrays' that are made of micro-sized electrodes. The number, size, and the composition of these sub-arrays can vary depending upon the nature and the scale of corrosion initiation process of interest. For instance, if the interest is to probe the effects of μm-sized IMPs such as those in Figure 7.14 on corrosion initiation, micro-sized electrodes would be used in these sub-arrays to specifically probe micro-scale processes as demonstrated in the use of micro-sized electrode arrays [3]. Such M3D array design is different from previous electrode arrays in

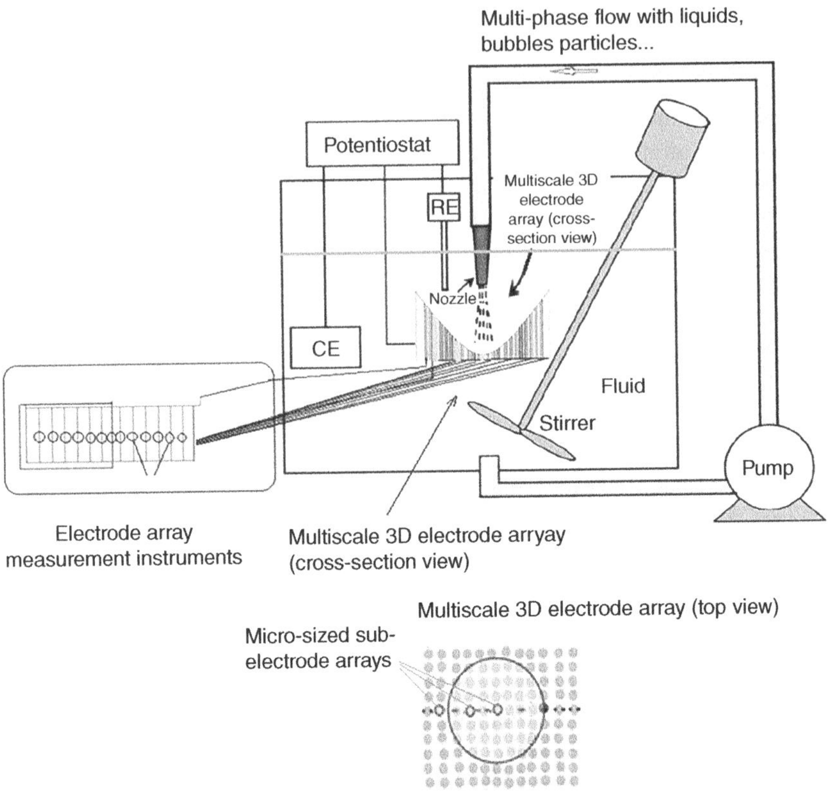

Figure 7.15 Illustrating direct impingement tests using a 3D electrode array for probing erosion corrosion over differently shaped M3D array surfaces for probing erosion corrosion of a pipe ebb.

various ways. Firstly, a combination of various sized sub-arrays in the M3D not only makes small-scale corrosion initiation process visible but also enable the probing of large-scale events because the overall M3D size can be large (e.g. cm scale). On the other hand, the three-dimensional surface of the M3D would enable the simulation of complex unevenly shaped surface such as an ebb of a pipeline corrosion environment. This effectively extends the electrode array method from a conventional mm to cm scale localized corrosion measurement method into a method that is able to measure corrosion data from complex localized corrosion surfaces of multi-length scales. This system will be able to show corrosion maps as videos using an electrode array measuring instrument system that has been developed and demonstrated previously (a sample video can be found from reference [91]), permitting *in situ* probing and monitoring of dynamic localized corrosion initiation and propagation. It should be noted however that the fabrication of multi-sized electrode arrays such as the multiscale 3D electrode array shown in Figure 7.15 is a manufacturing challenge. An idea to overcome this challenge is to fabricate such a complex electrode array using additive manufacturing techniques. Microelectrode array technology has also been described as a new electrochemical test method between macroscopic classical electrochemical technology and micro-area scanning probe technology for measuring not only the overall average information of the large-area electrode of the alloy but also for detecting the characteristics of potential and current distribution in the local micro-area of the alloy [92].

7.2.2 Probing Corrosion of Multi-Metal Structures and Weldments

Most work reported in the literature on multiscale corrosion have been on the study of galvanic corrosion behavior of multiple metals/alloys. For instance, three metal alloys commonly used in water desalination plants was investigated using coupled multi-electrode array probes (WBE probes) consisting of aluminum-brass (HAl77-2), titanium alloy (TA2), and 316L stainless steel (316L SS) [93]. The three electrode types were coupled galvanically and arranged in different geometric configurations. The potential and current distributions of the three-electrode coupling systems displayed electrochemical heterogeneities. Generally, the aluminum–brass wires are anodic versus the titanium alloy and stainless steel. The titanium alloy acts as a primary cathode, and the 316L SS acts as a secondary cathode. When the anodic HAl77-2 wires were located on the edge and connected to the 316L SS wires in the coupling system, the main anodic area enlarged, especially in the area adjacent to the 316L SS wires [93]. Electrode arrays composed of zinc and mild steel wire sensors were also used to measure electrochemical inhomogeneity due to variations of distances between the zinc and mild steel wire sensors over the electrode array, as well as their surface electrochemical status [94]. Similar devices were used to study the galvanic corrosion behavior of copper and copper alloys in chloride solution under different temperature conditions [95]. It was found that the average galvanic current of the Cu and Cu–Sn alloy, which shifted from the cathode at 35 °C to the anode at 80 °C, increased as temperature increased. In contrast, the average galvanic current of Cu–Zn alloy was ranked as 35 °C > 55 °C > 80 °C [95]. Differently designed WBE probes composed of zinc and mild steel wires were used to simulate hot-dip galvanized steel with spot coating defect at different dimensions in order to obtain the spatial distributions of potential and current density and their variations with time during the galvanic corrosion. The results show that the zinc wires within WBE probes could provide enough cathodic protection to steel wires after immersed in seawater, with surface area ratio of zinc wires to steel wires as 1 : 120, 9 : 112, and 25 : 96, respectively [96].

Galvanic corrosion of weldment metals is a major issue that has also been studied using variously designed electrode array probes. The composition, microstructures and properties of the welded joints of structures such as pipelines are highly heterogeneous, resulting in corrosion such as galvanic corrosion between different parts. An eight-channel multi-electrode array sensor system was used to study the corrosion process of carbon steel weld and the inhibition performance of nitrite on weldment. The galvanic corrosion among weld metal, heat affected zone and base metal and the inhibition performance of nitrite were studied by measuring the individual open circuit potential, coupled potential and galvanic current [97]. In another work, a welded joint was reconstructed by the WBE probe to measure the electrochemical corrosion behavior of X80 pipeline steel welded joint. The weld metal acted as the main anode, while the heat affected zone acts as the main cathode, and the current polarity of base metal electrode changed alternatively with the distance between the micro-electrode and the weld metal zone during the immersion test [98]. The WBE was also designed to simulate and measure corrosion of a X80 steel welded joint [99] and it was found that the corrosion tendency of the isolated electrodes that constituted the X80 steel welded joint increased in the following order: fine grain heat affected zone, inter-critical heat affected zone, base metal, coarse grain heat affected zone, weld metal. These results revealed that weld metal and coarse grain heat affected zone were the weak links in the simulated X80 pipeline steel welded joints during its long-term service [99]. The corrosion behavior of various zones of the aluminum alloy 7050-T76 friction stir welding joint and their interaction were also studied using the WBE [100], showing that the heat-affected zone adjacent to nugget zone receives the most intense negative current and thus presented the intensified susceptibility to corrosion. Both pitting and intergranular corrosion were aggravated greatly in heat-affected zone while suppressed in base material when the whole joint was put into NaCl solution, indicating the dominating influence from macro-galvanic effect [100].

Figure 7.16a shows an example of designing an electrode array probe for simulating and measuring three critical forms of corrosion at different regions of a pipeline: pitting corrosion at welds along segments of pipe, crevice corrosion at bolts on flanges between sections of the pipe, and crevice corrosion between flanges bolted together. Figure 7.16b shows the schematic representation of the designed electrode array probe to simulate the industry conditions shown in Figure 7.16a. Figure 7.16c shows a typical corrosion probe designed and made to acquire data on the initiation and propagation of these three types of localized corrosion that occur over different temporal and spatial scales. Figure 7.17 shows sample corrosion data obtained using the corrosion probe in a tank containing tap water simulating river conditions. Corrosion data in Figure 7.17 clearly shows weldment corrosion in tap water. These results confirm that the corrosion probe is

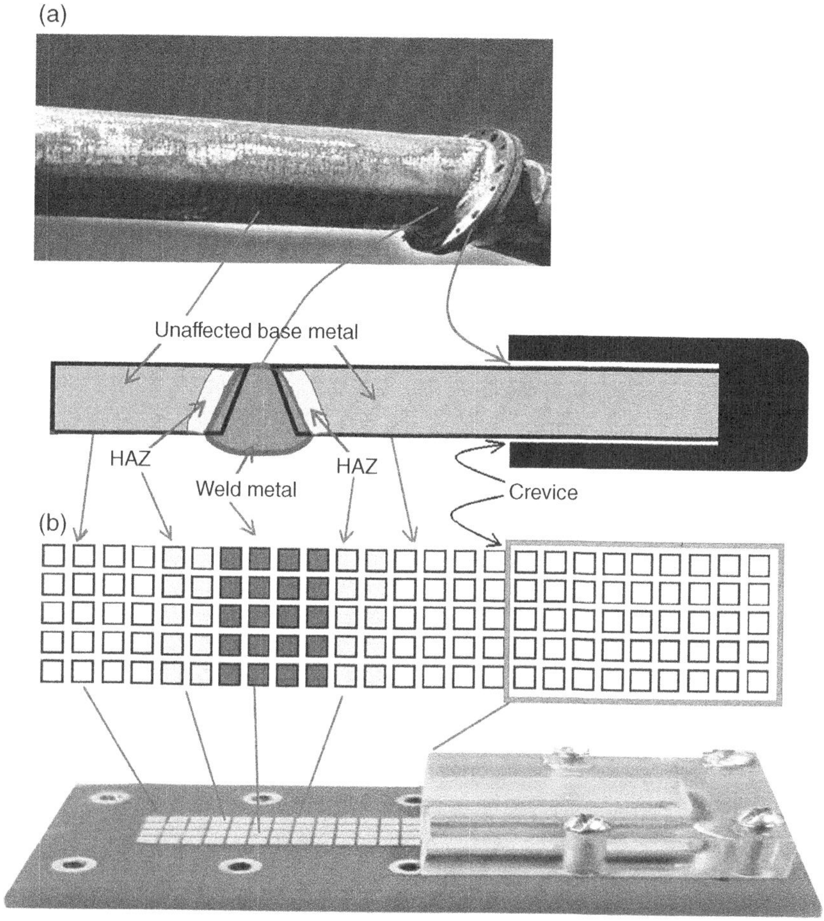

Figure 7.16 (a) Shows the corrosion at different regions of a pipeline. The region in yellow rectangle indicates the co-existence of three types of critical corrosion, i.e. pitting corrosion at welds along spools (spools = segments of pipe), crevice corrosion at bolts on flanges between spools, and crevice corrosion between flanges bolted together. (b) The schematic representation of the designed electrode array probe to simulate the industry conditions shown in (a).

sensitive in acquiring corrosion data (anodic current of 3200 nA and cathodic current of −2400 nA were measured from major anodes and cathodes of the electrode array). In tap water, corrosion occurred mainly in heat-affected zones and little corrosion was observed under the crevice area. Sample measurement data in Figure 7.17 clearly show corrosion at heat-affected zones with the weldment as the cathodic area in tap water, with peak localized corrosion rate of approximately 1 mm/y. Little corrosion occurred in the crevice area – this is expected because of

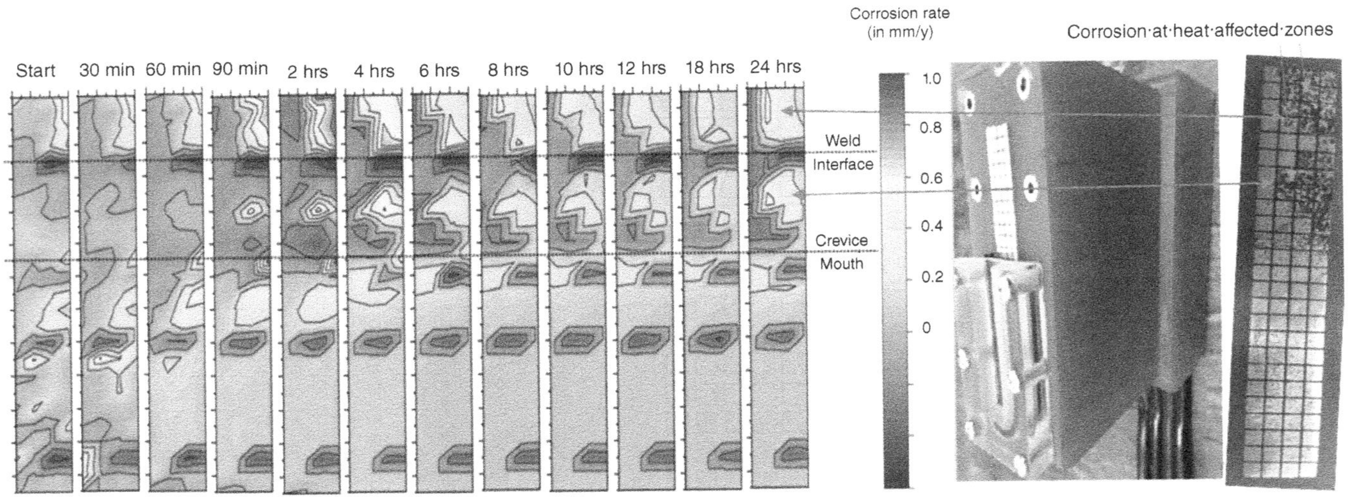

Figure 7.17 Sample measurement data that clearly shows weldment corrosion in tap water. The left images are corrosion reaction rates maps measured after various periods of exposure (scale bars: red/positive indicate corrosion anode zones with corrosion rates in mm/y, while blue/negative indicate cathodic reaction zones). The corroded probe surface was taken after a test duration of 24 hours in tap water open to the air. The distance of the sensor to the water surface was approximately 55 cm.

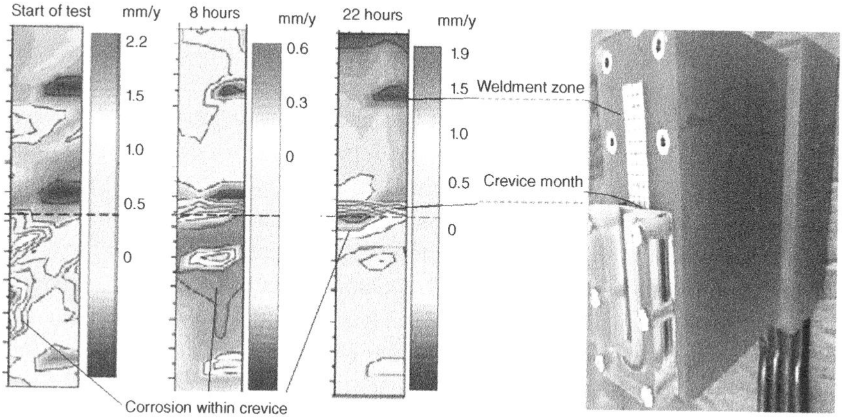

Figure 7.18 Sample measurement data that clearly shows crevice corrosion in 3.5%NaCl solution simulating seawater conditions. The images on the left are corrosion reaction rate maps after various periods of exposure (scale bars: red/positive indicate corrosion anode zones with corrosion rates in mm/y, while blue/negative indicate cathodic reaction zones).

the high resistivity of the tap water prevented the interaction inside and outside the crevice areas, therefore the main corrosion occurred over the heat-affected zones. Figure 7.18 shows results from the corrosion probe in 3.5%NaCl solution simulating seawater conditions. In this solution, corrosion is mainly in the crevice zone, with peak localized corrosion rate of approximately 2.2 mm/y. This suggests that in highly conductive solution, corrosion would be mainly controlled by galvanic interactions inside and outside the crevice area. Areas outside the crevice turned cathodic because of the more sufficient oxygen supply. Crevice corrosion occurred inside the crevice and gradually concentrated at the crevice month. These results are within expectations of corrosion of a steel pipeline as shown in Figure 7.16a in river water and seawater environmental conditions.

7.2.3 Simplified Method for Probing Multi-Scale Weldment Corrosion

In industry practice high temporal and spatial resolution corrosion measurements such as those shown in Figures 7.17 and 7.18 are often not required, instead simplified measurements are often sufficient for probing the corrosion behavior of weldments. An example is the study of corrosion of 316 stainless steel (316 SS) welded joints using a simple method of electrochemical measurement of selected zones on a welded joint of 316 stainless steel specimen [101]. Stainless steels are susceptible to localized corrosion attacks in chloride-containing environments [101–103]. Welding often makes this situation even worse because it can result in changes of microstructures, formation of thermal oxide (heat tint), and residual stresses [104, 105]. Due to the nonuniform temperature distribution during the thermal cycle, the changes in microstructures and surface conditions are nonuniform and therefore different zones of a welded joint could show different corrosion behavior. Numerous studies of localized corrosion processes on passive metal weldments have been carried out [106–109]. Standard cyclic polarization method (ASTM G61 – 86 -2009) is commonly used to evaluate the relative localized corrosion susceptibility of stainless steels by determining the breakdown potential; however, in principle this method cannot been used to welded joints where different heat-affected zones exist. Techniques such as local electrochemical cells have also been developed to perform local potential or polarization measurements; however, these local cell techniques often experience difficulties in the control of the standard testing conditions. Microelectrochemical methods were also used to study localized corrosion processes on small areas of passive metals [110–113] based on microcapillary-based droplet cells due to the advantage of the small size of the exposed working area [114]. However, in these cases, the electrolyte cannot flow through the capillary tip and causes some difficulties in its use [115]. A specific small-scale electrochemical cell called "minicell," different from the capillary-based microcell, was applied to study different forms of localized corrosion in welded joints of austenitic stainless steels [116, 117]. Nevertheless, it is still limited to extremely small volume of testing solution and it is difficult to apply standard testing conditions for the local cell tests.

A simple method of carrying out electrochemical measurement on selected welding zones has been reported and used to perform standard cyclic polarization tests on selected zones on welded stainless steel joints exposed to sodium chloride solution [22–24, 101]. Figure 7.19 shows the way to perform local electrochemical measurement on designated different zones in the heat tint area of a welded joint of 316 stainless steel (sample 1 [101], sample 2 [23], and the cross section of weldment of chromized SS316L [25]). To measure a selected zone in the heat tint area, as shown in Figure 7.19, all the other areas of the sample were coated by an insulating clear nail oil. Pitting corrosion behavior of each selected zone was investigated by standard cyclic potentiodynamic polarization tests and analysis of the microstructure and surface of the selected zones. The cyclic potentiodynamic polarization measurements were carried out according to ASTM standard G-61. Along with the electrochemical measurement, surface chemistry and morphology of the selected zones were also studied. Cyclic potentiodynamic polarization test results are shown in Figure 7.20. It can be seen in Figure 7.20a that zone Z1 behaved as an active metal, while zone Z2 behaved as passive metal. Pitting corrosion occurred at zone Z2 and the pitting potential (E_{pit}) was approximately 0.62 V. SEM and EDS analyses were performed on each of the zones and found that the surface of zone Z1 was rough and the high magnification image showed the surface was loose and powdery. In contrast, surface of zone Z2 was smooth

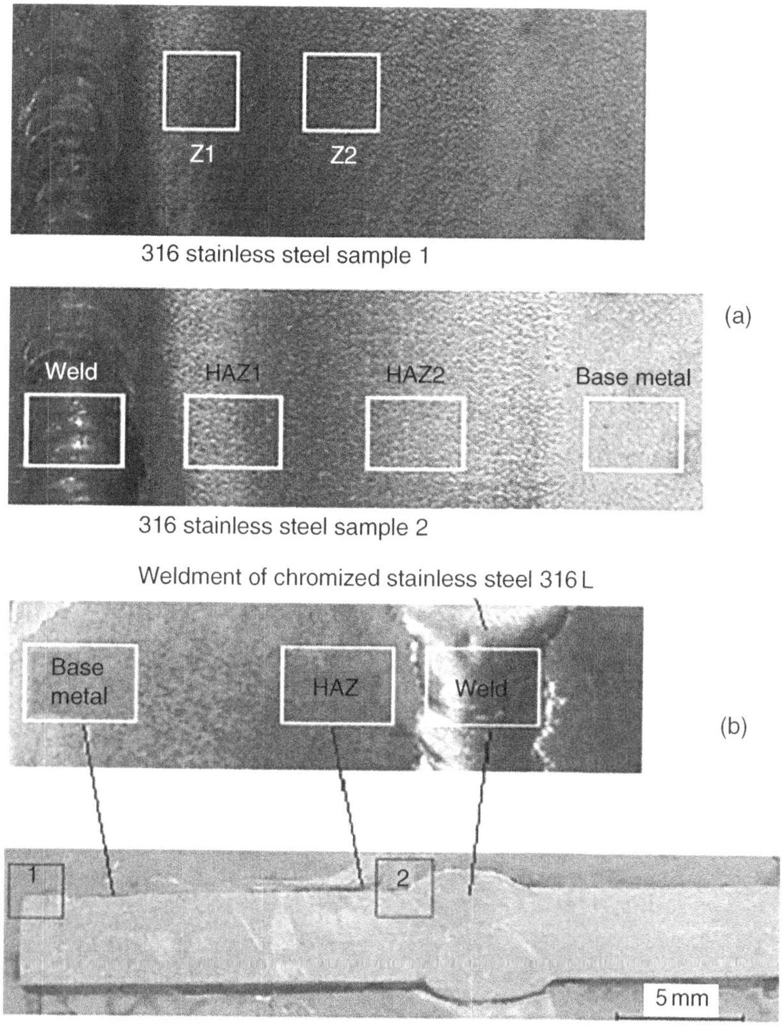

Figure 7.19 (a) Designated different zones Z1 and Z2 in the heat tint area of a welded joint of 316 stainless steel (sample 1 [101] and sample 2 [23]). (b) Cross section of weldment of chromized SS316L. *Source:* Xiong [25], Taylor & Francis Group.

and dense. The rough oxide layer at zone Z1 could not act as a protecting film. Based on the above discussion, the different corrosion behaviors of zone Z1 and Z2 can be related to the difference in the surface chemistry and morphology. Similar observations can be made in Figure 7.20b.

This simple method can be used in various ways. For instance, the localized polarisation measurement shown in Figure 7.7 obtained polarization curves

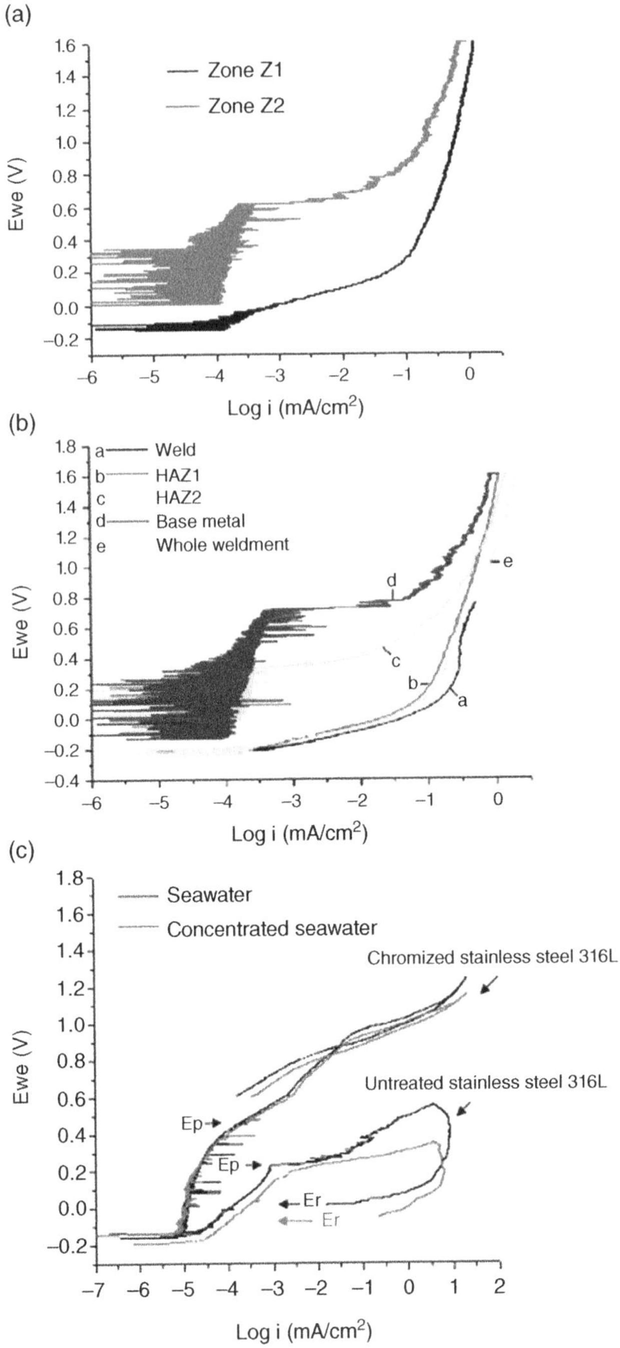

Figure 7.20 (a) CPP test results for selected zones Z1 and Z2 on a typical stainless steel 316 welded joint shown in Figure 7.19 SS sample 1. *Source:* Adapted from Xiong et al. [101]. (b) Potentiodynamic polarization test results for ((a) weld, (b) HAZ1, (c) HAZ2, (d) base metal, and (e) whole weldment of SS316 shown in Figure 7.19 SS sample 2). *Source:* Xiong et al. [23] / With permission of Elsevier. (c) CPP measurement results from untreated SS316L and chromized SS316L (chromized at 1050 °C) electrodes in seawater and concentrated seawater as shown in Figure 7.19b. *Source:* Xiong et al. [25] / With permission of Elsevier.

recorded in 0.6 M NaCl solution from different regions of SLM-produced 316L SS specimens. This method was also used to probe the corrosion susceptibility of chromized SS316L at the welding zones in order to improve the localized corrosion resistance of heat affect zones on a stainless steel weldment can suffer serious localized corrosion [23–25]. In a typical work, the SS316L chromized at 1050 °C was subjected to welding and the effects of the welding process on the Cr-enriched surface layer were investigated by performing electrochemical and surface analytical investigations. Figure 7.20c shows polarization curves that indicate that the electrochemical corrosion behavior for the HAZ area was similar to that at the base chromized SS316L area, with the CPP tests showing low current density values (below 10^{-3} mA/cm^2) over a larger potential range (to approximately 0.6 V). These results suggest that the Cr-enriched layer remained intact post welding of the surface-treated 316L alloy and that the HAZ remained resistant to corrosion. The localized corrosion behavior of the chromized stainless steel SS316L, chromized via fluid-bed reactor chemical vapor deposition, in simulated seawater reverse osmosis desalination environments was investigated using cyclic potentiodynamic polarization and crevice corrosion experiments in conjunction with surface analytical techniques. It is shown that the chromizing process could significantly improve the localized corrosion resistance of low-grade stainless steel SS316L, in particular, its resistance to crevice corrosion and corrosion at weldments. Chromium-enriched surface layers are found to be retained after the welding process, which leads to the improvement in corrosion resistance of weld zones [25].

References

1 Tan, Y. (2009). Sensing electrode inhomogeneity and electrochemical heterogeneity using an electrochemically integrated multielectrode array. *Journal of the Electrochemical Society* 156: C195–C208.

2 Tan, Y.J. (2013). *Heterogeneous Electrode Processes and Localized Corrosion*, 246. Hoboken, NJ, USA: John Wiley & Sons Inc.

3 Parvizi, R., Hughes, A.E., and Tan, Y. (2021). New approach to probing localized corrosion processes over wide length and time scales using integrated multi-scale electrode arrays. *Corrosion Science* 181: 109238.

4 Glenn, A.M., Hughes, A.E., Torpy, A. et al. (2015). Defect density associated with constituent particles in AA2024-T3 and its role in corrosion. *Surface and Interface Analysis* 48: 780–788.

5 Fonseca, I.T.E., Lima, N., Rodrigues, J.A. et al. (2002). Passivity breakdown of Al 2024-T3 alloy in chloride solutions: a test of the point defect model. *Electrochemistry Communications* 4: 353–357.

6 Zhou, X., Luo, C., Ma, Y. et al. (2013). Grain-stored energy and the propagation of intergranular corrosion in AA2xxx aluminium alloys. *Surface and Interface Analysis* 45: 1543–1547.
7 Luo, C., Zhou, X., Thompson, G.E., and Hughes, A.E. (2012). Observations of intergranular corrosion in AA2024-T351: the influence of grain stored energy. *Corrosion Science* 61: 35–44.
8 Parvizi, R., Hughes, A.E., Glenn, A.M. et al. (2018). Role of microstructure in corrosion initiation of a highly-deformed AA2024 wire. *Corrosion Science* 144: 184–197.
9 Parvizi, R., Hughes, A.E., Tan, M.Y. et al. (2017). Probing corrosion initiation at interfacial nanostructures of AA2024-T3. *Corrosion Science* 116: 98–109.
10 Knight, S.P., Salagaras, M., Wythe, A.M. et al. (2010). In situ X-ray tomography of intergranular corrosion of 2024 and 7050 aluminium alloys. *Corrosion Science* 52: 3855–3860.
11 Knight, S.P., Salagaras, M., and Trueman, A.R. (2011). The study of intergranular corrosion in aircraft aluminium alloys using X-ray tomography. *Corrosion Science* 53: 727–734.
12 Lim, M.L.C., Scully, J.R., and Kelly, R.G. (2013). Intergranular corrosion penetration in an Al-Mg alloy as a function of electrochemical and metallurgical conditions. *Corrosion* 69: 35–47.
13 Augustin, C., Andrieu, E., Blanc, C. et al. Intergranular corrosion of 2024 alloy in chloride solutions. *Journal of the Electrochemical Society* 154 (2007): C637–C644.
14 Ralston, K.D., Birbilis, N., Cavanaugh, M.K. et al. (2010). Role of nanostructure in pitting of Al-Cu-Mg alloys. *Electrochimica Acta* 55: 7834–7842.
15 Glenn, A.M., Hughes, A.E., MacRae, C.M. et al. (2020). Observations on the early stages of corrosion on AA2099-T83. *Microscopy and Microanalysis* 26: 821.
16 Zhang, W. and Frankel, G.S. (2003). Transitions between pitting and intergranular corrosion in AA2024. *Electrochimica Acta* 48: 1193–1210.
17 Li, J.F., Ziqiao, Z., Na, J., and Chengyu, T. (2005). Localized corrosion mechanism of 2xxx-series Al alloy containing S(Al2CuMg) and θ′(Al2Cu) precipitates in 4.0% NaCl solution at pH 6.1. *Materials Chemistry and Physics* 91: 325–329.
18 Wang, J., Zhang, B., Zhou, Y.T., and Ma, X.L. (2015). Multiple twins of a decagonal approximant embedded in S-Al2CuMg phase resulting in pitting initiation of a 2024Al alloy. *Acta Materialia* 82: 22–31.
19 Na, K.-H. and Pyun, S.-I. (2008). Comparison of susceptibility to pitting corrosion of AA2024-T4, AA7075-T651 and AA7475-T761 aluminium alloys in neutral chloride solutions using electrochemical noise analysis. *Corrosion Science* 50: 248–258.
20 Ren, J. and Zuo, Y. (2004). Study of electrochemical behavior and morphology of pitting on anodized 2024 aluminum alloy. *Surface and Coatings Technology* 182: 237–241.

21 Shao, M., Fu, Y., Hu, R., and Lin, C. (2003). A study on pitting corrosion of aluminum alloy 2024-T3 by scanning microreference electrode technique. *Materials Science and Engineering A* 344: 323–327.

22 Leblanc, P. and Frankel, G.S. (2002). A study of corrosion and pitting initiation of AA2024-T3 using atomic force microscopy. *Journal of the Electrochemical Society* 149: B239–B247.

23 Xiong, J., Tan, M., and Forsyth, M. (2013). The corrosion behaviors of stainless steel weldments in sodium chloride solution observed using a novel electrochemical measurement approach. *Desalination* 327: 39–45.

24 Xiong, J., Agarwala, S., Tan, M.Y., and Forsyth, M. (2015). The restoration of the passivity of stainless steel Weldments in pickling solutions observed using electrochemical and surface analytical methods. *Corrosion* 71 (10): 1248–1256.

25 Xiong, J., Manjaiah, G., Fabijanic, D. et al. (2018). Enhancing the localized corrosion resistance of 316L stainless steel via FBR-CVD chromising treatment. *Corrosion Engineering, Science and Technology* 53 (2): 114–121.

26 Sephton, M. and Pistorius, P.C. (2000). Localized corrosion of carbon steel weldments. *Corrosion-NACE* 56 (12): 1272. https://doi.org/10.5006/1.3280516.

27 Pagotto, J.F., Montemor, M.F., Recio, F.J. et al. (2015). Visualisation of the galvanic effects at welds on carbon steel. *Journal of the Brazilian Chemical Society* 26 (4): 667–675.

28 Choucri, J., Zanotto, F., Grassi, V. et al. (2019). Corrosion behavior of different brass alloys for drinking water distribution systems. *Metals* 9: 649. https://doi.org/10.3390/met9060649.

29 Gunasegaram, D.R., Venkatraman, M.S., and Cole, I.S. (2014). Towards multiscale modelling of localized corrosion. *International Materials Reviews* 59: 84–114.

30 Isaacs, H.S., Cho, J.H., Rivers, M.L., and Sutton, S.R. (1995). In situ X-ray microprobe study of salt layers during anodic dissolution of stainless steel in chloride solution. *Journal of the Electrochemical Society 142*: 1111–1117.

31 Ryan, M.P., Williams, D.E., Chater, R.J. et al. (2002). Why stainless steel corrodes. *Nature* 415: 770–774.

32 Schmuki, P., Hildebrand, H., Friedrich, A., and Virtanen, S. (2005). 'The composition of the boundary region of MnS inclusions in stainless steel and its relevance in triggering pitting corrosion. *Corrosion Science* 47: 1239–1250.

33 Schmutz, P. and Frankel, G.S. (1998). Characterization of AA2024-T3 by scanning Kelvin probe force microscopy. *Journal of the Electrochemical Society* 145: 2285–2295.

34 de Wit, J.H.W. (2004). Local potential measurements with the SKPFM on aluminum alloys. *Electrochimica Acta* 49: 2841–2850.

35 Davoodi, A., Pan, J., Leygraf, C., and Norgren, S. (2007). Integrated AFM and SECM for in situ studies of pitting corrosion of Al alloys. *Electrochimica Acta* 52: 7697–7705.

36 Jönsson, M., Thierry, D., and LeBozec, N. (2006). The influence of microstructure on the corrosion behavior of AZ91D studied by scanning Kelvin probe force microscopy and scanning Kelvin probe. *Corrosion Science* 48: 1193–1208.

37 Ben-Haroush, M., Ben-Hamu, G., Eliezer, D., and Wagner, L. (2008). The relation between microstructure and corrosion behavior of AZ80 Mg alloy following different extrusion temperatures. *Corrosion Science* 50: 1766–1778.

38 Apachitei, I., Fratila-Apachitei, L.E., and Duszczyk, J. (2007). Microgalvanic activity of an Mg–Al–Ca-based alloy studied by scanning Kelvin probe force microscopy. *Scripta Materialia* 57: 1012–1015.

39 Dasilva, M.F., Shimizu, K., Kobayashi, K. et al. (1995). On the nature of the mechanically polished aluminum surface. *Corrosion Science* 37: 1511–1514.

40 Tanem, B.S., Svenningsen, G., and Mardalen, J. (2005). Relations between sample preparation and SKPFM Volta potential maps on an EN AW-6005 aluminum alloy. *Corrosion Science* 47: 1506–1519.

41 Turley, M. and Samuels, L.E. (1981). The nature of mechanically polished surfaces of copper. *Metallography* 14: 275–294.

42 Raicheva, S.N. (1984). The effect of the surface state on the electrochemical behavior of copper electrodes. *Electrochimica Acta* 29: 1067–1073.

43 Williams, D.E., Kilburn, M.R., Cliff, J., and Waterhouse, G.I.N. (2010). Composition changes around sulphide inclusions in stainless steels, and implications for the initiation of pitting corrosion. *Corrosion Science* 52: 3702–3716.

44 Szklarska-Smialowska, Z. (1999). Pitting corrosion of aluminum. *Corrosion Science* 41: 1743–1767.

45 Pardo, A., Merino, M.C., Coy, A.E. et al. (2008). Influence of microstructure and composition on the corrosion behavior of Mg/Al alloys in chloride media. *Electrochimica Acta* 53: 7890–7902.

46 Burstein, G.T., Liu, C., Souto, R.M., and Vines, S.P. (2004). The origins of pitting corrosion. *Corrosion Engineering, Science and Technology* 39: 25–30.

47 Zuo, Y., Wang, H., Zhao, J., and Xiong, J. (2002). The effects of some anions on metastable pitting of 316L stainless steel. *Corrosion Science 44*: 13–24.

48 Garfias-Mesias, L.F., Alodon, M., James, P.I., and Smyrl, W.H. (1998). Determination of precursor sites for pitting corrosion of polycrystalline titanium by using different techniques. *Journal of the Electrochemical Society* 145: 2005–2010.

49 Isaacs, H.S. and Vyas, B. (1981). Scanning reference electrode technique in pitting corrosion. In: *Electrochemical Corrosion Testing* (ed. F. Mansfeld and U. Bertocci). ASTM STP 727, 1981.

50 Isaacs, H.S., Ryan, M.P., and Oblonsky, L.J. (1997). *Proc. Research Topic Symposium, NACE Corrosion 97*: 65.

51 Allah-karam, S.R., Vasantasree, V., and Hocking, M.G. (2001). Electrochemical potential mapping of a rapidly solidified processed light alloy. *Corrosion Science* 43: 1645–1656.

52 McMurry, H.N., Magill, S.R., and Jeffs, B.D. (1996). Scanning reference electrode technique as tool for investigating localized corrosion phenomena in galvanized steels. *Ironmaking and Steelmaking* 23: 183–188.

53 McMurry, H.N. (2001). Localized corrosion behaviour in aluminium-zinc alloy coatings investigated using the scanning reference electrode technique. *Corrosion* 57: 313–322.

54 Isaacs, H.S. (1987). The use of the scanning vibrating electrode technique for detecting defects in ion vapour-deposited aluminium on steel. *Corrosion* 43: 594–598.

55 Lillard, R.S., Moran, P.J., and Isaacs, H.S. (1992). A novel method for generating quantitative local electrochemical impedance spectroscopy. *Journal of the Electrochemical Society* 139: 1007–1012.

56 Murray, J.N. (1997). Electrochemical test methods for evaluating organic coatings on metals: an update. Part III: Multiple test parameter measurements. *Progress in Organic Coatings* 31: 375–391.

57 Barranco, V., Carmona, N., Galvan, J.C. et al. (2010). Electrochemical study of tailored sol–gel thin films as pre-treatment prior to organic coating for AZ91 magnesium alloy. *Progress in Organic Coatings* 68: 347.

58 Wipf, D.O. and Bard, A.J. (1991). Scanning electrochemical microscopy 10. High-resolution imaging of active-sites on an electrode surface. *Journal of the Electrochemical Society* 138: L4–L6.

59 Wipf, D.O. and Bard, A.J. (1991). Scanning electrochemical microscopy 7. Effect of heterogeneous electron-transfer rate at the substrate on the tip feedback current. *Journal of the Electrochemical Society* 138: 469–474.

60 Zhu, Y.Y. and Williams, D.E. (1997). Scanning electrochemical microscopic observation of a precursor state to pitting corrosion of stainless steel. *Journal of the Electrochemical Society* 144: L43–L45.

61 Hladky, K. and Dawson, J.L. (1981). The measurement of pitting corrosion using electrochemical noise. *Corrosion Science* 21: 317–322.

62 Hladky, K. and Dawson, J.L. (1982). The measurement of corrosion using electrochemical 1/f noise. *Corrosion Science* 22: 231–237.

63 Cottis, R.A. (2001). Interpretation of electrochemical noise data. *Corrosion* 57: 265–285.

64 Tan, Y.J. (2009). Sensing localized corrosion by means of electrochemical noise detection and analysis. *Sensors and Actuators B: Chemical* 139: 688–698.

65 Laleh, M., Hughes, A.E., Yang, S. et al. (2020). Two and three-dimensional characterisation of localized corrosion affected by lack-of-fusion pores in 316L stainless steel produced by selective laser melting. *Corrosion Science* 165 (1): 108394.

66 Laleh, M., Hughes, A.E., Xu, W. et al. (2019). Unexpected erosion-corrosion behaviour of 316L stainless steel produced by selective laser melting. *Corrosion Science* 155: 67–74.

67 Chao, Q., Cruz, V., Thomas, S. et al. (2017). On the enhanced corrosion resistance of a selective laser melted austenitic stainless steel. *Scripta Materialia* 141: 94–98.

68 Sander, G., Thomas, S., Cruz, V. et al. (2017). On the corrosion and metastable pitting characteristics of 316L stainless steel produced by selective laser melting. *Journal of the Electrochemical Society* 164 (6): C250–C257.

69 Kong, D., Ni, X., Dong, C. et al. (2018). Bio-functional and anti-corrosive 3D printing 316L stainless steel fabricated by selective laser melting. *Materials & Design* 152: 88–101.

70 Eklund, G.S. (1974). Initiation of pitting at sulfide inclusions in stainless steel. *Journal of the Electrochemical Society* 121 (4): 467–473.

71 Stewart, J. and Williams, D. (1992). The initiation of pitting corrosion on austenitic stainless steel: on the role and importance of sulphide inclusions. *Corrosion Science* 33 (3): 457465–463474.

72 Baker, M. and Castle, J. (1993). The initiation of pitting corrosion at MnS inclusions. *Corrosion Science* 34 (4): 667–682.

73 Scotto, V., Ventura, G., and Traverso, E. (1979). The influence of non-metallic inclusion nature and shape on the pitting corrosion susceptibility of 18Cr9Ni and 17Cr11Ni2Mo austenitic stainless steels. *Corrosion Science* 19 (4): 237247–245250.

74 Suter, T., Webb, E., Böhni, H., and Alkire, R. (2001). Pit initiation on stainless steels in 1 M NaCl with and without mechanical stress. *Journal of the Electrochemical Society* 148 (5): B174–B185.

75 Sun, Y., Moroz, A., and Alrbaey, K. (2014). Sliding wear characteristics and corrosion behaviour of selective laser melted 316L stainless steel. *Journal of Materials Engineering and Performance* 23 (2): 518–526.

76 Kong, D., Ni, X., Dong, C. et al. (2018). Heat treatment effect on the microstructure and corrosion behavior of 316L stainless steel fabricated by selective laser melting for proton exchange membrane fuel cells. *Electrochimica Acta* 276: 293–303.

77 Geenen, K., Röttger, A., and Theisen, W. (2017). Corrosion behavior of 316L austenitic steel processed by selective laser melting, hot-isostatic pressing, and casting. *Materials and Corrosion* 68 (7): 764–775.

78 Khairallah, S.A., Anderson, A.T., Rubenchik, A., and King, W.E. (2016). Laser powder-bed fusion additive manufacturing: Physics of complex melt flow and formation mechanisms of pores, spatter, and denudation zones. *Acta Materialia* 108: 36–45.

79 Bauereiß, A., Scharowsky, T., and Körner, C. (2014). Defect generation and propagation mechanism during additive manufacturing by selective beam melting. *Journal of Materials Processing Technology* 214 (11): 2522–2528.

80 Zhang, B., Li, Y., and Bai, Q. (2017). Defect formation mechanisms in selective laser melting: a review. *Chinese Journal of Mechanical Engineering* 30 (3): 515–527.

81 Sander, G., Tan, J., Balan, P. et al. (2018). Corrosion of additively manufactured alloys: a review. *Corrosion* 74 (12): 1318–1350.

82 Schaller, R.F., Mishra, A., Rodelas, J.M. et al. (2018). The role of microstructure and surface finish on the corrosion of selective laser melted 304L. *Journal of the Electrochemical Society* 165 (5): C234–C242.

83 Frankel, G. (1998). Pitting corrosion of metals a review of the critical factors. *Journal of the Electrochemical Society* 145 (6): 2186–2198.

84 Ernst, P., Laycock, N., Moayed, M.H., and Newman, R. (1997). The mechanism of lacy cover formation in pitting. *Corrosion Science* 39 (6): 1133–1136.

85 Pan, C., Liu, L., Li, Y., and Wang, F. (2013). Pitting corrosion of 304ss nanocrystalline thin film. *Corrosion Science* 73: 32–43.

86 Yang, Y., Gureyev, T., Tulloh, A. et al. (2010). Feasibility of a data-constrained prediction of hydrocarbon reservoir sandstone microstructures. *Measurement Science and Technology* 21 (4): 047001.

87 Burstein, G., Liu, C., Souto, R., and Vines, S. (2004). Origins of pitting corrosion. *Corrosion Engineering, Science and Technology* 39 (1): 25–30.

88 Otero, E., Pardo, A., Utrilla, M. et al. (1998). Corrosion behaviour of AISI 304L and 316L stainless steels prepared by powder metallurgy in the presence of sulphuric and phosphoric acid. *Corrosion Science* 40 (8): 1421–1434.

89 Tan, Y.J., Aung, N.N., Liu, T. et al. (2006). Novel corrosion experiments using the wire beam electrode (I-IV). *Corrosion Science* 48: 23–78.

90 Tan, M.Y.J. (2020). A critical overview of monitoring infrastructural health using corrosion probes. *Corrosion Engineering, Science and Technology* 55: 103–117. https://doi.org/10.1080/1478422X.2019.1695390.

91 Varela, F., Tan, M.Y.J., and Forsyth, M. (2015). An electrochemical method for measuring localized corrosion under cathodic protection. *ECS Electrochemistry Letters* 4: C1–C4.

92 Bu, H., Li, X., Qi, J., and Li, Y. (2019). Microelectrode array technology used for investigating the corrosion of alloys. *Cailiao Daobao/Materials Reports* 33 (12): 3963–3970.

93 Ju, H., Duan, J., Yang, Y. et al. (2018). Mapping the galvanic corrosion of three coupled metal alloys using coupled multielectrode array: Influence of chloride ion concentration. *Materials* 11 (4): 634.

94 Zhang, D.-L., Wang, W., and Li, Y. (2010). An electrode array study of electrochemical inhomogeneity of zinc in zinc/steel couple during galvanic corrosion. *Corrosion Science* 52 (4): 1277–1284.

95 Chen, C., Xu, Y.-X., Zhong, Q.-D. et al. (2018). Correlation of temperature with galvanic corrosion behaviour of copper alloys based on wire beam electrode. *Corrosion Engineering, Science and Technology* 53 (5): 331–339.

96 Zhang, D.-L., Wang, W., Jin, Y.-H. et al. (2011). Wire beam electrode technique for investigating galvanic corrosion behavior of galvanized steel-spot defect. *Zhongguo Youse Jinshu Xuebao/Chinese Journal of Nonferrous Metals* 21 (9): 2168–2174.

97 Liu, L., Xu, Y., Wang, X. et al. (2017). Preferential corrosion behavior of carbon steel weld in simulation pore solution and the inhibition performance of nitrite. *Cailiao Daobao/Materials Review* 31 (9): 119-124 and 130.

98 Li, Y., Li, Y., Yang, R., and Tang, X. (2018). Reconstruction of X80 pipeline steel welded joints and corrosion behavior in NACE solution. *Zhongguo Shiyou Daxue Xuebao (Ziran Kexue Ban)/Journal of China University of Petroleum (Edition of Natural Science)* 42 (6): 153–160.

99 Li, Y., Li, Q., Tang, X., and Li, Y. (2019). Reconstruction and characterization of galvanic corrosion behavior of X80 pipeline steelwelded joints. *Jinshu Xuebao/Acta Metallurgica Sinica* 55 (6): 801–810.

100 Chen, Y., Wang, Y., zhou, L. et al. (2020). Macro-galvanic effect and its influence on corrosion behaviors of friction stir welding joint of 7050-T76 Al alloy. *Corrosion Science* 164: 108360.

101 Xiong, J., Tan, Y., Nam, N., and Forsyth, M. (2013). A new method of studying the corrosion behavior of selected heat affected zones on UNS S31600 stainless steel welded joints. *NACE Corrosion Conference and Expo (2013: Orlando, Florida)*, Orlando, Florida (17–21 March 2013). CORROSION 2013: Proceedings of the 2013 NACE Corrosion Conference and Expo. Curran Associates, Red Hook, NY. 2444–2448.

102 Lu, B.T., Chen, Z.K., Luo, J.L. et al. (2005). Pitting and stress corrosion cracking behavior in welded austenitic stainless steel. *Electrochimica Acta* 50 (6): 1391–1403.

103 Gooch, T.G. (1996). Corrosion behavior of welded stainless steel. *Welding Journal* 75 (5): 135–154.

104 Lancaster, J.F. (1999). *Metallurgy of Welding*, 6e. Cambridge: Woodhead Publishing Limited.

105 Lamb, S. (2002). *Stainless steels and nickel alloys*, 2e. Canada: Casti Publishing Inc.

106 Meadows, C. and Fritz, J.D. (2005). Understanding stainless steel heat-affected zones. *Welding Journal* 84 (7): 26–30.

107 Tsai, W.T., Yu, C.L., and Lee, J.I. (2005). Effect of heat treatment on the sensitization of Alloy 182 weld. *Scripta Materialia* 53 (5): 505–509.

108 de Lima-Neto, P., Farias, J.P., Flávio, L. et al. (2008). Determination of the sensitized zone extension in welded AISIS 304 stainless steel using non-destructive electrochemical techniques. *Materials and Design* 50 (4): 1149–1155.

109 Park, S.H.C., Sato, Y.S., Kokawa, H. et al. (2004). Corrosion resistance of friction stir welded 304 stainless steel. *Scripta Materialia* 51 (2): 101–105.

110 Suter, T. and Böhni, H. (1998). Microelectrodes for studies of localized corrosion processes. *Electrochimica Acta* 43 (19-20): 2843–2849.

111 Perren, R.A., Suter, T.A., Uggowitzer, P.J. et al. (2001). Corrosion resistance of super duplex stainless steels in chloride ion containing environments: investigations by means of a new microelectrochemical method I. Precipitation-free states. *Corrosion Science* 43 (4): 707–726.

112 Wloka, J., Laukant, H., Glatzel, U., and Virtanen, S. (2007). Corrosion properties of laser beam joints of aluminium with zinc-coated steel. *Corrosion Science* 49 (7): 4243–4258.

113 Park, K. and Kwon, H. (2010). Effects of Mn on the localized corrosion behavior of Fe–18Cr alloys. *Electrochimica Acta* 55 (9): 3421–3427.

114 Lohrengel, M.M., Moehring, A., and Pilaski, M. (2001). Capillary-based droplet cells: limits and new aspects. *Electrochimica Acta* 47 (1–2): 137–141.

115 Birbilis, N., Padgett, B.N., and Buchheit, R.G. (2005). Limitations in microelectrochemical capillary cell testing and transformation of electrochemical transients for acquisition of microcell impedance data. *Electrochimica Acta* 50 (16–17): 3536–3544.

116 Garcia, C., de Tiedra, M.P., Blanco, Y. et al. (2008). Intergranular corrosion of welded joints of austenitic stainless steels studied by using an electrochemical minicell. *Corrosion Science* 50 (8): 2390–2397.

117 Garcia, C., Martin, F., de Tiedra, M.P. et al. (2008). Pitting corrosion of welded joints of austenitic stainless steels studied by using an electrochemical minicell. *Corrosion Science* 50 (4): 1184–1194.

8

Challenges and Opportunities in Managing Complex Localized Corrosion

Although corrosion science and engineering has made significant progresses over the past century, it has not yet achieved its goals. The ultimate goal of corrosion science and engineering should be to prevent the premature failure of engineering materials and to extend the safe operational life of engineering structures through detecting, predicting, mitigating, and managing corrosion, especially those occurring in complex and varying environments. A prerequisite for achieving this is the capability to effectively sense, locate, and monitor complex forms of localized corrosion – the prime threat to the integrity of metallic structures. Currently there are knowledge and technological gaps that limit the ability of engineers to effectively, reliably, and proactively detect and control localized corrosion. Many of these gaps, challenges, and needs have been extensively discussed in various occasions over the past decades [1–5]; however some critical issues still remain especially in complex engineering structures such as underground, multiphase and shore crossing pipelines. New challenges have also emerged with new needs for clean "greener" energy that requires significant advancements in materials suitable for application in infrastructures for generating, transporting and using solar, geothermal, nuclear, or other types of energy [1, 2]. Higher safety and environment protection requirements have also posed new demands for more reliable corrosion prediction and warning capabilities, as well as corrosion control capabilities and for more environmentally friendly anticorrosion materials and methods.

This chapter discusses unsolved localized corrosion issues and new challenges that have arisen from, for instance, new demand for eco-informed materials selection and environmentally friendly anticorrosion materials. Particular focus is on two specific challenges that are considered to be critical for making substantial progresses in corrosion science and engineering: (i) the timely and reliable detection and prediction of complex forms of localized corrosion in variable environmental conditions using corrosion probes and predictive models, and (ii) the

Localized Corrosion in Complex Environments, First Edition. Mike Yongjun Tan.

development of smart anticorrosion materials and methods in order to meet the new requirement for environmental protection and the effective control of localized corrosion in harsh engineering conditions.

8.1 Future Perspectives in Corrosion Monitoring Tools and Predictive Models

The management and protection of engineering structures could be considered to be comparable with the management of human health. Human body has many "sensors," the eyes, ears, nose, skin, and tongue that provide disease information through vision, hearing, smell, touch, and taste. This information provides warning to patients and assist medical doctors in diagnosing and preventing diseases although testing, monitoring, and analysis. Unfortunately engineering structures do not have such natural "sensors" on them, artificial corrosion sensing and monitoring devices are needed in order to provide engineers with structural health data required for corrosion detection, diagnosis, and prediction. For this reason, corrosion monitoring tools, data analytics, and predictive models are critical for corrosion management. An "ideal" corrosion monitoring and control system should be one that not only provides *in situ* and site-specific corrosion data required to visualize localized corrosion in variable corrosion environments, but also to use such data to inform corrosion predictive modeling, mitigation, and management actions that may need to be adjusted smartly and dynamically based on the prevailing corrosion condition and mechanism. For instance, corrosion data are needed to guide local coating repair and to regulate local cathodic protection potential and corrosion inhibitors injection. In this manner, the threat of localized corrosion to the integrity and safety of engineering structures would be minimized and the safe operational life of infrastructure would be maximized.

A prerequisite for achieving such effective and reliable corrosion management is timely knowledge about the initiation, propagation, and seriousness of localized corrosion occurring over an engineering structure. However currently industrial knowledge of localized corrosion is still generally from time-based routine inspections using various condition assessment and in-line inspection tools. Although corrosion data from such inspection are useful for identifying longer-term corrosion trends, they often do not have sufficient temporal and spatial resolutions required for predicting and responding to dynamic changes in complex localized corrosion. Nevertheless these technological limitations, challenges, and knowledge gaps provide opportunities to future corrosion research and development. Great opportunities exist to use "engineering tools to achieve quantitative life prediction, to incorporate state-of-the-art sensing approaches into experimentation and materials architectures, and to introduce environmental degradation factors into these capabilities" [4].

8.1.1 Opportunities in Developing More Reliable Monitoring Tools and Models

Corrosion mechanisms and kinetics in complex engineering structures are environment sensitive and can change with time and location. Even if corrosion systems have identical initial environments, corrosion conditions in these systems can still differ significantly with time due to the localization of corrosion reactions such as local cathodic reduction of oxygen, local dissolution and hydrolysis of metals, and local deposition of corrosion products. This is especially true for aged engineering structures where material and environmental conditions could have varied significantly over time, therefore the prediction of corrosion especially localized corrosion is a significant challenge under such changing local conditions. Effective and reliable corrosion monitoring and prediction tools are important for acquiring corrosion data that are critical for providing warning and prediction of safety issues and providing timely advice of maintenance needs especially for aged engineering structures hidden in soil, concrete, and deep sea.

Despite of progresses in developing corrosion inspection, detection, and monitoring tools and sensors over the past decades, as described in Chapter 2, the temporal and spatial resolutions of corrosion data still need improvement in order to enhance the reliable and timely warning of corrosion. It is expected that widespread industrial application of various effective corrosion monitoring tools will lead to the availability of real-time corrosion data that in conjunction with data analytics and artificial intelligence technologies will enable more reliable corrosion predictive modeling. There are opportunities in making significant progresses in the development and especially the industrial application of effective and reliable corrosion monitoring tools and predictive models.

The prediction of long-term corrosion behavior would need to be based upon the accumulation and analysis of long-term and reliable corrosion data with appropriate temporal and spatial resolutions. This is achievable through the development of new technologies such as smart corrosion monitoring probes and computational corrosion modeling and analysis. As discussed in Chapter 2, currently the prime challenge in corrosion monitoring is difficulties in simulating localized corrosion mechanisms on corrosion probe surfaces, especially when corrosion mechanisms are complex and changing with the extension of corrosion. A major reason that could lead to reporting of false corrosion rates and patterns has been identified to be the inappropriate design and use of corrosion probes that overlook critical corrosion mechanisms and possible changes in corrosion mechanisms with the extension of corrosion processes. Therefore, several progresses will need to be made in order to significantly enhance the reliability and applicability of corrosion probes as a structural health monitoring tool. Successful corrosion probe application in a structural health monitoring system requires the probe surface to simulate not only the corrosion environment but also the critical corrosion mechanisms occurring on the infrastructure. Significant attention should be given to addressing challenges in understanding localized corrosion mechanisms and in correctly monitoring them. It is essential to ensure that the corrosion probes are able to effectively simulate corrosion behavior in actual service environments and reliably evaluate the effects of various factors on corrosion processes, rates, and mechanisms. It is well appreciated that corrosion probes need to simulate the actual service exposure environment; however relatively less considerations have been given to the effects of environmental parameters on corrosion patterns and mechanisms. This challenge is more acute when corrosion is affected by many inter-related variables such as nonuniform temperature and pressure, heterogeneous metallurgy, inhomogeneous soil or solution chemistry and thermo-mechanical conditions, local mechanical stress, coating defects, and cathodic potential and excursions. Some more details on these challenges have already discussed in Section 2.3.3.

Another issue is that currently the practical and successful application of corrosion probes as structural health monitoring tools in industry is limited. This situation needs to be changed since the imitation of the environment in the laboratory is practically difficult, real-time, and site-specific corrosion monitoring using corrosion probes as structural health monitoring tools is almost essential for acquiring data for understanding the corrosion processes and mechanisms without disturbing the actual engineering environment. Further developments are needed to enhance the reliability of corrosion monitoring by improving the design and application of corrosion probes. Corrosion probes should become reliable tools for early detection and diagnosis of corrosion, for providing industrial system "health" alarm, for forecasting maintenance requirements, and for generating data for integrated and automated corrosion management system. It is expected that corrosion monitoring probes will become an important and much more widely applicable tool in structural health monitoring systems to acquire, visualize, and interpret industrial system "health" information, enabling early detection, warning, diagnosis, and prediction of infrastructural corrosion and failure.

Predictive models are probably the most widely used tools for predicting and managing corrosion with various mathematical or empirical models developed covering different length scales. Macroscopic models are designed to predict corrosion cells over centimeter to meter length scales; meso-scale models are developed to predict corrosion activities at sub-micrometer to millimeter scales; while atomistic models describe the motion of individual atoms involved in the fundamental corrosion mechanisms of bonding and charge transfer processes. However, in engineering practice, accurate modeling of localized corrosion processes in complex and variable environment can be very difficult and the prediction of corrosion patterns and locations in an engineering structure is deemed to be a major challenge in corrosion management. It is widely believed that corrosion behavior in the real world is nearly unpredictable because there are too many varieties and changes of environmental factors that can affect corrosion, including temperature, pressure, chemical composition, constituent concentration, pH value, electrical or thermal conductivity, viscosity, etc.. The changing nature of corrosion in real engineering systems suggests that a corrosion predictive model needs to be dynamically varying with changes in the corrosion environment and mechanism. This is challenging but it is not impossible if in-situ and site-specific corrosion data become available. Opportunities should exist in the development of "dynamic" corrosion predictive models and corrosion predictive networks supported by corrosion data, data analytics, and artificial intelligence.

The availability of corrosion data with good temporal and spatial resolutions from effective and reliable corrosion monitoring would also enable the establishment of a long-term corrosion database. Such database will enable big data, computational data analytics, and artificial intelligence to be applied as new approaches for the corrosion engineering community to make breakthrough in corrosion prediction, control, and management. An ideal corrosion monitoring and forecast system for managing complex localized corrosion in practical engineering structures should include a corrosion data collection system including various corrosion inspection and monitoring probes and tools to collect data of required temporal and spatial resolutions, a big data and data analytic system for data integration and analysis, as well as an artificial intelligence system for smart pattern recognition and prediction. Generally speaking, the future for corrosion monitoring and predictive modeling is promising. As pointed out by a National Research Council study, Research Opportunities in Corrosion Science and Engineering "the field of corrosion science and engineering is on the threshold of important advances" and "Corrosion science is on the threshold of great advances in high fidelity, in-operando experimental methods that can probe the chemical, electronic, and structural nature of materials on a local scale, in computational modeling from the atomic scale to application, in theory, data analytics, as well as their integration. Given the fact that corrosion is a complex process involving coupled chemical, electrochemical, biological, and solid-state reaction dynamics, materials science, surface science, and environmental chemistry. The multiscale nature of corrosion processes and the need to connect across length (and time) scales to advance our knowledge of the factors governing corrosion are daunting. High-resolution studies reveal that processes that seemed uniform at the mesoscale are often extremely heterogeneous at the local scale. Phenomena at each length scale are often regulated by underlying elementary processes at a smaller scale that in concert produce damage of engineering consequence. Thus, processes at the materials science length scale must be factored into atomic scale analysis to gain a complete picture. Drivers, dependencies, and triggers for all of these processes must be understood, including mechanical driving forces in the case of environmental cracking or film rupture" [1, 4].

8.1.2 Obstacles in Probing Multiscale Corrosion

Although various approaches of probing multiscale corrosion have been discussed in previous chapters, especially in Chapter 7, the detection and prediction of multiscale corrosion remain a practical challenge in industry. Corrosion research and inspection in laboratories and in the field are mainly based on one or two length scales in isolation without much attention to connection or integration of multiscale processes. It is considered that "the biggest difficulty in corrosion research is to gain a deep understanding of the multiscale degradation process *in situ* on a corroding surface that is covered by corrosive media" [5]. At present gaining understanding of multiscale corrosion is generally through "look and see" forensic type characterization by firstly probing corrosion at each scale and then connecting data and analysis from each scale to inform the next level. Practical industry studies that produce empirical macroscopic field data are sometimes lacking understanding of the root cause of corrosion that could be in microscopic scales. Similarly, laboratory surface science and atomic scale studies using various advanced surface analytical technologies are often weakly connected to engineering field applications. New approach to probing corrosion over multiscales is desirable because corrosion processes can interact over different scales that is often beyond the capability of "look and see" type of characterization.

To overcome difficulties in probing multiscale corrosion, technological progresses are needed. Currently there are technological difficulties, for instance, most of the advanced surface analytical instruments and tools used in characterizing the microstructure and composition of metal surfaces at a nano level must operate in high vacuum, they cannot produce real *in situ* data from an undisturbed corroding surface in aggressive media. It would be desirable if new emerging high-resolution *in situ* methods and tools such as corrosion electrolyte cell that can be installed in scanning electron microscopy (SEM) or transmission electron microscopy (TEM) could become available for investigating changes in the microstructure and composition of a corroding surface in liquid media [5]. Three-dimensional electrode designs such as those illustrated in Figures 7.15 and 7.16 could also create opportunities in probing multiscale corrosion. There are also opportunities that corrosion field could advance significantly by further connecting and integrating laboratory and industry information at each scale [1] through the use of data analytics and artificial intelligence technologies.

Better understanding of multiscale corrosion will not only fill knowledge gaps in corrosion science and engineering but also will enable better models for corrosion prediction and prognosis, as well as more effective application of corrosion control technologies such as cathodic protection, protective coatings, and corrosion inhibitors. Monitoring, probing, and understanding of local corrosive environment, its multiscale evolution on engineering structures and its multiscale interaction with engineering materials, especially in confined spaces such as pits, crevice, and cracks and under disbonded coatings, would enable progresses in the control of localized corrosion [1].

8.1.3 Challenges in Quantifying Mechanical Impact on Material Behavior

A factor that can significantly affect localized corrosion and materials degradation behavior is local mechanical impacts. Mechanical stresses can induce materials damage and corrosion failure as a result of complicated interactions between stress –materials–electrochemical reactions. There are many well-known examples in which specific combinations of metals and environments cause major problems, such as stress corrosion cracking (SCC) of brass in solutions with ammonia; steel SCC in high pH caustic and amine solutions; stainless steels and aluminum alloys SCC in solutions containing chlorides; Ti-alloys SCC in nitric acid or methanol; as well as high nickel alloys SCC in aerated water under high pressure and temperature. Local mechanical stresses could also cause damages to surface protective films and coatings, dramatically facilitating the loss of passivity and electrochemical corrosion processes. Because of the interdependence of corrosion and degradation on mechanical stresses and environmental conditions, probing and understanding these processes would need *in situ* testing of mechanical – materials – corrosion interactions. *Ex situ* measurements may not generate essential information of the interacting corrosion – cracking processes or may distort the information. However this has been a relatively weak area in corrosion studies that needs more attention.

The study of mechanical impact on localized corrosion and local materials degradation requires *in situ* techniques with suitably high resolution to reveal the interaction of mechanical stress and corrosion processes and mechanisms. Over recent decade, efforts have been made to develop and apply variously designed experimental devices for probing and quantifying mechanical impacts on localized materials damage. An example is a new experimental method for studying the effects of coating mechanical cracking based on in-situ electrochemical monitoring of stresses coating specimens and X-ray computed tomography (CT) observations [6–8]. Coatings on pipelines are subjected to mechanical stress during pipeline installation processes such as hydrostatic testing and field bending. For instance, pipeline coatings are exposed to the influence of various forms of mechanical stresses during the pipeline installation and operation, as a result of horizontal directional drilling, hydrostatic testing, the internal fluid pressure, the soil stress, and other forms of mechanical effects. Additionally, the external soil loads and the internal pipeline pressure are the continual sources of mechanical stress during the pipeline operation. At the same time, coatings on the buried pipelines are also under the influence of the underground soil environment and cathodic protection. In order to probe the synergistic influence of multiple factors on coating degradation in such a highly complex system, there is a requirement for an experimental method that is able to simulate such a complex condition and measure its effects on coating degradation.

As illustrated in Figure 8.1(i), a novel experimental method has been developed by Ranade et al. [6–8] for simulating and measuring coating degradation and cracking under the combined effects of three major factors affecting buried pipeline coatings: mechanical stress, corrosive environment, and cathodic protection (CP). This new testing assembly consisting of a specially designed tensile rig and a standard flat cell has been constructed to measure and study the effects of these factors on coating degradation based on electrochemical impedance spectroscopy (EIS) analysis. The tensile testing rig is employed to impose mechanical strains on coated steel samples, which are monitored using strain gauges. The corrosive environment and cathodic protection have been achieved using an electrochemical flat cell. The degradation of coatings has been measured based on changes in the amplitude of electrochemical impedance. Results show a continuous decrease in the amplitude of coating's impedance values with the increase in applied strain level and the extension of exposure [6–8]. These are consistent with coating degradation observed visually using various surface analytical methods including SEM and X-ray CT [6–8]. Figure 8.1(ii) shows a typical image of coating cracking, degradation and corrosion after being exposed to corrosive medium under 5% mechanical strain, showing cracks in the coating, accompanied by brownish rust coloring. Figure 8.1(iii) shows that when the strain is increased to 3.5%, there is a rapid drop in the impedance at a frequency of 0.05 Hz of almost two orders of magnitude, indicating rapid cracking of the coating. *In situ* and *ex situ* scanning electron microscopy was also employed to trace and characterize coating defects from the very early stages of micro shear bands initiation to the formation of full coating cracks [7] in conjunction with finite element analysis of the influence of strain on the initiation and propagation of coating defects. The strain distribution and shear stress distribution patterns obtained using the finite element analysis were used to explain typical features observable in EIS data based on the formation of electrolytic pathways through unstrained and strained coatings. This approach has been consolidated into a two-dimensional model proposed to explain the electrolyte movement in a coating impacted by applied mechanical strain and environmental exposure. Figure 8.2 shows typical FEA models (c, d) for samples under strains of 3.5% to present possible electrolytic path for typical SEM image of the coating surface (a) and cross-section (b) [7].

The impact of mechanical strain and the environment on pipeline coatings was further visualized from a three-dimensional perspective using the data-constrained modeling (DCM) technique with multi-energy X-ray computed tomography (X-ray CT), in conjunction with three dimensional (3D) finite element analysis (FEA) modeling that provides a qualitative interpretation of the DCM microstructure data in terms of the concentration of stress at heterogeneous interfaces within

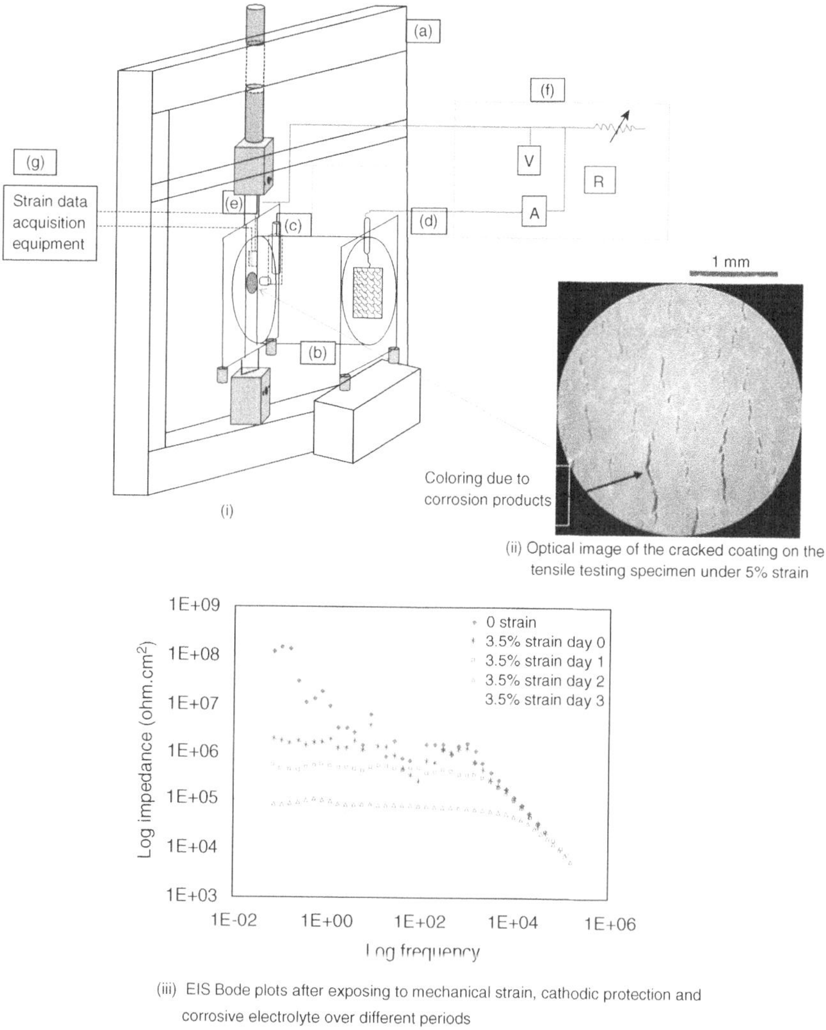

Figure 8.1 (i) a setup of the tensile testing rig in combination with an electrochemical flat cell, where (a) tensile testing rig; (b) flat cell filled with electrolyte; (c) reference electrode; (d) counter electrode; (e) coated sample as working electrode; (f) potentiostat; where V = voltmeter, A = ammeter, R = rheostat; and (g) strain data acquisition equipment connected to the strain gauge in yellow color at the back of the sample. (ii) The cracked coating surface image. (iii) EIS plots showing impedance changes with coating cracking. *Source:* Ranade et al. [6, 7], ELSEVIER.

coatings [8]. It has been found using the DCM technique that unstrained coating films are heterogeneous in nature, showing characteristic features similar to the D-type and I-type coating regions espoused in the historical literature. For strained coating films, the interfaces of heterogeneities were seen to provide preferential

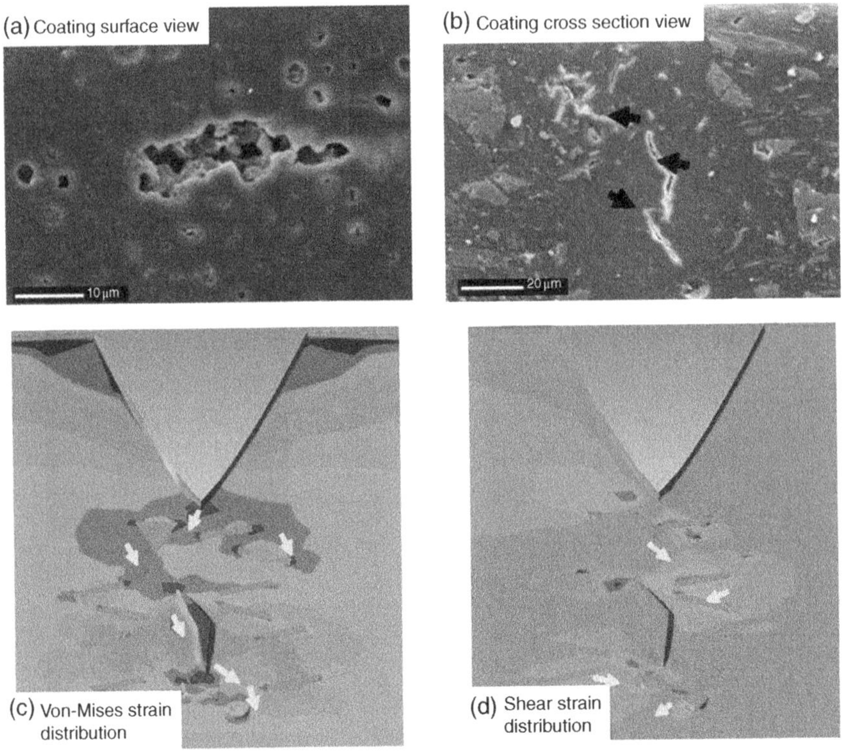

Figure 8.2 Typical SEM image of (a) the coating surface and (b) the cross-section; (c, d) sample FEA models for samples under strains of 3.5% to present possible electrolytic path for possible direction of electrolytic movement (indicated by bright arrows) [7]. *Source:* Ranade et al. [7], ELSEVIER.

sites to form voids by mechanical straining. Moreover, the void network was also found at the interfaces of the inorganic fillers and the organic polymers. These have been validated by simulation carried out using the FEA modeling, with a 3D model indicating the role of fillers in the formation of a tortuous void network in a heterogeneous coating subjected to both mechanical straining and corrosive environment. This approach therefore allows the determination of the spatial distribution of void, polymeric, and inorganic components within the coating. When such a heterogeneous polymer coating is subjected to mechanical straining, the polymer properties and structures are likely to influence the resulting morphological variations. As shown in Figure 8.3, there are two major void features including large void spaces denoted by an arrow in Figure 8.3b in between the higher and lower polymer compositions [8].

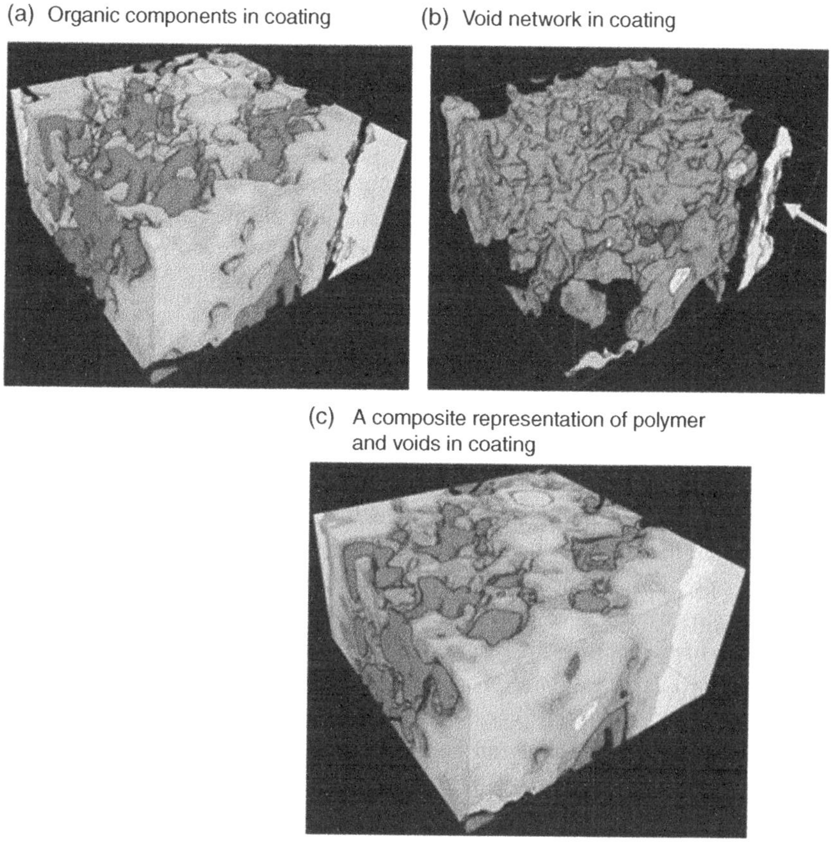

Figure 8.3 DCM representation for components in high build epoxy coating strained to 2.5%, where (a) organic, and (b) void network, and (c) void network in presence of organic component. The yellow arrow in (b) represents the straight rupture formed between different polymer compositions shown in (a). *Source:* [8]/With permission of Elsevier.

8.2 Opportunities in Developing Smart Anticorrosion Methods and Materials

8.2.1 Design and Select Materials for Severe Industrial Conditions

Corrosion control and management under some severe conditions also still remain challenges, for instance in the mining industry slurry pumps needs changing every few months due to severe erosion corrosion. These "old" tenacious issues remain unsolved partly because of the fact that corrosion resistance is often not among top materials selection considerations for many industrial structures.

Take oil and gas pipeline as an example, mechanical strength (e.g. yield strength), toughness (especially fracture toughness), and the impact toughness (to prevent dent induced safety problems) are often considered to be top three "design-limiting" materials properties considered in materials selection, while weldability and corrosion are often among secondary properties that are left to address through the use of corrosion prevention technologies such as surface coatings and cathodic protection. Another example is magnesium alloys that are selected as structural materials for the next generation of automobiles because of their desirable high strength–weight ratio despite of their poor corrosion resistance. This creates new challenges since these untraditional materials often have a corrosion behavior different from traditional alloys, and existing corrosion and protection knowledge and technology cannot be directly applied to them. There is a need to further develop the existing corrosion principles and techniques in order to cover these advanced engineering alloys [5]. The applied nature of engineering suggests that it needs to respond to changing industrial and social needs, for instance the need for hydrogen transportation and storage; the need for permanent storage of nuclear wastes; the need for offshore solar and wind farms etc., creating new challenges and opportunities for corrosion science, engineering, and technology.

The emerging hydrogen economy has created a need for steel pipelines such as X65 steel natural gas pipelines to be used to transport and store hydrogen-containing fuels as an economical methods of distributing and storing gaseous hydrogen. However, in order to ensure the reliability and safety of pipelines it is critical to understand the material behavior and interaction of these standard pipeline materials, aged or new, with hydrogen when they are used in pressurized hydrogen–natural gas mixtures. There have been many studies on the interaction of standard pipeline materials with hydrogen in the past, and it has been reported that a general reduction in the materials ability to plastically deform was noted in pipeline steels [9–12]. Research has confirm the need for caution when injecting significant levels of hydrogen into transmission pipelines, especially into higher grade pipelines operating at higher pressures because hydrogen additions can generally increase hydrogen embrittlement and the fatigue crack growth rate as the hydrogen content in the methane increases [11–13]. It has also been noted that there is a critical hydrogen concentration that causes the significant loss of local fracture resistance of material including three API grade pipeline steels, X52; X70 and X100 [10]. It has been suggested that existing high-pressure natural gas pipelines are not suitable for hydrogen transport because they are susceptible to embrittlement if the hydrogen gas pressure is higher than the threshold pressure [10]. Overall the long-term performance of materials in renewable gas environment, in particular hydrogen methane blends, is not fully understood.

A formidable obstacle to understanding of hydrogen-materials-surface interactions is the lack of experimental surface science tools capable of directly detecting hydrogen. Hydrogen can be measured globally, that is simply as an overall content from a bulk material specimen, or locally, from a region within the material. Furthermore, hydrogen can be measured using microstructure-specific methods that can assign the presence of hydrogen to microstructural features without any specific spatial information, such as thermal desorption spectrometry, or using more advanced, highly spatially specific techniques that can pin-point the exact location within a microstructure such as atom-probe tomography. Nevertheless materials development can be a useful approach to enhancing hydrogen embrittlement resistance of pipeline steels. For instance, a suitable microstructure for optimum hydrogen embrittlement resistance would contain maximum number density of small irreversible hydrogen trapping sites well distributed across the whole volume of the alloy.

Another example of materials development is to significantly enhance the repassivation ability and localized corrosion resistance of an additively manufactured duplex stainless steel through microstructural modification and control by post-processing heat treatment [14]. Additive manufacturing (AM) has attracted considerable attention during the past decade as a new means of producing engineering metallic materials, in particular components with intricate geometries. However, a major challenge associated with AM metallic materials is their non-equilibrium and sometimes anisotropic microstructures compared to their conventionally manufactured counterparts, which may affect materials properties and deteriorate their corrosion resistances [15–17]. This is even more significant in AM of materials where their microstructure evolution is highly dependent on the cooling rate during solidification, e.g. dual- and multi-phase materials. Duplex stainless steel 2205 produced by laser powder bed fusion (LPBF) additive manufacturing was found to possess poor repassivation ability and weak resistance to localized corrosion. This was due to the nonequilibrium phase composition and elemental distribution in the as-built LPBF DSS affected by the rapid solidification inherent to LPBF processing. A short-time post-processing heat treatment at 1000 °C for 10 minutes followed by water quenching to room temperature was shown to be able to establish the repassivation ability and localized corrosion resistance by retrieving the equilibrium dual-phase microstructure [14]. The complication of applying post-AM heat treatment on the corrosion properties of AM stainless steel has been comprehensively reviewed by Laleh et al. [18], indicating that the response to heat treatment depends on the initial microstructure, AM processing, and heat treatment conditions of a metal or alloy.

8.2.2 Design and Apply Environmentally Friendly Anticorrosion Materials

Over the recent decades, requirements for environmentally friendly engineering materials have significantly affected corrosion engineering. For instance, the development of environmentally friendly corrosion-resistant functional coatings has become an important need. Coatings have the ability to alter the properties of a metal surface, which opens up many new application fields for metals. Generally speaking, the corrosion resistance of a coating is a function of the stability of the coating material, the integrity and thickness of the coating layer, and the adhesion of the coating on metals. It also depends upon how the coating responds to temperature, stress, and abrasion in service. If a new coating can be made to be electrical or thermal insulating or conducting, anti-skidding or self-lubricant-releasing, super-hydrophobic or hydrophilic, selectively adsorptive and absorptive to certain species, etc., it will find a much wider application and also significantly extend the applications of metallic components/parts in industry. However, the development of such a coating is challenging because it need to satisfy many contradicting requirements [5]. Molecular design has been used as an approach to developing such multi-functional coatings that satisfy many contradicting requirements. In a typical study, copolymerization has been carried out to enhance the anticorrosion performance of multifunctional waterborne acrylic coatings [19, 20] in order to overcome drawbacks of waterborne coatings in poor water and corrosion resistance, low solid content, and so forth [21]. The acrylic-alkyd copolymers were found not only to possess combined advantages of acrylic and alkyd resins in coating properties such as stability, fast drying, and hardness, but also to possess several properties required for enhanced corrosion resistance, especially a higher hydrophobicity of the coating surface, a lower water diffusion coefficient, and a denser network of copolymer films. The higher hydrophobicity of the coating surface was achieved through the design of acrylic-alkyd copolymers with acrylic branches and hydrophobic alkyd backbones, leading to increase in the water contact angles of the copolymer films. The denser coating network was achieved by the branched structure and the flexibility of the copolymer chains. This allows the copolymer particle stacking during the curing process and builds a denser barrier film with lower water diffusion coefficients. These coating properties have led to significant enhancement in the corrosion resistance of these acrylic-alkyd copolymer films, which has been confirmed by the results from electrochemical impedance spectra (EIS) and corrosion exposure tests.

Similar approaches have also been taken to design and make waterborne coatings with intended multifunctionalities through manipulating acrylic-based nanocomposites with different nanostructures [20]. Taking advantage of the favorable structure of acrylic copolymers, three waterborne coatings with various desired properties have been created through molecular engineering either by copolymerizing with other components or through nanocomposite formation. This approach was demonstrated by synthesizing acrylic-based waterborne coatings with homogeneous, worm-like, and spherical-like nanostructures. The incorporation of 3-methacryloxypropyltrimethoxysilane (MPS)-modified nanoparticle

TiO_2 in acrylic base enabled the formation of a nanocomposite with nanoparticles uniformly distributed in the acrylic base. The coating film with this acrylic-TiO_2 nanocomposite has shown significantly better UV absorption performance than coating made by physical blending. The copolymerization of acrylic copolymers with an organic polymer (alkyd) led to a wormlike nanostructure of acrylic-alkyd composite, allowing generally uniform distribution of the acrylic-alkyd nanocomposite in coating films. The coating film with this acrylic-alkyd nanocomposite was found to be more closely packed, resulting in a dense film with enhanced barrier property and corrosion resistance. The copolymerization of acrylic monomers with an inorganic polymer [polydimethylsiloxane (PDMS)], which is with vinyl group led to a spherical-like nanostructure of acrylic-PDMS composite film, permitting the surface nanostructure of acrylic-PDMS nanocomposite to be more organized and incorporated, resulting in a coating film with PDMS functionalities such as low dirt-picking behavior [20].

Figure 8.4 shows atomic force microscopy (AFM) images of the acrylic-TiO_2 composite films that exhibited homogeneous nanostructure due to the chemical bond formation between nanoparticles and acrylic base, therefore the filling of TiO_2 nanoparticles by copolymerization could obtain uniform composite film. In general, the enhancement in UV absorbance of acrylic-TiO_2 composite film suggest that the uniform distribution of nano-TiO_2 particles into acrylic base plays an important role in improving UV protection of acrylic base. The anticorrosion performance of the acrylic-alkyd composite film enhanced significantly [20]. The significantly enhanced corrosion resistances of acrylic-alkyd composite film is believed to be due to the uniform distribution of alkyd in acrylic base and the formation of a dense coating film. In addition, it is clear that the acrylic-alkyd composite film formed a nanocomposite with uniform and wormlike nanostructure as shown in Figure 8.4. The wormlike nanostructure is more packed, and this creates better barrier property, which can reduce the penetration of salt solution into the composite film, therefore enhancing the anticorrosion performance of acrylic base [20].

Similarly multifunctional and environmentally friendly materials have also been developed from various other materials for various purposes. For instance, a triple-functional carbon-fiber-reinforced polymer was developed for strengthening and protecting reinforced concrete structures [22, 23]. Carbon-fiber-reinforced polymers (CFRP) are composite materials that have been applied for repairing and retrofitting civil engineering structures. Reinforced concrete (RC) has been widely used for constructing major civil and infrastructures such as bridges because of its stability, economy, versatility, and especially its resilience in extreme environmental conditions. Unfortunately, steel reinforcements embedded in concrete for improving the tensile strength of RC structures are subjected to corrosion

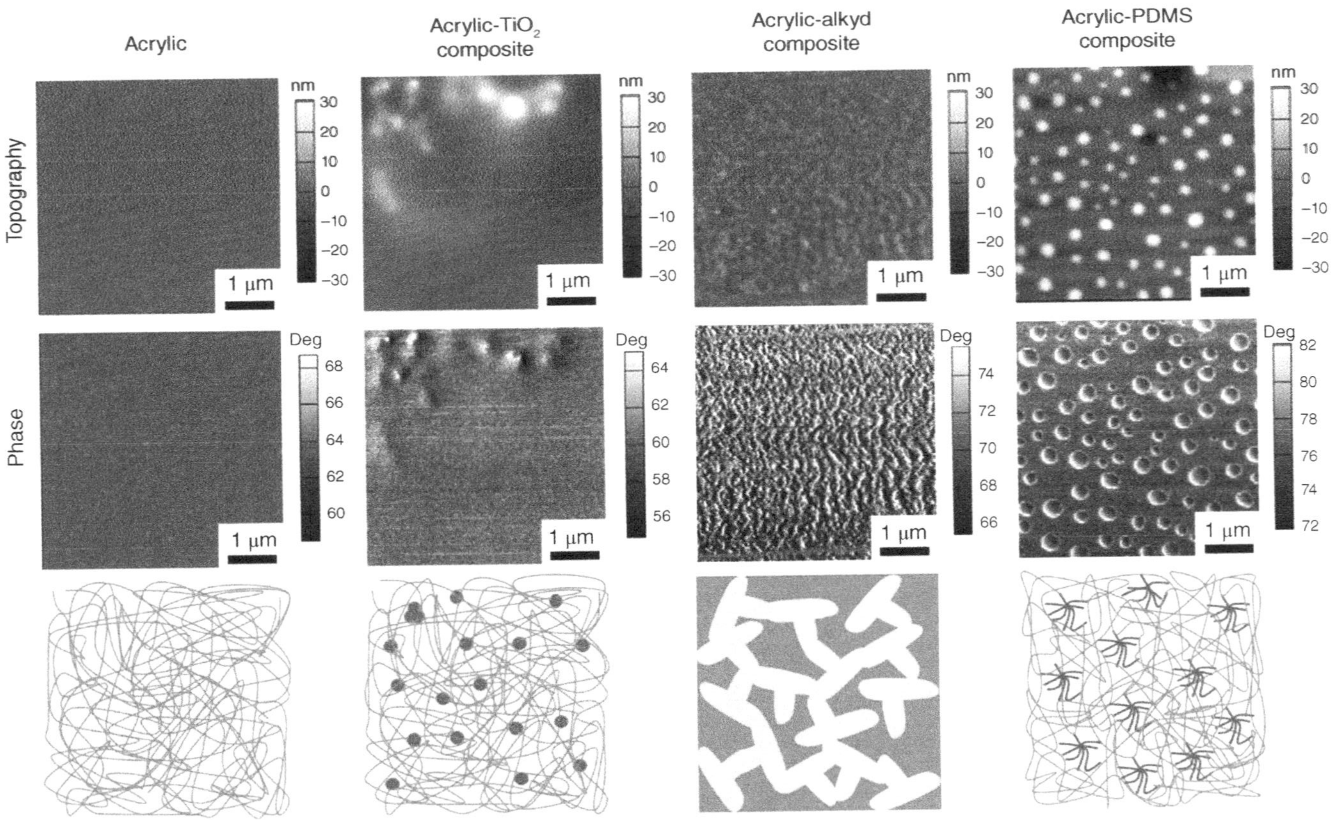

Figure 8.4 AFM images and models of acrylic-alkyd composite films formed as nanocomposite with uniform and wormlike nanostructures. *Source:* Ji et al. [20] / With permission of Elsevier.

when RC structures are exposed to harsh environments, inducing concrete cracks and reducing structural mechanical strength and stiffness. The most common method for repairing RC structures is concrete patching, in which deteriorated concrete is removed and then corroded steel reinforcements are cleaned with patching mortar, in conjunction with cathodic protection in which a cathodic current is applied for controlling steel rebar corrosion in RC structures. CFRPs, because of their high strength, chemical and corrosion resistance, have been applied for retrofitting and strengthening civil engineering structures by CFRP jacketing and CFRP wrapping RC structures such as columns, beams, and slabs. Recently long-term performance tests have demonstrated that the CFRP can not only perform as a strengthening material for RC, but also can simultaneously perform as a protective coating layer and an impressed current cathodic protection (ICCP) anode for protecting steel rebar from corrosion [22]. Figure 8.5 shows long-term experimental setup in a tank with simulated harsh marine environmental conditions for evaluating CFRP coating layer in its ability of reducing steel rebar corrosion and ICCP current [22]. Long-term immersion tests under sustained loading and simulated marine cyclic wet–dry conditions have been conducted to demonstrate triple functions of a CFRP-based coating layer for protecting steel rebar in RC specimens: as a protective coating, a strengthening material, and an ICCP anode. The coating layer was found to be protective and simultaneously reducing the current requirement for effective ICCP. At a reduced current density requirement of 20 mA/m^2, ICCP-induced CFRP damage was avoided, permitting long-term mechanical strengthening of RC structures [22].

Figure 8.5 The long-term experimental setup in a tank with simulated harsh marine environmental conditions. *Source:* Wei et al. [22], ELSEVIER.

References

1 Scully, J.R. (2018). Future frontiers in corrosion science and engineering, Part I. *Corrosion* 74 (1): 3–4. https://doi.org/10.5006/2734.

2 National Academy of Engineering (2017). *NAE Grand Challenges for Engineering*. Washington, DC: National Academy of Science http://www.engineeringchallenges.org/challenges.aspx (accessed September 2022).

3 Department of Energy (2017). Basic Research Needs and Priority Research Directions in Nuclear Power (August). https://www.osti.gov/biblio/1616270-basic-research-needs-future-nuclear-energy-report-basic-energy-sciences-workshop-future-nuclear-energy-august (accessed 4 September 2022)

4 National Research Council (2011). *Research Opportunities in Corrosion Science and Engineering*. Washington, DC: National Academies Press https://doi.org/10.17226/13032, https://www.nap.edu/catalog/13032/research-opportunities-in-corrosion-science-and-engineering (4 September 2022).

5 Song, G.-L. (2014). The grand challenges in electrochemical corrosion research, SPECIALTY GRAND CHALLENGE article. *Front. Mater.* 10: https://doi.org/10.3389/fmats.2014.00002, https://www.frontiersin.org/articles/10.3389/fmats.2014.00002/full (accessed 4 September 2022).

6 Ranade, S., Forsyth, M., and Tan, M.Y.J. (2016). In situ measurement of pipeline coating integrity and corrosion resistance losses under simulated mechanical strains and cathodic protection. *Progress in Organic Coatings* 101: 111–121 (11 pages).

7 Ranade, S., Forsyth, M., and Tan, M.Y.J. (2017). The initiation and propagation of coating morphological and structural defects under mechanical strain and their effects on the electrochemical behavior of pipeline coatings. *Progress in Organic Coatings* 110: 62–77 (16 pages).

8 Ranade, S.D., Hughes, T., Yang, S. et al. (2018). Visualizing the impact of mechanical strain and the environment on pipeline coatings from a three dimensional perspective. *Progress in Organic Coatings* 122: 45–55.

9 Smith, N., Byrne, N., Coates, M. et al. (2017). *Research Report on identifying the commercial, technical and regulatory issues for injecting renewable gas in Australian distribution gas networks*. Energy Networks Australia, Final Report.

10 Dodds, P.E. and Demoullin, S. (2013). Conversion of the UK gas system to transport hydrogen. *International Journal of Hydrogen Energy* 38: 7189–7200.

11 Moro, I., Briottet, L., Lemoinea, P. et al. (2010). Hydrogen embrittlement susceptibility of a high strength steel X80. *Materials Science and Engineering A* 527: 7252–7260.

12 Briottet, L., Moro, I., and Lemoine, P. (2012). Quantifying the hydrogen embrittlement of pipeline steels for safety considerations. *International Journal of Hydrogen Energy* 37: 17616–17623.

13 Traidia, A., Chatzidouros, E., Jouiad, M. et al. (2018). Review of hydrogen-assisted cracking models for the oil and gas industry. *Corrosion Reviews* 36: 323–347.

14 Laleh, M., Haghdadi, N., Hughes, A.E. et al. (2022). Enhancing the repassivation ability and localised corrosion resistance of an additively manufactured duplex stainless steel by post-processing heat treatment. *Corrosion Science* 198: 12. pages Article number 110106.

15 Laleh, M., Hughes, A.E., Yang, S. et al. (2021). A critical insight into lack-of-fusion pore structures in additively manufactured stainless steel. *Additive Manufacturing* 38: 101762.

16 Laleh, M., Hughes, A.E., Yang, S. et al. (2020). Two and three-dimensional characterisation of localised corrosion affected by lack-of-fusion pores in 316L stainless steel produced by selective laser melting. *Corrosion Science* 165: 108394.

17 Laleh, M., Hughes, A.E., Xu, W. et al. (2019). Unexpected erosion-corrosion behaviour of 316L stainless steel produced by selective laser melting. *Corros. Sci.* 155: 67–74.

18 Laleh, M., Hughes, A.E., Xu, W. et al. (2021). A critical review of corrosion characteristics of additively manufactured stainless steels. *Int. Mater. Rev.* 66: 563–599.

19 Ji, S., Gui, H., Guan, G. et al. (2021). Molecular design and copolymerization to enhance the anti-corrosion performance of waterborne acrylic coatings. *Progress in Organic Coatings* 153: 1–12. Article number 106140.

20 Ji, S., Gui, H., Guan, G. et al. (2021). A multi-functional coating based on acrylic copolymer modified with PDMS through copolymerization. *Progress in Organic Coatings* 156: 1–9. Article number 106254.

21 Panda, S.S., Panda, B.P., Nayak, S.K., and Mohanty, S. (2018). A review on waterborne thermosetting polyurethane coatings based on castor oil: synthesis, characterization, and application. *PPTEn* 57 (6): 500–522.

22 Wei, A., Al-Ameri, R., Koay, Y.C., and Tan, M.Y.J. (2021). Triple-functional carbon fibre reinforced polymer for strengthening and protecting reinforced concrete structures. *Composites communications* 24: 1–6. Article number 100648.

23 Wei, A., Tan, Y., Koay, Y.-C. et al. (2021). Effect of carbon fiber waste on steel corrosion of reinforced concrete structures exposed to the marine environment. *Journal of cleaner production* 316: 1–12. Article number 128356.

Index

Localized Corrosion in Complex Environments, First Edition. Mike Yongjun Tan.

Printed and bound by CPI Group (UK) Ltd, Croydon, CR0 4YY
19/09/2026
14965095-0002